Texts in Statistical

The BUGS Book

A Practical Introduction to Bayesian Analysis

David Lunn
Christopher Jackson
Nicky Best
Andrew Thomas
David Spiegelhalter

CRC Press
Taylor & Francis Group
Boca Raton London New York

CRC Press is an imprint of the
Taylor & Francis Group an **informa** business

A CHAPMAN & HALL BOOK

CHAPMAN & HALL/CRC
Texts in Statistical Science Series

Series Editors

Francesca Dominici, *Harvard School of Public Health, USA*
Julian J. Faraway, *University of Bath, UK*
Martin Tanner, *Northwestern University, USA*
Jim Zidek, *University of British Columbia, Canada*

CRC Press
Taylor & Francis Group
6000 Broken Sound Parkway NW, Suite 300
Boca Raton, FL 33487-2742

© 2013 by Taylor & Francis Group, LLC
CRC Press is an imprint of Taylor & Francis Group, an Informa business

No claim to original U.S. Government works

Printed in the United States of America on acid-free paper
Version Date: 20120808

International Standard Book Number: 978-1-58488-849-9 (Paperback)

**Visit the Taylor & Francis Web site at
http://www.taylorandfrancis.com**

**and the CRC Press Web site at
http://www.crcpress.com**

Contents

Preface

History Markov chain Monte Carlo (MCMC) methods, in which plausible values for unknown quantities are simulated from their appropriate probability distribution, have revolutionised the practice of statistics. For more than 20 years the BUGS project has been at the forefront of this movement. The BUGS project began in Cambridge, United Kingdom, in 1989, just as Alan Gelfand and Adrian Smith were working 80 miles away in Nottingham on their classic Gibbs sampler paper (Gelfand and Smith, 1990) that kicked off the revolution. But we never communicated (except through the intermediate node of David Clayton) and whereas the Gelfand–Smith approach used image processing as inspiration, the philosophy behind BUGS was rooted more in techniques for handling uncertainty in artificial intelligence using directed graphical models and what came to be called Bayesian networks (Pearl, 1988). Lunn et al. (2009b) lay out all this history in greater detail.

Some people have accused Markov chain Monte Carlo methods of being slow, but nothing could compare with the time it has taken for this book to be written! The first proposal dates from 1995, but things got in the way, as they do, and it needed a vigorous new generation of researchers to finally get it finished. It is slightly galling that much of the current book could have been written in the mid-1990s, since the basic ideas of the software, the language for model description, and indeed some of the examples are unchanged. Nevertheless there have been important developments in the extended gestational period of the book, for example, techniques for model criticism and comparison, implementation of differential equations and nonparametric techniques, and the ability to run BUGS code within a range of alternative programs.

The BUGS project is rooted in the idea of generic reusable components that can be put together as desired, like a child's construction set but not quite as colourful. In this book we typically tackle each of these components one by one using deliberately simplified examples, but hopefully it will be clear that they can be easily assembled into arbitrarily complex models. This flexibility has enabled BUGS to be applied in areas that we had never dreamed about, which is gratifying. But it is also important to note that in many situations BUGS may not be the most efficient method, and there are many things it cannot do. Yet...

What's in the book? Perhaps we should start by saying what is *not* in the book. First, there is minimal statistical theory, neither of statistical inference nor of Markov chain Monte Carlo methods (although a presumption of some familiarity with probability theory is made). This is partly to keep the book to a manageable length, but also because the very way in which BUGS works removes the need for much of the theory that is taught in standard Bayesian statistics courses. Second, we do not cover decision theory, as BUGS has been designed for handling Bayesian inferences expressed as an appropriate posterior distribution. Finally, we take it for granted that a Bayesian approach is desired, and so barely bother to lay out the reasons why this may be appropriate.

A glance at the chapter contents will reveal that we introduce regression models, techniques for criticism and comparison, and a wide range of modelling issues before going into the vital and traditional Bayesian area of hierarchical models. This decision came after considerable thought and experimentation, and was based on the wish to deal with the essentials of modelling without getting bogged down in complexity. Our aim is to bring to the forefront model criticism, model comparison, sensitivity analysis to alternative priors, and thoughtful choice of prior distributions — all those aspects of the "art" of modelling that are easily overlooked in more theoretical expositions. But we have also really enjoyed going systematically through the large range of "tricks" that reveal the real power of the BUGS software: for example, dealing with missing data, censoring, grouped data, prediction, ranking, parameter constraints, and so on.

Our professional background has meant that many of the examples are biostatistical, but they do not require domain knowledge and hopefully it will be clear that they are generalisable to a wide range of other application areas. Full code and data for the examples, exercises, and some solutions can all be found on the book website: www.mrc-bsu.cam.ac.uk/bugs/thebugsbook.

The BUGS approach clearly separates the model description from the "engine," or algorithms and software, used to actually do the simulations. A brief introduction to WinBUGS is given in Chapter 2, but fully detailed instructions of how to run WinBUGS and similar software have been deferred to the final chapter, 12, and a reference guide to the modelling language is given in the appendices. Since BUGS now comes in a variety of flavours, we have tried to ensure that the book works for WinBUGS, OpenBUGS, and JAGS, and any differences have been highlighted. Nevertheless the software is constantly improving, and so in some areas the book is not completely prescriptive but tries to communicate possible developments.

Finally, we acknowledge there are many shades of Bayesianism: our own philosophy is more pragmatic than ideological and doubtless there will be some who will continue to spurn our rather informal attitude. An example of this informality is our use of the term 'likelihood', which is sometimes used when referring to a sampling distribution. We doubt this will lead to confusion.

How to use the book. Our intended audience comprises anyone who would like to apply Bayesian methods to real-world problems. These might be practising statisticians, or scientists with a good statistical background, say familiarity with classical statistics and some calculus-based probability and mathematical statistics. We do not assume familiarity with Bayesian methods or MCMC. The book could be used for self-learning, for short courses, and for longer courses, either by itself or in combination with a textbook such as Gelman et al. (2004) or Carlin and Louis (2008).

Chapters 1 to 6 provide a basic introduction up to regression modelling, which should be a review for those with some experience with Bayesian methods and BUGS. Beyond that there should be new material, even for experienced users. For a one-semester course we would recommend Chapters 1 to 6, most of Chapter 8 on model criticism and comparison, and Chapter 10 on hierarchical models. A longer course could select from the wide range of issues and models outlined in Chapters 7, 9 and 11, depending on what is most relevant for the audience.

Whether studying on your own or as part of a course, instructions for running the WinBUGS software are given briefly in Chapter 2 and fully in Chapter 12. A full explanation of BUGS model syntax and a list of functions and distributions are given in the appendices. Chapter 12 explains how Open-BUGS and JAGS differ from WinBUGS and gives examples of how all varieties of BUGS can be conveniently run from other software, in particular from R.

Other sources. If an accompanying text on the underlying theory of Bayesian inference is required, possibilities include Gelman et al. (2004), Carlin and Louis (2008) and Lee (2004), with Bernardo and Smith (1994) providing a deeper treatment. Other books focus explicitly on BUGS: Ntzoufras (2009) provides a detailed exposition of WinBUGS with accompanying theory, Gelman and Hill (2007) explore both standard and hierarchical regression models using both R and BUGS, while the texts by Congdon (2003, 2005, 2006, 2010) explore a staggering range of applications of BUGS that we could not hope to match. Jackman (2009) covers both theory and BUGS implementations within social science, ecology applications are covered by Kéry (2010) and Kéry and Schaub (2011), while Kruschke (2010) gives a tutorial in Bayesian analysis and BUGS with applications in psychology. Expositions on MCMC theory include Gamerman and Lopes (2006) and Brooks et al. (2011), while Gilks et al. (1996) is still relevant even after many years.

Finally, there are numerous websites that provide examples and teaching material, and when tackling a new problem we strongly recommend trying to find these using appropriate search terms and adapting someone else's code. We have always been impressed by the great generosity of BUGS users in sharing code and ideas, perhaps helped by the fact that the software has always been freely available.

A suggested strategy for inference and reporting. Rather than leaving it until later in the book, it seems appropriate to lay out at an early stage the approach to modelling and reporting that we have tried to exemplify. Bayesian analysis requires a specification of prior distributions and models for the sampling distribution for the data. For prior distributions, we emphasise that there is no such the thing as the "correct" prior, and instead recommend exploring a range of plausible assumptions and conducting sensitivity analysis. Regarding assumptions for the sampling distribution, throughout this book we try to exemplify a reasonably consistent approach to modelling based on an iterative cycle of fitting and checking. We recommend starting with fairly simple assumptions, cross-checking with graphics and informal checks of model fit which can then suggest plausible elaborations. A final list of candidate models can then be compared using more formal methods.

There have been limited "guidelines" for reporting Bayesian analyses, e.g., Spiegelhalter et al. (2004), Sung et al. (2005), and Johnson (2011) in a medical context, and also BaSiS (2001). Naturally the data have to be summarised numerically and graphically. We need to acknowledge that Bayesian methods tend to be inherently more complex than classical analyses, and thus there is an additional need for clarity with the aim that the analysis could be replicated by another investigator who has access to the full data, with perhaps full details of computational methods and code given online.

If "informative" priors are included, then the derivation of the prior from an elicitation process or empirical evidence should be detailed. If the prior assumptions are claimed to be "non-informative," then this claim should be justified and sensitivity analysis given. The idea of "inference robustness" (Box and Tiao, 1973) is crucial: it would be best if competing models with similar evidential support, or alternative prior distributions, gave similar conclusions, but if this is not the case then the alternative conclusions must be clearly reported. Where possible, full posterior distributions should be given for major conclusions, particularly for skewed distributions.

Finally. We would like to thank, and apologise to, our publishers for being so patient with the repeatedly deferred deadlines. Special thanks are extended to Martyn Plummer for his contributions to the book and for keeping us on our toes with his persistent efforts at doing everything better than us. Special thanks also to Simon White for his contribution, and to four reviewers, whose comments were extremely helpful. Thanks also to our friends, colleagues, and families for their support and words of encouragement, such as "Have you not finished that bloody book yet?" Many thanks to the (tens of) thousands of users out there, whose patience, enthusiasm, and sense of humour are all very much appreciated. And finally, we are deeply grateful to all those who have freely contributed their knowledge and insight to the BUGS project over the years. We shall be thinking of you when we get to share out whatever minimal royalties come our way!

All MATLAB® files found in the book are available for download from the publisher's Web site. MATLAB is a registered trademarks of The Mathworks, Inc. For product information please contact:

The Math Works, Inc.
3 Apple Hill Drive
Natick, MA 01760-2098 USA
Tel: 508-647-7000
Fax: 508-647-7001
E-mail: info@mathworks.com
Web: www.mathworks.com

1

Introduction: Probability and parameters

1.1 Probability

The Reverend Thomas Bayes (1702–1761) of Tunbridge Wells started his famous paper (Bayes, 1763) as shown in Figure 1.1: In modern language we

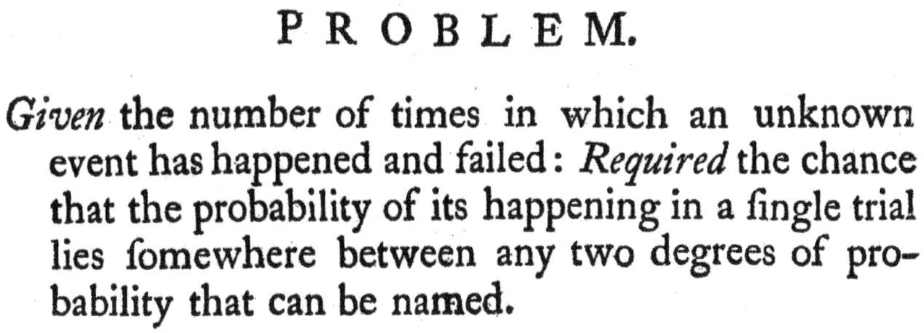

P R O B L E M.

Given the number of times in which an unknown event has happened and failed: *Required* the chance that the probability of its happening in a fingle trial lies fomewhere between any two degrees of probability that can be named.

FIGURE 1.1
Reproduction of part of the original printed version of Bayes (1763): note the font used for an 's' when starting a word.

might translate this into the following problem: suppose a random quantity has a binomial distribution depending on a true underlying 'failure' probability θ, and we observe r failures out of n observations, then what is the chance that θ lies between two specified values, say θ_1 and θ_2? We will return to Bayes' main achievement later, but first we should pay careful attention to his precise use of the terms to describe uncertainty. He uses 'probability' to define the underlying risk of the event occurring (which we have called θ), and this is standard usage for a fixed but currently unknown risk. However, he also describes the uncertainty concerning θ, using the term "chance." This is a vitally important component of his argument (although we shall revisit his use of specific terminology in the next section). Essentially he wants to make a direct numerical expression of uncertainty about an unknown parameter in a probability model: this usage appeared natural to Bayes but is still deeply controversial.

So what do we mean, in general, by "probability"? From a *mathematical* perspective there is no great problem: probabilities of events are numbers between 0 and 1, where 0 represents impossibility and 1 certainty, which obey certain rules of addition (for mutually exclusive events) and multiplication (for conditional events). A "random variable" Y is said to have a probability distribution $p(y)$ when sets of possible realisations y of Y are assigned probabilities, whether Y is discrete or continuous. If the set of possible probability distributions for Y can be limited to a family indexed by a parameter θ, then we may write $p(y|\theta)$ for the distribution, which now depends on some fixed but unknown θ (note that capital Roman letters are generally used for potentially observable quantities, lower case Roman for observed quantities, Greek letters for unobservable parameters).

Using standard statistical techniques we can derive estimates, confidence intervals, and hypothesis tests concerning θ. The particular procedures chosen are justified in terms of their properties when used in repeated similar circumstances. This is known as the "classical" or "frequentist" approach to statistical inference, since it is based on long-run frequency properties of the procedures under (hypothetical) repeated application. See § 3.6 for further discussion on classical procedures.

But Bayes' usage went beyond this. He wanted to express uncertainty about θ, which is not directly observable, as a probability distribution $p(\theta)$. Thus the crucial step taken in Bayesian analysis is to consider θ as a random variable (in principle we should therefore start using capital and lower case Greek letters, but this is not generally done and does not seem to lead to undue confusion). As we shall see later in Chapter 3, when a distribution $p(\theta)$ is directly specified it is known as a "prior" distribution, whereas if it arises as a result of conditioning on some observed data y, it is known as a "posterior" distribution and given the notation $p(\theta|y)$. Of course, parameters of interest may reflect different characteristics depending on the questions being asked: for example, the mean treatment effect in a population, the true variability across individuals, and so on.

Example 1.1.1. *Surgery: direct specification of a prior distribution*
Suppose we are going to start to monitor mortality rates for a high-risk operation in a new hospital. Experience in other hospitals indicates that the risk θ for each patient is expected to be around 10%, and it would be fairly surprising (all else being equal) if it were less than 3% or more than 20%. Figure 1.2 is seen to represent this opinion as a formal probability distribution – in fact this distribution has a specific mathematical form which will be explored in the next section.

We note that we are talking about the underlying risk/long-term rate, and not the actual observed proportion of deaths which would, of course, be subject to additional chance variability. We also note the common habit of referring to the

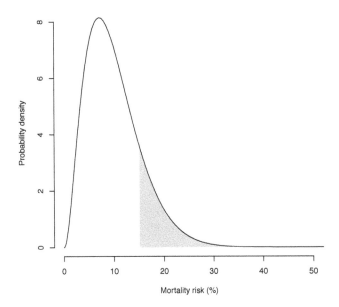

FIGURE 1.2
A prior distribution supporting risks of mortality between 3% and 20%, and expected to be around 10%. The shaded area indicates risks greater than 15%.

underlying risk of mortality both as, say, "10%" and a probability (which must lie between 0 and 1) of "0.1" – we hope the context will make the meaning clear.

For those used to standard statistical analysis, a distribution such as that shown in Figure 1.2 may be doubly suspect: first, it is treating an unknown parameter as a random variable, and, second, the distribution apparently expresses *opinion* rather than being solely based on formal data analysis. In answer to the first concern, what advantages are there to providing a direct probability distribution for such quantities of interest? We might summarise these as follows:

1. The analysis tells us precisely what we want to know: what are the plausible values for the parameter of interest? Presentation of conclusions is therefore intuitive to a general audience.

2. There is no need for p-values or α levels as measures of evidence, as we can directly provide the probability of hypotheses of interest: for example, the shaded tail area shown in Figure 1.2 expresses the probability

that the underlying mortality risk is greater than 15%: we shall see later that this is 0.17.

3. There are no (difficult to interpret) confidence intervals: we can just report that, say, a central range of 3% to 20% has 90% probability.

4. It is generally straightforward to make predictions (see §2.7, §3.2).

5. The process extends naturally to a theory of rational decision making (Berger, 1985; Bernardo and Smith, 1994), although we shall not be concerned with this topic in this book.

6. Importantly, there is a procedure for adapting the distribution in the light of additional evidence: i.e., *Bayes' theorem* allows us to learn from experience and turn a prior distribution into a posterior distribution.

And what about the potential disadvantages of this apparently intuitive approach? These may become more apparent later, but briefly we could list the following:

1. Bayes' theorem tells us how to learn from new evidence, but inevitably requires the specification of what we thought *before* that evidence was taken into account. Specification of one or more such "prior" distributions is an additional responsibility of the analyst.

2. There is an explicit allowance for quantitative subjective judgement in the analysis, which is controversial within a supposedly "objective" scientific setting (although of course one could argue that a standard statistical analysis rests on many assumptions that are not testable given the available data).

3. The analysis may be more complex than a traditional approach.

4. Computation may be more difficult (which is precisely why BUGS has been developed).

5. There are no established standards for Bayesian reporting (although some suggestions have been made; see the discussion in our Preface).

Most of these potential difficulties relate to *accountability*, in the sense of ensuring that the analysis is transparent, reproducible, and not unreasonably influenced by uncheckable assumptions that may not necessarily be generally agreed. These issues are having to be faced by journal editors and other bodies acting as "gatekeepers" for the dissemination of claims made using Bayesian analyses. For example, the Center for Devices and Radiological Health (CDER) of the U.S. Food and Drug Administration (FDA) has published guidelines for the use of Bayesian methods in submissions for the approval of new medical devices (U.S. Department of Health and Human Services, 2010). These guidelines emphasise the need to be explicit about the evidential basis for prior assumptions, conducting sensitivity analysis, providing software for reproducing the analysis, and so on, and we shall repeatedly return to these themes later.

1.2 Probability distributions

A wide range of parametric probability distributions is described in Appendix C, with BUGS notation. But how do we know which distribution to use in a particular context? Choice of the appropriate distribution inevitably depends on knowledge of the specific subject matter and a strong degree of judgement. We can distinguish between four different scenarios:

1. Choice of a fully specified sampling distribution, say $p(y)$: for example, if $Y =$ the number of "heads" when tossing a single fair coin, taking on the values 0 or 1, we may be happy to agree that $Y \sim \text{Bernoulli}(0.5)$ or, equivalently, $Y \sim \text{Binomial}(0.5, 1)$.

2. Choice of the functional form of a parametric sampling distribution for an observation, say $p(y|\theta)$: for example, if $Y =$ annual number of road accidents at a certain location, we may assume that $Y \sim \text{Poisson}(\theta)$ for some annual rate θ.

3. Choice of a fully specified "prior" distribution, say $p(\theta)$: for example, if θ is a proportion, we might assume a uniform distribution between 0 and 1, so that $\theta \sim \text{Uniform}(0, 1)$.

4. Choice of the functional form of a parametric distribution for parameters, say $p(\theta|\mu)$: in Chapter 10 we shall describe how "hierarchical" models include distributions for parameters that themselves contain unknown parameters, for example, when θ has a normal distribution with mean μ and variance ω^2, so that $\theta \sim \text{Normal}(\mu, \omega^2)$.

We shall use standard notation for dealing with probability distributions. For example, a binomial distribution representing the number of events in n trials, each with probability θ of occurring, will be written $\text{Binomial}(\theta, n)$: a variable R could be represented as having such a distribution either by $R|\theta, n \sim \text{Binomial}(\theta, n)$ or $p(r|\theta, n) = \text{Binomial}(\theta, n)$, where r is the realisation of R. $\Pr(A)$ is used to denote the probability of a general event A. Often we drop the explicit conditioning, and thus Bayes' original aim can be expressed formally as follows: if $R \sim \text{Binomial}(\theta, n)$, what is $\Pr(\theta_1 < \theta < \theta_2|r, n)$?

Random variables, whether denoted by Roman or Greek letters, have a range of standard properties arising from probability theory. Here we use notation suitable for continuous quantities, which can be easily translated to discrete quantities by substituting summation for integration.

Consider a generic probability distribution $p(\theta)$ for a single parameter θ. All the usual properties of probability distributions are defined, for example,

Distribution function: $F(\theta^*) = \Pr(\theta < \theta^*)$, sometimes referred to as the "tail area."

Expectation: $E[\theta] = \int \theta p(\theta)\, d\theta$, where the integral is replaced by a summation for discrete θ.

Variance, standard deviation and precision:
$Var[\theta] = \int (\theta - E[\theta])^2 p(\theta)\, d\theta = E[\theta^2] - E^2[\theta]$; standard deviation $= \sqrt{\text{variance}}$; precision $=1/\text{variance}$.

Percentiles: the $100q$th percentile is the value θ_q such that $F(\theta_q) = q$, in particular the median is the 50th percentile $\theta_{0.5}$.

% interval: A subset of values of θ with specified total probability: generally a $100q\%$ interval will be (θ_1, θ_2) such that $F(\theta_2) - F(\theta_1) = q$. Such an interval might be "equi-tailed," in that $F(\theta_2) = 1 - q/2, F(\theta_1) = q/2$, although for asymmetric distributions narrower intervals will be possible. The narrowest interval available is known as the Highest Posterior Density (HPD) interval: see below for an example.

Mode: the value of θ that maximises $p(\theta)$.

These properties extend naturally to multivariate distributions, although percentiles are not generally uniquely defined.

Example 1.2.1. *Surgery (continued): properties of a probability distribution*
The distribution shown in Figure 1.2 is actually a Beta$(3, 27)$, which, from Appendix C.3, we find has probability density proportional to $\theta^2(1 - \theta)^{26}$. From formulae in Appendix C.3 and standard software we can obtain the following properties: mean $= 3/(3+27) = 0.1$, standard deviation 0.054, variance 0.003, median 0.091, mode 0.071. An equi-tailed 90% interval is $(0.03, 0.20)$, which has width 0.17, but a narrower HPD interval is $(0.02, 0.18)$ with width 0.16.

Fitting parametric distributions to expressed subjective judgements will be discussed in Chapter 5.

Bayesian analysis is based on expressing uncertainty about unknown quantities as formal probability distributions. This provides an agreed mathematical framework, but still leaves the possibility for confusion arising from the use of terms such as "chance," "risk," "uncertainty" and so on. Some consistency in terminology may be useful. In Bayes' original aims shown in Figure 1.1, he used "probability" to refer to uncertainty concerning an observable event and "chance" to refer to uncertainty concerning that probability. We shall diverge from this usage: specifically, it seems more natural to use "chance" to refer to "frequentist" or agreed probabilities, say based on physical characteristics of a coin, while retaining the term "probability" for more subjective assessments. Furthermore, if we were being properly pedantic, we might say "*the* chance of this coin coming up heads is 0.5," with the understanding that this was an agreed probability based on physical assumptions about the symmetry of the coin, i.e., *it was a property of the coin itself*, while saying "*my* probability that someone will be killed by falling junk from space in the next 10 years is 0.2," clearly communicating that this is a subjective judgement on my part, perhaps expressing my willingness to bet on the outcome, and is *a property of my relationship with the event*, conditional on all the background evidence available to me, and not solely of the event itself. This essentially subjective interpretation of all probability statements arising in Bayesian analysis will be implicit in all subsequent discussion.

1.3 Calculating properties of probability distributions

Bayesian inference entirely rests on reporting properties of probability distributions for unknown parameters of interest, and therefore efficient calculation of tail areas, expectations, and so on is vital.

Options for calculating these quantities include:

Exact analytic: for example, when tail areas can be calculated exactly using algebraic formulae.

Exact numeric: where, although no closed-form algebraic formula is available, the quantity can be calculated to arbitrary precision, such as tail areas of a normal distribution.

Approximate analytic: for example, using normal approximations to distributions of random variables.

Physical experimentation: for example, by physically repeating an experiment many times to determine the empirical proportion of "successes."

Computer simulation: using appropriate functions of random numbers, generate a large sample of instances of the random variable and empirically estimate the property of interest based on the sample. This technique is popularly known as *Monte Carlo*, and this will be the focus of the methods used in this book.

1.4   Monte Carlo integration

Monte Carlo integration is a widely used technique in many branches of mathematics and engineering and is conceptually very simple. Suppose the random variable X has arbitrary probability distribution $p(x)$ and we have an algorithm for generating a large number of independent realisations $x^{(1)}, x^{(2)}, ..., x^{(T)}$ from this distribution. Then

$$E(X) = \int x p(x)\, dx \approx \frac{1}{T} \sum_{t=1}^{T} x^{(t)}.$$

In other words, the theoretical expectation of X may be approximated by the sample mean of a set of independent realisations drawn from $p(x)$. By the Strong Law of Large Numbers, the approximation becomes arbitrarily exact as $T \to \infty$. Monte Carlo integration extends straightforwardly to the evaluation of more complex integrals. For example, the expectation of any function of X, $g(X)$, can be calculated as

$$E(g(X)) = \int g(x) p(x)\, dx \approx \frac{1}{T} \sum_{t=1}^{T} g(x^{(t)}),$$

that is, the sample mean of the functions of the simulated values. In particular, since the variance of X is simply a function of the expectations of X and X^2, this too may be approximated in a natural way using Monte Carlo integration. Not surprisingly, this estimate turns out to be the sample variance of the realisations $x^{(1)}, x^{(2)}, ..., x^{(T)}$ from $p(x)$.

Another important function of X is the indicator function, $I(l < X < u)$, which takes value 1 if X lies in the interval (l, u) and 0 otherwise. The expectation of $I(l < X < u)$ with respect to $p(x)$ gives the probability that X lies within the specified interval, $\Pr(l < X < u)$, and may be approximated using Monte Carlo integration by taking the sample average of the value of the indicator function for each realisation $x^{(t)}$. It is straightforward to see that this gives

$$\Pr(l < X < u) \approx \frac{\text{number of realisations } x^{(t)} \in (l, u)}{T}. \qquad (1.1)$$

In general, any desired summary of $p(x)$ may be approximated by calculating the corresponding summary of the sampled values generated from $p(x)$, with the approximation becoming increasingly exact as the sample size increases. Hence the theoretical quantiles of $p(x)$ may be estimated using the equivalent empirical quantile in the sample, and the shape of the density $p(x)$ may be approximated by constructing a histogram (or alternatively a "kernel density estimate" which effectively "smooths" the histogram) of the sampled values.

Suppose we obtain an empirical mean $\widehat{E} = \widehat{E}(g(X))$ and variance $\widehat{V} = \widehat{Var}(g(X))$ based on T simulated values, and we consider $\widehat{E}$ as the estimate of interest. Then, since $\widehat{E}$ is a sample mean based on T independent samples, it has true sample variance $Var(g(X))/T$, which may be estimated by $\widehat{V}/T$.

Hence $\widehat{E}$ has an estimated standard error $\sqrt{\widehat{V}/T}$, which is known as the *Monte Carlo error*: see §4.5 for further discussion of this concept. We note that this may be reduced to any required degree of precision by increasing the number of simulated values.

Example 1.4.1. *Coins: a Monte Carlo approach to estimating tail areas*
Suppose we want to know the probability of getting 2 or fewer heads when we toss a fair coin 8 times. In formal terms, if $Y \sim \mathrm{Binomial}(\pi, n)$, $\pi = 0.5, n = 8$, then what is $\Pr(Y \leq 2)$? We can identify four methods:

1. An *exact analytic* approach uses knowledge of the first three terms of the binomial distribution to give

$$
\Pr(Y \leq 2) = \sum_{y=0}^{2} p\,(y|\pi = 0.5, n = 8)
$$

$$
= \binom{8}{0}\left(\frac{1}{2}\right)^{8}\left(\frac{1}{2}\right)^{0} + \binom{8}{1}\left(\frac{1}{2}\right)^{7}\left(\frac{1}{2}\right)^{1} + \binom{8}{2}\left(\frac{1}{2}\right)^{6}\left(\frac{1}{2}\right)^{2}
$$

$$
= 0.1445.
$$

2. An *approximate analytic* approach might use our knowledge that $E[Y] = n\pi = 4$ and $Var[Y] = n\pi(1-\pi) = 2$ to create an approximate distribution $p(y) \approx \mathrm{Normal}(4, 2)$, giving rise to an estimate of $\Pr(Y \leq 2) = \Phi((2 - 4)/\sqrt{2}) = 0.079$, or with a "continuity correction" $\Phi((2.5 - 4)/\sqrt{2}) = 0.144$; the latter is a remarkably good approximation.

3. A *physical* approach would be to repeatedly throw a set of 8 coins and count the proportion of trials where there were 2 or fewer heads. We did this 10 times, observed 0/10 cases of 2 or fewer heads, and then got bored!

4. A *simulation* approach uses a computer to toss the coins! Many programs have random number generators that produce an unstructured stream of numbers between 0 and 1. By checking whether each of these numbers lies above or below 0.5, we can simulate the toss of an individual fair coin, and by repeating in sets of 8 we can simulate the simultaneous toss of 8 coins. Figure 1.3 shows the empirical distributions after 100 and 10,000 trials and compares with the true binomial distribution. It is clear that extending the simulation improves the estimate of the required property of the underlying probability distribution.

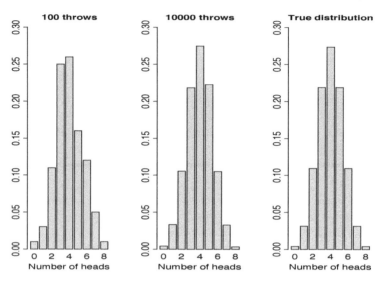

FIGURE 1.3

Distribution of the number of "heads" in trials of 8 tosses, from which we calculate the proportion with 2 or fewer heads: (a) after 100 trials (0.160); (b) after 10,000 trials (0.1450); (c) the true binomial distribution (0.1445).

Suppose we consider an indicator function $P2$ which takes on the value of 1 when there are 2 or fewer heads, 0 otherwise, so that $P2$ is a Bernoulli random quantity with expectation π, which we can calculate to be 0.1445, and true variance $\pi(1 - \pi) = 0.124$. The true Monte Carlo error for an estimate of π based on T simulated values is therefore $\sqrt{\pi(1 - \pi)/T}$, corresponding to the classical standard error of an estimate of π. Our estimates of π after 100 and 10,000 samples are 0.16 and 0.145, respectively, and so we can estimate Monte Carlo errors of 0.037 for $T = 100$ and 0.0035 for $T = 10,000$. If we took a classical statistical perspective we could therefore calculate approximate confidence intervals for π of $0.16 \pm 2 \times 0.037 = (0.09, 0.23)$ after 100 iterations, and $0.145 \pm 2 \times 0.0035 = (0.138, 0.152)$: both comfortably include the true value of 0.1445.

The above results are enormously useful, but to see the real beauty of Monte Carlo integration, suppose now that $\boldsymbol{X}$ is a random *vector* comprising k components, $X_1, ..., X_k$. Further suppose that $\boldsymbol{x}^{(1)}, \boldsymbol{x}^{(2)}, ..., \boldsymbol{x}^{(T)}$ are k-dimensional realisations, with elements denoted $x_j^{(i)}$ ($i = 1, ..., T$, $j = 1, ..., k$), from the joint distribution $p(\boldsymbol{x})$. Then for any $j \in \{1, ..., k\}$, $x_j^{(1)}, x_j^{(2)}, ..., x_j^{(T)}$ represents a sample from $p(x_j)$. In other words, we can make inferences with respect to any marginal distribution by simply using those realisations that pertain to the random variable(s) of interest, and ignoring all others. This result holds for all possible marginal distributions, including those of arbitrary subsets of $\boldsymbol{X}$.

Such marginalisation, for example, integrating out of "nuisance" parameters, is a key component of modern Bayesian inference.

One could argue that the whole development of Bayesian analysis was delayed for decades due to lack of suitable computational tools, which explains why recent availability of high-performance personal computers has led to a revolution in simulation-based Bayesian methods.

2

Monte Carlo simulations using BUGS

2.1 Introduction to BUGS

2.1.1 Background

BUGS stands for Bayesian inference Using Gibbs Sampling, reflecting the basic computational technique originally adopted (see Chapter 4). The BUGS project began in 1989 and from the start was strongly influenced by developments in artificial intelligence in the 1980s. Briefly, these featured an explicit attempt to separate what was known as the "knowledge base," encapsulating what was assumed about the state of the world, from the inference engine "used to draw conclusions" in specific circumstances. The knowledge base naturally makes use of a "declarative" form of programming, in which the structure of our "model" for the world is described using a series of local relationships that can often be conveniently expressed as a graph: see the next section for further discussion of interpretation and computation on graphs. As an essentially separate endeavour, one or more inference engines can be used to compute results on the basis of observations in particular contexts.

This philosophy has been retained within the BUGS project, with a clear separation between the BUGS *language* for specifying Bayesian models and the various programs that might be used for actually carrying out the computations. This book is primarily about the BUGS language and its power to describe almost arbitrarily complex models using a very limited syntax. This language has remained extremely stable over a long period. In contrast, programs to actually run BUGS models are in a state of constant development, and so are only described in the final chapter of this book.

2.1.2 Directed graphical models

The basic idea of a graphical representation is to express the joint relationship between all known and unknown quantities in a model through a series of simple local relationships. Such a decomposition not only allows a simple way of expressing and communicating the essential structure of the model, but also provides the basis for computation. However, this is only possible if substantial assumptions can be made about the qualitative structure of the model, and these assumptions concern *conditional independence*: we shall use the notation $X \perp\!\!\!\perp Y | Z$ to represent the assumption that X is independent of Y, conditional on fixing Z.

Suppose we have a set of quantities $\mathcal{G}$ arranged as a *directed acyclic graph* (DAG), in which each quantity $v \in \mathcal{G}$ is represented as a node in the graph, and arrows run into nodes from their direct influences or *parents*. Formally, such a model represents the assumption that, conditional on its parent nodes pa[v], each node v is independent of all other nodes in the graph except "descendants" of v, where descendant has the obvious definition.

In the context of probability models, these conditional independence assumptions imply that the full joint distribution of all the quantities $\mathcal{G}$ has a simple factorisation in terms of the conditional distribution $p(v|\text{pa}[v])$ of each node given its parents, so that

$$p(\mathcal{G}) = \prod_{v \in \mathcal{G}} p(v|\text{pa}[v]) \tag{2.1}$$

this conditional distribution may be "degenerate," in the sense that the child may be a logical function of its parents. Thus we only need to provide the parent–child relationships in order to fully specify the model: the crucial idea behind BUGS is that this factorisation forms the basis for both the model description and the computational methods (§4.2.2).

Example 2.1.1. *Family: a simple graphical model*
The language of familial relationships is extremely useful when discussing DAGs. For example, consider the graph shown in Figure 2.1, in which A, B, and D are termed "founders," as they have no parents, and A and B are parents of C, which is in turn a parent (with D) of E and F. Considered as random quantities, the conditional independence relationships exactly match those found in simple Mendelian genetics. For example, A, B, and D are marginally independent, E and F are conditionally independent given C and D, and C and D are also marginally independent. However, say we observe E. Then this will induce a dependency between C and D and between A and B, since two nodes without common parents are only independent given no descendants have been observed. Using a genetic analogy, once it is known a child has a particular gene, then the ancestors are no longer probabilistically independent, in the sense that knowing the gene was inherited from a particular ancestor reduces the chance that it came from any other source. From the graph we can see that the joint distribution of the set of quantities may be written

$$p(V) = p(A, B, C, D, E, F) = p(A)p(B)p(C|A, B)p(D)p(E|C, D)p(F|C, D) \tag{2.2}$$

To repeat, the crucial point is that we only need to specify these parent–child conditional relationships in order to express the full joint distribution.

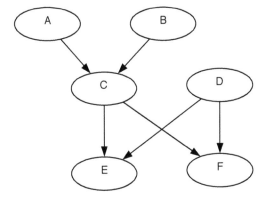

FIGURE 2.1
A typical directed graphical model. Nodes represent variables in the model and arrows show direct dependence between variables. If an arrow emanating from one node points to another node, then the former is said to be a "parent" of the latter; additionally, the latter is said to be a "child" of the former. For example, D is a parent of F; F is a child of D.

In this book we shall use graphs as an aid to communicating qualitative conditional independence structure, emphasising that the graphical representation allows us to reduce "globally" complex models into a set of fairly simple "local" components: furthermore we shall see this graphical structure not only underlies the language used to represent such models, but also directly leads to the computational procedures necessary to draw inferences in the light of any available data.

2.1.3   The BUGS language

The BUGS language comprises syntax for a limited (but extensible — see §12.4.8, §12.5, §12.6.1) list of functions and distributions which allow a series of logical or stochastic local relationships between a node and its parents to be expressed. By "chaining" these relationships together, a full joint distribution over all unknown quantities is expressed using the factorisation in (2.1). The ideas are very similar to a spreadsheet, in which local relationships are arranged into a directed graph so that when "founder" nodes are altered, the implications are propagated through the graph to the "child-less" nodes which form the conclusions. The BUGS language provides a similar representation, except allowing stochastic as well as logical connectives. When combined with a BUGS "engine," inferences can be made on any unknown quantities in the graph conditional on observed data, but instead of a spreadsheet that can only work "down" the graph following the direction of the arrows, BUGS allows you to fix the value of any node in the graph and establish plausible values for all other nodes. Quantities are specified to be constants or data by giving them values in a data file — see Chapter 12.

The BUGS language is "declarative," and so in particular it does not matter in which order the statements come (provided loop constraints are obeyed). This is in contrast to more traditional statistical packages, which tend to use a "procedural" language to execute commands in sequence. This can lead to some perceived difficulties in BUGS model descriptions: for example, there is no `if-then-else` construct.* So we can contrast two different approaches to statistical packages:

Traditional approach:

1. Start with data, in some appropriate format.
2. Apply different statistical techniques to the data using a sequence of commands.
3. Report estimates and intervals, and so on.

Graphical modelling approach:

1. Start with a model describing assumptions concerning the relationships in the world, thus providing a full joint probability model for all quantities, whether parameters or potentially observable data.
2. Offer up to the model whatever relevant data have been observed.
3. Use an appropriate engine to obtain inferences about all unobserved quantities conditional on the observed data.

The BUGS syntax will be introduced through examples, with extensive cross-references to a full listing in the appendices.

2.1.4 Running BUGS models

The currently available software applications for running BUGS models are described in Chapter 12. Each program has the same basic functionality:

1. Checking the syntax of the model specification.
2. Reading in any data provided.
3. "Compiling" the BUGS model, which means constructing an internal representation and working out the sampling methods to be used for each stochastic node.
4. Starting the simulation at an appropriate set of values for the unknown quantities.

---

*But the `step` or `equals` functions can be used to define nodes conditionally on the values of other nodes — see `step` or `equals` in the index for some examples.

5. In response to appropriate commands, simulate unknown quantities.

6. Report summary statistics and other tabular and graphical output.

The analyses in this book have been carried out using the currently most popular engine: WinBUGS 1.4.3. This can be run interactively, performing each of the above six (or more) steps one at a time; alternatively there is a "script" facility to run an entire analysis in batch mode (§12.4.5). Scripts also enable WinBUGS to be called from other software, and interfaces have been developed for a variety of other packages (§12.4.6).

New developments are now made in the OpenBUGS program. This provides a BUGS computation engine with a variety of interfaces, including one which is very similar to WinBUGS. Another program for implementing BUGS models, called JAGS, has been developed entirely independently (Plummer, 2003) and is more portable to different computing platforms. More details about these programs are provided in Chapter 12.

2.1.5 Running WinBUGS for a simple example

The following example illustrates the most basic use of BUGS.

Example 2.1.2. *Coins: running WinBUGS*
The model for Example 1.4.1 is

$$Y \sim \text{Binomial}(0.5, 8)$$

and we want to know $\Pr(Y \leq 2)$. This model is represented in the BUGS language as

```
model {
   Y    ~ dbin(0.5, 8)
   P2 <- step(2.5 - Y) # does Y = 2, 1 or 0?
}
```

P2 is a *step function* that will take on the value 1 if 2.5 − Y is ≥ 0, i.e., if Y is 2 or less, and 0 if Y is 3 or more: this corresponds to the indicator function used in Example 1.4.1.

The following steps are used to run a basic model interactively in WinBUGS (and in the graphical interface to OpenBUGS). This process is explained in more detail for a more complex example in §12.4 and in the WinBUGS user manual accessible from the Help menu.

1. Make a new document (New from the File menu) and type in the BUGS model code, or open a document containing code which has been written already (Open from the File menu).

2. Open Specification Tool from the Model menu. A dialog like the one in Figure 2.2 will appear.

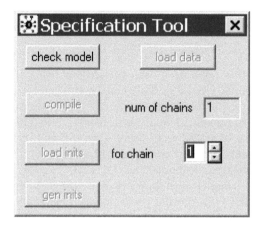

FIGURE 2.2
WinBUGS Model Specification Tool.

3. Highlight the word `model` in the BUGS code by double clicking on the word. Click on `check model`. Any error messages are shown on the bottom left of the screen, or `model is syntactically correct` will appear if there are no errors.

4. There are no observed data in this model; therefore we can ignore `load data`. See Example 3.3.2 for a simple example with observed data.

5. In this example it is sufficient to leave the number of parallel chains to run (`num of chains`) at 1, but see §4.4.2 for an example of where running more than one chain is helpful.

6. Click on `compile`. Again check for any error messages at the bottom left.

7. We can ignore `load inits` in this simple example. See §4.3 for an example which needs initial values to be supplied by the user.

8. Click on `gen inits`. A message `initial values generated, model initialized` should appear.

9. Open `Update...` from the `Model` menu (Figure 2.3) and `Samples...` from the `Inference` menu (Figure 2.4).

10. Specify the nodes we want to *monitor* or record the sampled values for. In this case, type `P2` into the node box in the `Sample Monitor Tool`, and click `set`. Similarly, type `Y` and click `set`.

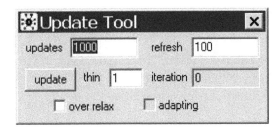

FIGURE 2.3
WinBUGS Update Tool.

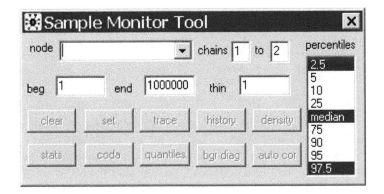

FIGURE 2.4
WinBUGS Sample Monitor Tool.

11. Type * into the node box in the `Sample Monitor Tool`, which means "all monitored nodes," and click `trace` to open a window where the sampled values will appear as they are generated.[†]

12. Go to the `Update Tool` and type the number of samples to be generated in `updates`. 10,000 are sufficient in this example. Click on `Update` to generate the samples. See §12.4.3 for more information about the `Update Tool`.

13. Type * in the `Sample Monitor Tool` again. Click `stats` to see summary statistics for all monitored nodes, and `density` to see plots of their empirical distributions.

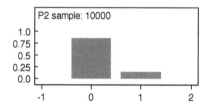 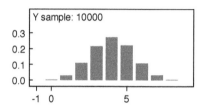

FIGURE 2.5

Empirical distributions for Y and P2 based on 10,000 simulations: output from WinBUGS 1.4.3.

These plots are shown in Figure 2.5. Taking the empirical mean of P2 gives the estimated probability that Y will be 2 or fewer. The summary statistics provided by WinBUGS are

```
node   mean    sd       MC error  2.5%  median  97.5%  start  sample
P2     0.1448  0.3519   0.003317  0.0   0.0     1.0    1      10000
Y      4.004   1.417    0.01291   1.0   4.0     7.0    1      10000
```

The `mean` and `sd` are simply the empirical average and standard deviation of the sampled values while, as described in §1.4, the `MC error` (Monte Carlo error; see §4.5.1) provides an assessment of the sampling error on the mean attributable to the limited number of iterations performed: we note that the MC error calculated for P2 matches that obtained in Example 1.4.1. The 2.5%, median, and 97.5% values are the empirical percentiles, while `start` is the iteration at which monitoring began, and `sample` indicates the total number of iterations contributing to the summary statistics.

[†]Note this will not work in the current version of OpenBUGS, which requires at least one update to have been performed before opening the trace window.

This example illustrates a number of aspects of the BUGS syntax. First, the entire model description is enclosed in `model{...}`. Second, there are two types of connective corresponding to different parent–child relationships:

- `<-` represents logical dependence. The left-hand side of a logical statement comprises a *logical node*, and the right-hand side comprises an expression formed from the logical functions listed in Appendix B applied to a set of stochastic or logical nodes, e.g., `m <- a + b*x`.

- `~` represents stochastic dependence. The left-hand side of a stochastic statement comprises a *stochastic node*, and the right-hand side comprises a distribution from the list in Appendix C, e.g., `r ~ dunif(a,b)` for a variable *r* that is uniformly distributed between *a* and *b*. Note that in WinBUGS and OpenBUGS, logical expressions are not permitted as parameters of distributions, so a statement such as `r ~ dunif(2*a,b)` is not permitted[‡].

- `#` is a *comment* character used to annotate the modelling code. Everything after `#` on the same line is ignored by BUGS. Clear and concise comments can be helpful when reading and maintaining models, particularly if it is not immediately clear what a piece of code does.

In general, each node in a model (apart from constants) should appear once and only once on the left-hand-side of a statement (although see §A.7 for exceptions to this rule).

2.2   DoodleBUGS

WinBUGS (and OpenBUGS) allow models to be specified by drawing a picture of the directed acyclic graph represented by the model. WinBUGS calls this picture a *Doodle*. Nodes in the graph are of three types.

1. *Constants* are fixed by the design of the study: they are always founder nodes (i.e., do not have parents) and are here denoted as rectangles in the graph.

2. *Stochastic nodes* are variables that are given a distribution and are denoted as ellipses in the graph; they may be parents or children (or both). Stochastic nodes may be observed and so be *data*, or may be unobserved and hence be *parameters*, which may be unknown quantities underlying

---

[‡]But this is permitted in JAGS.

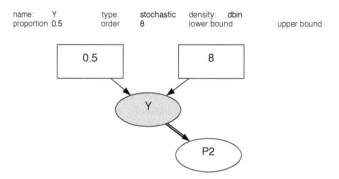

FIGURE 2.6

Doodle for coins example: Y is generated from a binomial distribution with parameters 0.5 and 8 represented by constant nodes, while P2 is a logical indicator function taking on the value 1 if Y is 2 or less, and 0 otherwise. The Y node has been highlighted by the user, whereby the underlying detail is shown above the Doodle: lower and upper bounds can also be specified for prior distributions (see Appendix A.2.2 and § 9.6 for discussion of the appropriate use of bounded distributions).

> a model, censored (partially observed) observations, or simply missing data.

  3. *Deterministic nodes* are logical functions of their parent nodes, again denoted by ellipses but with a double arrow from their parents.

Figure 2.6 shows a Doodle representation of the coins example. If we correctly compose a Doodle, then we can run the program directly from the Doodle or print out the equivalent BUGS code, though we cannot automatically draw a Doodle for a given piece of code. See the WinBUGS or OpenBUGS User Manual for more details on drawing Doodles. Although we feel such graphs are extremely useful for explanation of the assumptions in a complex model, they can be tricky to set up and we would not generally recommend using them to specify complex models.

2.3 Using BUGS to simulate from distributions

We can use BUGS to simulate samples from any of the built-in distributions: a sample of size n can be obtained either as n iterations or as a single iteration of an array of size n.

Example 2.3.1. *Simulating from a Student's t distribution*
Suppose we wanted a sample of size 1000 from a Student's *t* distribution with
mean 10, precision parameter 2 (Appendix C.1), and 4 degrees of freedom. This
could be obtained by either running the code

```
model {
   Y ~ dt(10, 2, 4)
}
```

and saving 1000 iterations (coda from the Sample Monitor Tool, Figure 2.4), or
by running the code

```
for (i in 1:1000) {Y[i] ~ dt(10, 2, 4)}
```

for a single iteration and saving the current state of Y (see, e.g., §12.4.3). Using
the former approach we obtain the following statistics:

node	mean	sd	MC error	2.5%	median	97.5%	start	sample
Y	10.04	0.9893	0.031	8.094	10.04	11.92	1	1000

From Appendix C.1, a t distribution with "precision" parameter r and d degrees
of freedom has variance $d/((d-2)r)$. Hence for $r = 2$ and $d = 4$ the exact
standard deviation is $\sqrt{1} = 1$. The density curve is shown in Figure 2.7, showing
the characteristic heavy tails of the t distribution.

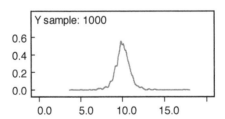

FIGURE 2.7
Kernel density plot of 1000 simulated values from a t distribution with mean 10,
precision parameter 2, and 4 degrees of freedom.

2.4   Transformations of random variables

Suppose we have a distribution $p_\Theta(\theta)$, where for clarity we introduce the Θ subscript to indicate the subject of the distribution. We wish to make inferences about a 1–1 transformation of θ, say $\phi = g(\theta)$ with inverse function $\theta = g^{-1}(\phi)$. If $p_\Theta(\theta)$ is discrete, then we have $p_\Phi(\phi) = p_\Theta(g^{-1}(\phi))$, so that the probability of a particular value of ϕ is obtained simply by making the appropriate transformation back to θ. For example, if $Y \sim \text{Bernoulli}(p)$, then the distribution of $X = 2Y + 1$ is simply a discrete distribution $\Pr(X = 1) = 1 - p, \Pr(X = 3) = p$.

If $p_\Theta(\theta)$ is continuous, then standard results from probability theory show that

$$p_\Phi(\phi) = p_\Theta(g^{-1}(\phi)) \left| \frac{d\theta}{d\phi} \right|,$$

where the final term is called the "Jacobian" and is required when transforming quantities with continuous probability distributions. The difficulty of computing these terms can make inferences on transformations of parameters complex to handle, particularly in multivariate situations.

However, transformations are straightforward when using a simulation approach. If we have a sample $\theta^{(1)}, \ldots, \theta^{(T)}$ from $p_\Theta(\theta)$, then we just need to create the transformed simulated values $\phi^{(1)}, \ldots, \phi^{(T)} = g(\theta^{(1)}), \ldots, g(\theta^{(T)})$ and treat them as a simulated sample from $p_\Phi(\phi)$. This trivial result has strong implications for ease of making inferences on measures, such as ranks, that can be extremely difficult using an exact or approximate analytic approach, whether classical or Bayesian.

Example 2.4.1. *Cube*
Take a standard normal Z with mean 0 and SD 1. Double it, add 1, and cube. What is the distribution of the resulting random quantity Y, and what is Y's expectation and the probability of Y exceeding 10?

We want to find the distribution of $Y = (2Z + 1)^3$ where $Z \sim \text{Normal}(0, 1)$, or equivalently the distribution of $Y = X^3$ where $X \sim \text{Normal}(1, 2^2)$. Analytically, we can show that

$$p(y) = \frac{1}{2\sqrt{\pi}} e^{-\frac{1}{2} \left(\frac{\text{sign}(y)|y|^{1/3} - 1}{2} \right)^2} \frac{1}{6} |y|^{-2/3},$$

which has an infinite mode at $y = 0$. This distribution is plotted in Figure 2.8. To calculate its expectation it is best to return to the original transformation to obtain

$$E[(2Z + 1)^3] = E[8Z^3 + 12Z^2 + 6Z + 1] = 8 \times 0 + 12 \times 1 + 6 \times 0 + 1 = 13,$$

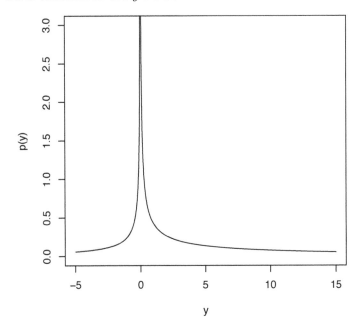

FIGURE 2.8

Exact distribution of $Y = (2Z + 1)^3$ where $Z \sim \text{Normal}(0, 1)$.

since $E[Z^3] = E[Z] = 0$. This expectation may appear surprisingly high, but reflects the remarkably long right-hand tail shown in Figure 2.8. For the tail area above 10, we obtain $\Pr(Y > 10) = \Pr(Z > (10^{1/3} - 1)/2) = \Pr(Z > 0.577) = 0.28$.

In BUGS code, the model is written:

```
Z    ~ dnorm(0, 1)
Y    <- pow(2*Z + 1, 3)
P10 <- step(Y - 10)
```

We note the use of the normal distribution dnorm and the pow function for powers. Note that dnorm is parameterised in terms of mean and *precision* (inverse variance) as opposed to the more conventional mean and variance, although in this example, the precision and variance are equal. Running 100,000 iterations gives estimates for $E[Y]$ of 12.83 and $\Pr(Y > 10)$ of 0.28.

2.5 Complex calculations using Monte Carlo

Since arbitrary functions can be calculated at each iteration, we may be able to use some ingenuity with the BUGS language to solve some otherwise intractable problems.

Example 2.5.1. *Repairs: the "how many" trick*

Suppose costs of a repair have a gamma distribution with mean £100 and standard deviation £50: how many items will I be able to repair for £1000? From Appendix C.2 we can work out that a Gamma$(4, 0.04)$ has mean 100 and sd 50. The "how many trick" is then, at each iteration, to simulate costs Y_i, $i = 1, ..., I$, from this distribution for a sufficiently large I, find the empirical cumulative distribution, and then find the value M which is the highest i such that the total cost does not exceed £1000. We do this by creating a new vector cum.step $= 1, 2, ..., M, 0, 0, ...$, where M is the largest integer such that $\sum_{i=1}^{M} Y_i < 1000$, and using the ranked function to find the maximum of the elements of cum.step.

```
for (i in 1:20) {Y[i] ~ dgamma(4, 0.04)}
cum[1]                  <- Y[1]
for (i in 2:20) {
  cum[i]                <- cum[i - 1] + Y[i]
}
for (i in 1:20) {
  cum.step[i]           <- i*step(1000 - cum[i])
}
number <- ranked(cum.step[], 20)  # maximum number in cum.step
check  <- equals(cum.step[20], 0) # always 1 if I=20 big enough
```

Running 10,000 iterations in WinBUGS produces the following summary statistics; the empirical distributions are shown in Figure 2.9.

node	mean	sd	MC error	2.5%	median	97.5%	start	sample
number	9.631	1.636	0.01503	7.0	10.0	13.0	1	10000
Y[1]	101.1	50.02	0.5408	28.51	93.23	222.2	1	10000

Therefore the median number we can repair is 10, with a 95% predictive interval 7 to 13. Note that each Y has mean 100 and sd 50, as required.

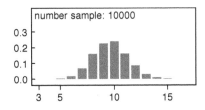
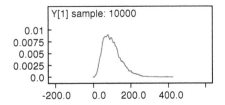

FIGURE 2.9
Distribution of number of items which can be repaired for 1000, given a random item repair cost Y.

2.6 Multivariate Monte Carlo analysis

Our examples up to now have comprised calculating samples of a single random variable. General Monte Carlo analysis extends this in two ways: first, simulating multiple random variables, and second, setting up "chains" of variables so that the parameters of distributions themselves depend on random quantities. It may help to think of this process as "adding uncertainty" to a spreadsheet, and indeed in domains such as risk analysis and health economics such analyses are often conducted using spreadsheets with additional macros or add-on programs that allow cell values to be generated randomly from probability distributions.

The following example illustrates multivariate Monte Carlo analysis.

Example 2.6.1. *Heart transplant cost effectiveness: risks assumed known*
Suppose a patient with heart failure has a survival time s_N, which is assumed exponential with mean $\theta_N = 2$ years, corresponding to a constant monthly mortality risk of 6.25%. A heart transplant has a $\theta_T = 80\%$ operative survival rate, and if a patient survives the operation their survival s_P is exponential with mean $\theta_P = 5$ years. Assume the operation costs £20,000, and each post-operative year costs $\theta_C = £3,000$ in immunosuppressive drugs. What is the expected additional cost per year of life gained by having a transplant?

We assume $s_N \sim$ Exponential(0.5), using the "rate" parameterisation for the exponential distribution given in Appendix C.2. Suppose $o_T = 1$ if the patient survives the transplant operation, $o_T = 0$ otherwise. Then the total survival for a patient receiving a heart transplant is $s_T = o_T s_P$, where $s_P \sim$ Exponential(0.2). The additional survival is $I_s = s_T - s_N$, at additional cost $I_c = 20,000 + 3000 s_T$, so that the additional cost per unit year of life gained is $r = I_c/I_s$. Obtaining the distribution of r analytically is difficult if not impossible. The BUGS code is as follows:

```
sN  ~ dexp(0.5)         # life without transplant (mean 2)
```

```
oT  ~ dbern(0.8)          # survive operation (prob 0.8)
sP  ~ dexp(0.2)           # life if survive transplant (mean 5)
sT <- oT*sP               # total life time if choose transplant
Ic <- 20000 + 3000*sT     # total additional cost of transplant
Is <- sT - sN             # total additional survival
r  <- Ic/Is               # individual cost per additional year
```

1,000,000 iterations provides the following summary statistics:

node	mean	sd	MC error	2.5%	median	97.5%	start	sample
Ic	3.2E+4	14690.0	14.56	2.0E+4	27050.0	71990.0	1	1000000
Is	2.002	5.287	0.005213	-5.68	0.6658	15.64	1	1000000
oT	0.7999	0.4	3.952E-4	0.0	1.0	1.0	1	1000000
r	-5885.0	7.948E+6	7904.0	-184600.0	5278.0	111100.0	1	1000000
sN	1.998	2.001	0.001936	0.05045	1.382	7.386	1	1000000
sP	4.995	4.992	0.005088	0.1261	3.469	18.46	1	1000000
sT	4.0	4.896	0.004855	0.0	2.351	17.33	1	1000000

The predictive distributions for I_s and r are shown in Figure 2.10. We note the

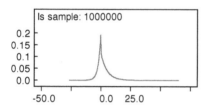
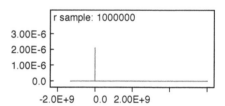

FIGURE 2.10
Empirical distributions from 1,000,000 samples of the incremental survival I_s and
the cost per additional life-year r.

huge standard deviation for the simulated values of r: this occurs because the
survival advantage I_s is often near 0 and leads to an MC error for r that is so
large that we cannot be confident whether r is expected to be positive. Indeed
after 10,000,000 iterations the estimate is -37730 with an MC error 30,780: the
fact that the MC error is not reducing indicates that the distribution for r does
not have a finite variance.

Fortunately this individual-level analysis is not appropriate when deciding on a
public policy of whether to fund a heart transplant programme for such patients.
When making policy decisions we would really like to know the total cost of the
programme compared to the total benefit, which depends on what is known as
the ICER (incremental cost-effectiveness ratio), which is the ratio of the expected
additional cost to the expected additional benefit. In this example this is simply
$E[I_c]/E[I_s]$, rather than $E[r] = E[I_c/I_s]$, which we were trying to estimate

above. In this case

$$E[I_c] = 20,000 + \theta_C E[o_T s_P] = 20,000 + \theta_C E[o_T] E[s_P] = 20,000 + \theta_C \theta_T \theta_P$$

and

$$E[I_s] = E[o_T s_P] - E[s_N] = \theta_T \theta_P - \theta_N,$$

from which we readily obtain that ICER $= (20,000+3000\times0.8\times5)/(0.8\times5-2) = 32,000/2 = 16,000$, with no uncertainty as it is a direct function of assumed parameters of the distribution. The next section will deal with the more interesting situation where there is uncertainty in the parameters. Whether £16,000 per additional year of life is a good investment depends, of course, on the willingness to pay of the healthcare funders.

2.7 Predictions with unknown parameters

Suppose we assume a parametric sampling distribution $p(y|\theta)$ and we are willing to express our uncertainty about the parameter θ as a distribution $p(\theta)$. Then before observing a future quantity Y, we can (in principle) integrate out the unknown parameter to produce a predictive distribution

$$p(y) = \int p(y|\theta) p(\theta) \, d\theta:$$

for discrete parameter distributions this takes the form

$$p(y) = \sum_i p(y|\theta_i) p(\theta_i).$$

Such predictions are useful in, for example, cost-effectiveness models, design of studies, checking whether observed data are compatible with expectations, and so on.

In some cases, such as when Y has a binomial distribution with chance of "success" θ and sample size n, and our uncertainty about θ is expressed in the form of a beta distribution, we can carry out such integration analytically. The reader is referred to Gelman et al. (2004) and Carlin and Louis (2008) for mathematical detail/background, and to Table 3.1 for closed-form expressions for predictive distributions in cases where they are available. In general, however, such analytic integration is not possible. In contrast, to make such predictions in BUGS we can just write

```
theta ~ dbeta(a, b)
Y     ~ dbin(theta, n)
```

and the integration is automatically carried out without requiring any algebraic manipulations.

Example 2.7.1. *Surgery (continued): prediction*
Suppose our hospital in Example 1.1.1 and Example 1.2.1 was going to do 20 operations next year — how many deaths might we expect, and what is the chance there will be at least 6 deaths?

If Y is the number of deaths next year, then since $\theta \sim$ Beta$(3, 27)$ and $Y \sim$ Binomial$(\theta, 20)$, we have from Table 3.1 that Y is *beta-binomial* with mean $0.1 \times 20 = 2$ and standard deviation 1.70. We can also calculate $\Pr(Y \geq 6) = 0.04$. In BUGS code we have:

```
theta  ~ dbeta(3, 27)      # prior distribution
Y      ~ dbin(theta, 20)   # sampling distribution
P6     <- step(Y - 5.5)    # =1 if y >= 6, 0 otherwise
```

We obtain the following WinBUGS output based on 100,000 iterations:

node	mean	sd	MC error	2.5%	median	97.5%	start	sample
P6	0.04058	0.1973	6.578E-4	0.0	0.0	1.0	1	100000
Y	1.998	1.708	0.005216	0.0	2.0	6.0	1	100000

The simulation-based estimates of $E[Y]$ and $\Pr(Y \geq 6)$ are within MC error (see §4.5.1) of the true values.

The underlying process here is actually very straightforward: we simply simulate from the assumed prior distributions for the unknown parameters, and then simulate future events conditional on the current values of the parameters. In contexts such as cost-effectiveness analysis of healthcare interventions this process is termed *probabilistic sensitivity analysis*, where it is necessary to simulate expected outcomes for populations using distributions that depend, say, on uncertain rates of disease progression and treatment effectiveness.

Example 2.7.2. *Heart transplant cost effectiveness (continued)*
In Example 2.6.1 we assumed all the input parameters were known, and we now relax that assumption. First, the operative survival θ_T, previously assumed to be 0.8, is now given a Beta$(8, 2)$ distribution which has mean 0.8, and could be considered equivalent to having observed 8 survivors and 2 deaths in the last 10 operations — see §3.3.1, §5.3.1. Second, the mean survival θ_P following a successful transplant operation, which we had assumed to be fixed at 5 years, is now given a normal distribution with mean 5 and standard deviation 1, corresponding to mean survival being between 3 and 7 years. Finally, the annual cost θ_C of transplant survivors, previously assumed to be £3000, is now given a normal distribution with mean £3000 and standard deviation £1000, representing considerable between-patient variability in drug requirements.

We recall that the crucial quantities of interest are the *expected incremental effectiveness*, which we shall denote E_e, where $E_e = \theta_T\theta_P - \theta_N$, and the *expected incremental cost* denoted E_c, where $E_c = 20,000 + \theta_C\theta_T\theta_P$. The incremental cost-effectiveness ratio is ICER = E_c/E_e. The essential BUGS code and results are shown below. Note that in BUGS the normal distribution is parameterised in terms of mean and precision (inverse variance), as opposed to mean and variance.

```
thetaN <- 2           # expected lifetime without transplant
thetaT  ~ dbeta(8,2)  # probability of surviving operation
thetaP  ~ dnorm(5,1)  # expected survival post-transplant (mean 5, sd 1)
thetaC  ~ dnorm(3000,0.000001)
                      # expected cost per year (mean 3000, sd 1000)
E.c    <- (20000 + thetaC*thetaT*thetaP)/1000
                      # expected additional cost of transplant
                      # in thousands of pounds
E.e    <- thetaT*thetaP - thetaN
                      # expected total additional survival
ICER   <- E.c/E.e     # incremental cost-effectiveness ratio
```

node	mean	sd	MC error	2.5%	median	97.5%	start	sample
E.c	31.98	5.097	0.01578	23.5	31.47	43.35	1	100000
E.e	1.995	1.007	0.003094	0.09311	1.972	4.027	1	100000
ICER	15.05	1113.0	3.702	7.189	15.98	84.68	1	100000

The ICER has a median of about £16,000, 95% interval £7200 to £84,700, and yet has a massive standard deviation. This is because the expected incremental benefit E_e in the denominator of ICER can plausibly be around 0, which creates occasional massive positive or negative values for ICER. Rather than focusing on the ICER alone in such circumstances, it is clearer to carry out a sensitivity analysis to different values of the "willingness to pay," denoted K, for a unit of benefit, which in this case is an expected additional year of life. For fixed K, the *incremental net benefit* (INB) is defined as

$$\text{INB}(K) = KE_e - E_c,$$

and is the expected benefit (in cost terms) for a single patient being given the intervention. Of course $\text{INB}(K)$ is an uncertain quantity which can be calculated and monitored, and of particular interest is the probability that the incremental net benefit is positive, denoted $Q(K) = \Pr(\text{INB}(K) > 0)$. Plotting $Q(K)$ for a range of values of K yields what is known as the *cost-effectiveness acceptability curve* (CEAC).

These quantities are trivial to calculate within the BUGS language. If we wished to conduct a sensitivity analysis for values of K between 0 and 100,000, in steps of 5000, we add the following code:

```
for (i in 1:21) {
  K[i]    <- (i-1)*5
  INB[i] <- E.e*K[i] - E.c
```

```
    Q[i]    <- step(INB[i])
    }
```

We note that this is generic code that can be added to any cost-effectiveness model. In the UK such assessments are carried out by the National Institute for Health and Clinical Effectiveness (NICE), and values of K around £20,000–£30,000 are considered as boundary cases for funding under the National Health Service. Figure 2.11 shows the expected incremental benefits and costs, the incremental net benefit for $K = 30,000$, and the CEAC. The probability of cost effectiveness for $K = 30,000$ is 0.84, so there is fairly convincing evidence that at this threshold the intervention is cost effective.

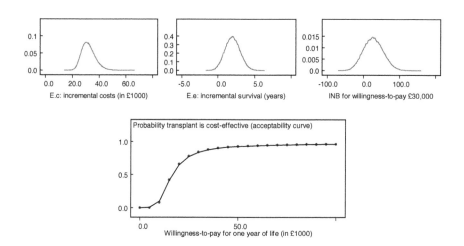

FIGURE 2.11
Expected incremental benefits and costs, the incremental net benefit for $K = 30,000$, and the cost-effectiveness acceptability curve (CEAC) for values of K (willingness to pay for an additional expected year of survival) from 0 to £100,000.

3

Introduction to Bayesian inference

3.1 Bayesian learning

The problem stated in Bayes' famous paper (Figure 1.1) involves two key ingredients. One is the use of probability as a means of expressing uncertainty about an unknown quantity of interest. The other is the conditional nature of the problem: what Bayes was interested in evaluating was the *conditional probability* of failure in a single trial, given some data on the previous number of failures. Put another way, he wanted to *learn* about the failure probability on the basis of observed data. In modern language, this translates to requiring $p(\theta|y, n)$ where θ is the unknown failure probability and we have observed data on y failures out of n binomial trials. Bayes proposed a theorem (easily provable from the axioms of probability) relating conditional and marginal probabilities of random variables which he used to calculate the required conditional probability for his problem.

3.1.1 Bayes' theorem for observable quantities

Bayes' theorem is usually stated in terms of probabilities for observable events. Let A and B be events; then

$$p(A|B) = \frac{p(B|A)p(A)}{p(B)}. \tag{3.1}$$

- $p(A)$ is the *marginal* probability of A, often referred to as the *prior* probability of A — where "prior" indicates "before taking account of the information in B." If the complement (not A) is denoted $\overline{A}$, then $p(\overline{A}) = 1 - p(A)$.

- $p(A|B)$ is the *conditional* probability of A given B, often referred to as the *posterior* probability of A after taking account of the value of B.

- $p(B|A)$ is the *conditional* probability of B given A — we will see later that this corresponds to the *likelihood function* when Bayes' theorem is applied in a statistical modelling context.

- $p(B)$ is the marginal probability of B and acts as a normalising constant to ensure that the value of $p(A|B)$ is a valid probability, i.e., a number between 0 and 1. $p(B)$ may be written as $p(B|A)p(A) + p(B|\overline{A})p(\overline{A})$, a process sometimes known as "extending the conversation."

Example 3.1.1. *Use of Bayes' theorem in diagnostic testing*
This example is taken from Spiegelhalter et al. (2004). Suppose a new HIV test is claimed to have "95% sensitivity and 98% specificity." In a population with an HIV prevalence of $1/1000$, what is the probability that a patient testing positive actually has HIV? We can use Bayes' theorem (3.1) to evaluate this.

Let $A = 1$ if the patient is truly HIV positive and $A = 0$ if they are truly HIV negative. Further, let $B = 1$ if they test positive and $B = 0$ if they test negative. The required probability is then $p(A = 1|B = 1)$. Now, "95% sensitivity" means that $p(B = 1|A = 1) = 0.95$, and "98% specificity" means that $p(B = 1|A = 0) = 1 - 0.98 = 0.02$. Writing $p(B = 1) = p(B = 1|A = 1)p(A = 1) + p(B = 1|A = 0)p(A = 0)$ and applying Bayes' theorem gives

$$p(A = 1|B = 1) = \frac{0.95 \times 0.001}{0.95 \times 0.001 + 0.02 \times 0.999} = 0.045.$$

Thus over 95% of those testing positive will, in fact, *not* have HIV!

This result generally comes as a surprise and illustrates that intuition is often poor when processing probabilistic evidence. The key issue is *how should this test result change our belief that a patient is HIV positive?* The disease prevalence can be thought of as the *prior* probability ($p = 0.001$) of having HIV; observing a positive result causes us to modify or update this to obtain a *posterior probability* of having HIV of $p = 0.045$ — hence the patient is 45 times more likely to have HIV after recording a positive test, but the absolute risk of HIV is still very small.

This result is perhaps better communicated by considering the expected status of a large population, say 100,000 people, in which 100 people are expected to be HIV+, of which 95 will test positive, and 99,900 will be HIV−, of which 1998 (2%) will also (erroneously) test positive. Thus, out of 2093 positive tests, only 95 (4.5%) will be truly HIV+ (Figure 3.1).

3.1.2   Bayesian inference for parameters

Bayes' theorem applied to *observable* random variables (as in the diagnostic testing example) is uncontroversial and established. More controversial is the use of Bayes' theorem in general statistical analysis, where *parameters* are the unknown quantities, and their prior distribution needs to be specified. As discussed in Chapter 1, frequentist and Bayesian interpretations disagree about what sort of things probabilities should be assigned to. In frequentist statistics only the data are assumed to be random variables with associated probability distributions; parameters are fixed but unknown quantities and their associated p-values and confidence intervals are based on long-run frequency

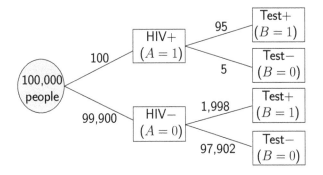

FIGURE 3.1
Bayes' theorem in HIV testing.

properties under repeated sampling of the data. From a Bayesian perspective, both data and parameters can have probability distributions, and so Bayes' theorem can be used to learn about probabilities of unobservable parameters as well as observable events: see §3.6.2 for discussion of the intersection between these viewpoints. Using the same notation as Chapter 1 to distinguish observed and unobservable quantities, Bayes' theorem for inference about parameters can be expressed as

$$p(\theta|y) = \frac{p(y|\theta)p(\theta)}{p(y)}$$

where $p()$ now denotes a probability density rather than a simple probability of an event. The interpretation is analogous to before: $p(\theta)$ is the prior distribution for θ and expresses our uncertainty about the values of θ before taking account of the observed data; $p(\theta|y)$ is the posterior distribution for θ and represents the uncertainty about θ after conditioning on the data y. The conditional distribution $p(y|\theta)$ describes how the data depend on the parameter values. The normalising constant, $p(y)$, simply ensures that $p(\theta|y)$ is a valid probability distribution that integrates to 1. It turns out that it is usually not necessary to calculate $p(y)$ to evaluate properties of the posterior, and so Bayes' theorem in this context is often expressed simply as

$$p(\theta|y) \propto p(y|\theta)\, p(\theta)$$

where the proportionality is considered with relation to θ.

While the form of $p(y|\theta)$ arises from an assumed sampling distribution for the data, it is clear that our only interest in $p(y|\theta)$ is as a function of θ for fixed y. This means that any function of θ, say $L(\theta; y)$, such that $L(\theta; y) \propto p(y|\theta)$, can be used in Bayes' theorem, so that

$$p(\theta|y) \propto L(\theta; y)\, p(\theta).$$

$L(\theta; y)$ is termed the *likelihood* and is the basis for standard likelihood-based statistical models. Hence Bayes' theorem essentially states that

$$\text{posterior} \propto \text{likelihood} \times \text{prior.}$$

In this book we will generally not use the notation $L(\theta; y)$ and indeed will often rather loosely refer to $p(y|\theta)$ as the likelihood, where it is clear that it should be interpreted as a function of θ for fixed y.

3.2   Posterior predictive distributions

In the same way as we are able to make predictions about future quantities based on our prior distribution for some parameter of interest θ (§2.7), we can also make predictions based on the posterior distribution of θ, that is, after learning about θ from observed data y. Denoting our future quantity of interest by $\tilde{Y}$, we derive the posterior-predictive distribution

$$p(\tilde{y}|y) = \int p(\tilde{y}, \theta|y) \, d\theta = \int p(\tilde{y}|\theta, y)p(\theta|y) \, d\theta.$$

Assuming past and future observations are conditionally independent given θ, this simplifies to

$$p(\tilde{y}|y) = \int p(\tilde{y}|\theta)p(\theta|y) \, d\theta. \qquad (3.2)$$

The posterior-predictive expectation is

$$E[\tilde{Y}|y] = \int E[\tilde{Y}|\theta]p(\theta|y) \, d\theta.$$

As we will see in the following section, in some cases, that is, when a particular prior distribution is chosen, such integrals are analytically tractable. This is not the case, in general, however, but we will see that a simulation approach again allows arbitrarily exact approximations to be derived in a straightforward manner.

3.3   Conjugate Bayesian inference

The following are some simple examples of Bayesian inference for continuous-valued parameters. In each case, we use what is known as a *conjugate prior* distribution for the parameter of interest, in order to make the calculations tractable.

3.3.1 Binomial data

Suppose we observe y responses out of n binomial trials. Assuming the trials are independent, with common unknown response probability θ, leads to a binomial sampling distribution

$$p(y|n, \theta) = \binom{n}{y} \theta^y (1 - \theta)^{n-y}.$$

When considered as a function of θ, we obtain a likelihood

$$p(y|n, \theta) \propto \theta^y (1 - \theta)^{n-y}.$$

Suppose that, before taking account of the evidence from our trials, we believe all values for θ are equally likely. This implies a uniform prior distribution for θ

$$\theta \sim \text{Unif}(0, 1).$$

The posterior is then proportional to likelihood $\times$ prior, or

$$p(\theta|y, n) \propto \theta^y (1 - \theta)^{n-y} \times 1.$$

From Appendix C.3, we know that

$$\theta \sim \text{Beta}(a, b) \;\Rightarrow\; p(\theta) = \frac{\Gamma(a + b)}{\Gamma(a)\Gamma(b)} \theta^{a-1} (1 - \theta)^{b-1}$$

and so $p(\theta|y, n)$ has the form of a $\text{Beta}(y + 1, n - y + 1)$ *kernel*.

To represent external evidence that some response rates are more plausible than others, it is mathematically convenient to use a $\text{Beta}(a, b)$ prior distribution for θ. Combining this with the binomial likelihood gives a posterior distribution

$$
\begin{aligned}
p(\theta|y, n) &\propto p(y|\theta, n)\, p(\theta) \\
&\propto \theta^y (1 - \theta)^{n-y} \theta^{a-1} (1 - \theta)^{b-1} \\
&= \theta^{y+a-1} (1 - \theta)^{n-y+b-1} \\
&\propto \text{Beta}(y + a,\, n - y + b).
\end{aligned}
\tag{3.3}
$$

The posterior mean of θ may thus be written as

$$E[\theta|y, n] = (y + a)/(n + a + b) = w \frac{a}{a + b} + (1 - w) \frac{y}{n}$$

where $w = (a + b)/(a + b + n)$: the posterior mean is a weighted average of the prior mean and y/n, the standard maximum-likelihood estimator, where the weight w reflects the relative contribution of the prior "effective sample size" $a + b$. Hence the prior parameters a and b can be interpreted as equivalent to observing a events in $a + b$ trials — see §5.3.1.

Suppose we return to a uniform prior on θ by setting $a = b = 1$ and consider the case $y = n$, i.e., the event has happened at every opportunity! What is the chance it will happen next time? The posterior-predictive expectation is given by the posterior mean for θ:

$$E[\tilde{Y}|y,n] = p(\tilde{Y} = 1|y,n) = \int \theta p(\theta|y,n)\, d\theta = \frac{n+1}{n+2}.$$

This is known as *Laplace's law of succession* and assumes "exchangeable events" (see §3.6.2): i.e., the same (unknown) θ applies to each. Laplace originally applied this to the problem of whether the sun will rise tomorrow. But he recognised that the background knowledge should overwhelm simplistic assumptions. *"But this number [the probability that the sun will rise tomorrow] is far greater for him who, seeing in the totality of phenomena the principle regulating the days and seasons, realises that nothing at the present moment can arrest the course of it."* (Stigler, 1986.)

More generally, with fixed a and b, as y and n increase, $E(\theta|y,n) \to y/n$ and the variance tends to zero. This is a general phenomenon: as n increases, the posterior distribution gets more concentrated and the likelihood dominates the prior.

Example 3.3.1. *Surgery (continued): conjugate analysis*

In Example 1.2.1 we used a Beta(3, 27) as a prior distribution for a mortality rate. Suppose we now operate on $n = 10$ patients and observe $y = 0$ deaths. What is the current posterior distribution, what is the probability that the next patient will survive the operation, and what is the probability that there are 2 or more deaths in the next 20 operations?

Plugging in the relevant values of $a = 3$, $b = 27$, $y = 0$ and $n = 10$ into (3.3) we obtain a posterior distribution for the mortality rate θ of $p(\theta|y,n) = \text{Beta}(3, 37)$. The prior, likelihood, and posterior are shown in Figure 3.2.

The probability of a death at the next operation is simply $E[\theta|y,n] = (y + a)/(n + a + b) = 3/40 = 0.075$. When considering the number $\tilde{Y}$ of deaths in the next 20 operations, from the beta-binomial predictive distribution (Appendix C.5), we can calculate $\Pr(\tilde{Y} \geq 2) = 0.42$.

Example 3.3.1 uses the closed-form solution to the beta-binomial analysis. Alternatively, we could have used simulation methods for inference. In the case of a conjugate model, we could sample directly from the closed-form posterior. However, the whole point of using BUGS is to avoid having to perform derivations of the type illustrated above for the beta-binomial model. All the software requires is specification of the likelihood (or more precisely, the sampling distribution) and prior distribution. From these it can usually derive the posterior in closed form when a closed form is available, or, more generally, it can sample indirectly from the posterior using Markov chain Monte Carlo (MCMC, see Chapter 4). Such specification of sampling distribution

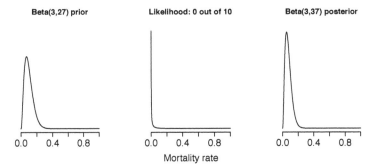

FIGURE 3.2
Prior, likelihood, and posterior distributions for Example 3.3.1.

and prior alone, as opposed to a closed form for the posterior, is illustrated in the following example.

Example 3.3.2. *Surgery (continued): beta-binomial analysis using BUGS*
Assuming we derive the closed form by hand, the BUGS syntax for direct sampling of the surgery mortality rate θ is simply

```
theta ~ dbeta(3, 37)
```

Alternatively, the BUGS syntax for direct sampling of the surgery posterior and predictive distribution for the next 20 patients is shown below. Note that we now need to specify some data along with our model, since we have observed $y = 0$. In this simple example we can just include the statement y <- 0 in the model code. Note also that in such examples we will separate observed data from modelling assumptions by a row of #s.

```
y <- 0
##################################################
theta   ~ dbeta(3, 27)       # prior distribution
y       ~ dbin(theta, 10)    # sampling distribution
Y.pred  ~ dbin(theta, 20)    # predictive distribution
P.crit <- step(Y.pred - 1.5) # =1 if Y.pred >= 2, 0 otherwise
```

Language notes. We note that y appears *twice* on the left-hand side of a statement — once as a logical node and once as a stochastic node. Strictly speaking, this goes against the declarative structure of the model specification, with the accompanying exhortation to construct a directed graph and then to make sure that each node appears once and only once on the left-hand-side of a statement. However, a check has been built in so that, when finding a logical node which also features as a stochastic node (such as y above), a stochastic node is created with the calculated values as fixed data; see §A.7.
In more generic code we could write

```
a <- 3; b <- 27; y <- 0; n <- 10; n.pred <- 20; n.crit <- 2
############################################################
theta    ~ dbeta(a, b)
y        ~ dbin(theta, n)
Y.pred   ~ dbin(theta, n.pred)
P.crit <- step(Y.pred - n.crit + 0.5)
```

Alternatively, the data and prior parameters could be included in a list of data kept separate from the model code:

```
list(a=3, b=27, y=0, n=10, n.pred=20, n.crit=2)
```

Recall the basic steps in Example 2.1.2 for running a model in WinBUGS. In this example, we would now need to load this list of data in Step 4. We previously ignored this step when there were no observed data. To do this,

- highlight the word `list` by double-clicking on it, and click `load data` in the `Specification Tool`.

See §12.4.2 for a full discussion of supplying data to WinBUGS and OpenBUGS and §12.6.3 for the different data format in JAGS.

Estimated posterior distributions for θ and the predicted number of deaths in 20 future operations are shown in Figure 3.3. Posterior summary statistics for the three unknowns are

```
node    mean     sd       MC error 2.5%     median  97.5%   start sample
P.crit 0.4175   0.4931   0.001496 0.0      0.0     1.0     1     100000
Y.pred 1.499    1.427    0.004347 0.0      1.0     5.0     1     100000
theta  0.07514 0.04134 1.322E-4 0.01611 0.06794 0.1739 1     100000
```

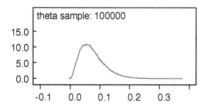
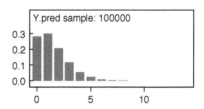

FIGURE 3.3
Posterior and predictive distributions for surgery mortality, calculated by simulation.

In this case, BUGS is able to derive the posterior in closed form and so is sampling `theta` directly from a Beta$(3, 37)$ distribution. Note the empirical mean and standard deviation are within Monte Carlo error of the true values, 0.075 and 0.04113, respectively. The estimated probability of at least two deaths in 20 future operations ($E[\texttt{P.crit}|\texttt{y},\texttt{n}]$) is 0.4175, which also agrees with the analytic result.

A Doodle or directed graph of this model is shown in Figure 3.4: this shows how the observed and future data are assumed conditionally independent given θ.

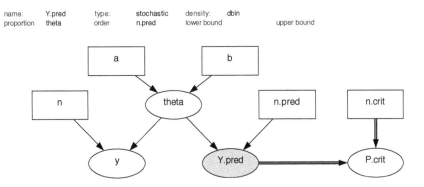

FIGURE 3.4

Graphical representation of model for surgery mortality. The observed number of deaths y is generated from a binomial distribution with probability theta. Information "flows down the arrows" from the prior parameters a and b, and "up the arrow" from the data y, to provide the posterior of the mortality rate theta. The posterior is used to predict the outcome Y.pred of the next n.pred patients, which is conditionally independent of Y given theta.

3.3.2 Normal data with unknown mean, known variance

Suppose we have an independent sample of normal data

$$y_i \sim \text{Normal}(\mu, \sigma^2), \quad i = 1 \ldots n, \tag{3.4}$$

where σ^2 is known and μ is unknown. The conjugate prior for the normal mean μ is also normal:

$$\mu \sim \text{Normal}(\gamma, \omega^2)$$

where γ and ω^2 are assumed specified. It is convenient to write ω^2 as σ^2/n_0, where n_0 represents the "effective number of observations" in the prior distribution. Then the posterior distribution for μ is given by

$$p(\mu|y) \propto p(\mu) \prod_{i=1}^{n} p(y_i|\mu)$$

$$\propto \exp\left[-\frac{1}{2}\left\{\frac{(\mu-\gamma)^2}{\sigma^2/n_0}\right\}\right]\exp\left[-\frac{1}{2}\left\{\frac{\sum(y_i-\mu)^2}{\sigma^2}\right\}\right]. \qquad (3.5)$$

Note that any terms in the normal sampling distribution or prior that do not depend on μ can be ignored as they are simply absorbed into the proportionality constant. By expanding the quadratics, collecting powers of μ together, and then completing the square, it is straightforward to show that (3.5) has the form of another normal density and so we can write the posterior for μ as

$$p(\mu|y) = \text{Normal}(\gamma_n, \omega_n^2), \quad \gamma_n = \frac{n_0\gamma + n\bar{y}}{n_0 + n}, \quad \omega_n^2 = \frac{\sigma^2}{n_0 + n}. \qquad (3.6)$$

There are three other equivalent expressions for the posterior mean:

$$\gamma_n = w\gamma + (1-w)\bar{y}, \quad w = \frac{n_0}{n_0 + n}; \qquad (3.7)$$

$$\gamma_n = \gamma + (\bar{y} - \gamma)\frac{n}{n_0 + n}; \qquad (3.8)$$

$$\gamma_n = \bar{y} - (\bar{y} - \gamma)\frac{n_0}{n_0 + n}. \qquad (3.9)$$

Expression (3.7) shows that the posterior mean is a weighted average of the prior mean and the sample mean, (3.8) emphasises the interpretation of the posterior mean as the prior mean adjusted towards the data mean, while (3.9) shows the data mean being "shrunk" towards the prior mean. It is also clear from the symmetry of the observed sample size n and the prior constant n_0 in these expressions that n_0 can be interpreted as a "prior sample size" — that is, the information content of the prior is equivalent to having observed an additional n_0 "data" points. All three expressions highlight the compromise between the prior and data means, with weights proportional to their relative "sample sizes" or precisions.

The posterior variance is best interpreted on the inverse (i.e., precision) scale, by re-writing (3.6) as $1/\omega_n^2 = n_0/\sigma^2 + n/\sigma^2$. This shows that the posterior precision is the sum of the prior precision and the data precision. Alternatively, we can write $\omega_n^2 = \frac{\sigma^2}{n}(1-w)$ where $(1-w) = n/(n_0 + n)$, emphasising that the posterior variance is the data variance shrunk by a factor proportional to the relative sample size of the data as a fraction of the total effective "sample size" (or precision) of the data plus prior.

We may consider a future observation $\hat{Y}$ as being equal to the sum of two independent normal quantities, $\epsilon \sim \text{Normal}(0, \sigma^2)$ and $\mu|y \sim \text{Normal}(\gamma_n, \omega_n^2)$ and hence the posterior predictive distribution is

$$p(\tilde{y}|y) = \text{Normal}\left(\gamma_n, \sigma^2 + \omega_n^2\right). \qquad (3.10)$$

So the predictive distribution is centered at the posterior mean of μ with variance equal to the sum of the posterior variance and the data (residual) variance.

Example 3.3.3. *Trihalomethanes in tap water*

Regional water companies in the UK are required to take routine measurements of trihalomethane (THM) concentrations in tap water samples for regulatory purposes. Samples are tested throughout the year in each water supply zone and analysed using an assay with known measurement error having standard deviation $\sigma = 5\,\mu g/L$. Suppose we want to estimate the average THM concentration in a particular water zone. Two independent measurements are taken, with values $y_1 = 128\,\mu g/L$ and $y_2 = 132\,\mu g/L$; hence their mean, $\bar{y}$, is $130\,\mu g/L$. What is the true mean THM concentration in this water zone?

Denote the true mean THM concentration by μ. A standard analysis would use the sample mean $\bar{y} = 130\,\mu g/L$ as an estimate of μ, with standard error $\sigma/\sqrt{n} = 5/\sqrt{2} = 3.5\,\mu g/L$. A 95% confidence interval is then $\bar{y} \pm 1.96 \times \sigma/\sqrt{n}$, i.e., 123.1 to 136.9 $\mu g/L$.

Suppose historical data on THM levels in other zones supplied from the same water source showed that the mean THM concentration was $120\,\mu g/L$ with standard deviation $10\,\mu g/L$. This suggests a Normal$(120, 10^2)$ prior for μ. If we express the prior standard deviation as $\sigma/\sqrt{n_0}$, we can solve to find $n_0 = (\sigma/10)^2 = 0.25$ (hence the information content of this prior is equivalent to one quarter of an observation). Our prior can thus be written as $\mu \sim$ Normal$(120, \sigma^2/0.25)$.

Substituting the relevant values above into (3.6), the posterior for μ is then

$$p(\mu|y) = \text{Normal}\left(\frac{0.25 \times 120 + 2 \times 130}{0.25 + 2},\ \frac{5^2}{0.25 + 2}\right)$$

$$= \text{Normal}(128.9, 3.33^2),$$

giving a 95% credible interval for μ of 122.4 to 135.4 $\mu g/L$. The prior, likelihood, and posterior are shown in Figure 3.5. Note how the informative prior distribution "pulls" the posterior to the left (away from the likelihood). The effect is only small, however, because the likelihood contains eight times $(2/0.25)$ as much information as the prior. Note also that the posterior is narrower than both likelihood and prior, due to the combination of both sources of evidence.

Suppose the water company will be fined if observed THM levels in the water supply exceed $145\,\mu g/L$. From (3.10) the predictive distribution for the THM concentration in a future sample taken from the water zone is Normal$(128.9, 3.33^2 + 5^2) = $ Normal$(128.9, 36.1)$ (see Figure 3.6). Hence the probability that the THM concentration in a future sample exceeds $145\,\mu g/L$ is $1 - \Phi[(145 - 128.9)/\sqrt{36.1}] = 0.0037$, which is very low.

BUGS code (and data) for the above analysis is as follows, and posterior summaries for the unknowns are presented in the table beneath. (Remember that in BUGS the normal distribution is parameterised in terms of mean and *precision* = $1/\text{variance}$.) The results are very close to the theoretical values derived above.

```
for (i in 1:n) {
    y[i]                 ~ dnorm(mu, inv.sigma.squared)
```

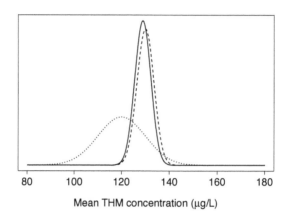

FIGURE 3.5
Likelihood (−−) for Example 3.3.3 with prior (···) and posterior (—) distributions.

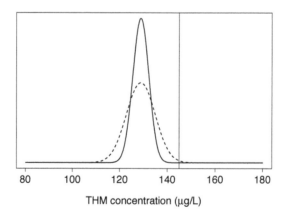

FIGURE 3.6
Posterior (—) and predictive (−−) distributions for Example 3.3.3. The vertical line represents the concentration at which water companies are fined.

```
}
mu                      ~ dnorm(gamma, inv.omega.squared)
inv.omega.squared <- n0/sigma.squared
inv.sigma.squared <- 1/sigma.squared
y.pred                  ~ dnorm(mu, inv.sigma.squared)
P.crit                  <- step(y.pred - y.crit)

list(n=2, y=c(128, 132), gamma=120, n0=0.25,
      sigma.squared=25, y.crit=145)
```

node	mean	sd	MC error	2.5%	median	97.5%	start	sample
P.crit	0.00363	0.06014	1.856E-4	0.0	0.0	0.0	1001	100000
mu	128.9	3.328	0.01017	122.3	128.9	135.4	1001	100000
y.pred	128.9	5.999	0.01936	117.1	128.9	140.6	1001	100000

More generally, a *conjugate* prior is one that is "compatible" with the likelihood, in the sense that they share the same functional form when the likelihood is viewed as a function of the parameter of interest; the posterior is then also of the same form, and hence has the same (closed) distributional form as the prior but with modified parameters. Table 3.1 shows examples of conjugacy in cases where the likelihood is a function of one continuous-valued parameter. The corresponding predictive distributions for future observations are also given. These are obtained from (3.2) by noting that since the posterior has the same form as the prior, and $p(\tilde{y}|\theta)$ has the same form as the likelihood, then the integrand must be proportional to a closed-form distribution (of the same distributional form as the prior and posterior). The predictive distribution is thus given by the proportionality constant.

3.4   Inference about a discrete parameter

In cases where the prior distribution has support on a finite set of discrete values, the posterior is derived trivially, by multiplying prior and likelihood for each possible value of the parameter of interest, and by then normalising each such product by their sum, as illustrated in the example below.

Example 3.4.1. *Three coins*
Suppose I have 3 coins in my pocket. The coins may be either fair, biased 3:1 in favour of heads, or biased 3:1 in favour of tails, but I do not know how many of each type there are among the 3 coins. I randomly select 1 coin and toss it once, observing a head. What is the posterior distribution of the probability of a head?

TABLE 3.1

Univariate conjugate prior distributions for various one-parameter likelihoods from a sample of size n. Also given are the corresponding posterior parameters and the predictive distribution for a single new observation $\tilde{y}^\dagger$. See Appendix C and/or Bernardo and Smith (1994), pp. 427–435, for definitions of distributions.

Sampling distribution	Conjugate prior	Posterior parameters	Predictive distribution
$y\|\theta \sim \text{Binomial}(\theta, n)$ including Bernoulli ($n = 1$)	$\theta \sim \text{Beta}(a, b)$	$a_n = a + y,$ $b_n = b + n - y$	Beta-Binomial(a_n, b_n, n)
$y\|\mu \sim \prod_{i=1}^n \text{Normal}(\mu, \sigma^2)$	$\mu \sim \text{Normal}(\gamma, \omega^2 = \frac{\sigma^2}{n_0})$	$\gamma_n = \frac{n_0\gamma + n\bar{y}}{n_0 + n},$ $\omega_n^2 = \frac{\sigma^2}{n_0 + n}$	Normal$(\gamma_n, \omega_n^2 + \sigma^2)^\ddagger$
$y\|\sigma^2 \sim \prod_{i=1}^n \text{Normal}(\mu, \sigma^2)$	$\sigma^{-2} \sim \text{Gamma}(a, b)$	$a_n = a + \frac{n}{2},$ $b_n = b + \frac{1}{2}\sum_i (y_i - \mu)^2$	Student-$t(\mu, \frac{b_n}{a_n}, 2a_n)^\S$
$y\|\theta \sim \prod_{i=1}^n \text{Poisson}(\theta)$	$\theta \sim \text{Gamma}(a, b)$	$a_n = a + n\bar{y},$ $b_n = b + n$	NegBin$(\frac{b_n}{b_n+1}, a_n)$
$y\|\theta \sim \prod_{i=1}^n \text{Gamma}(\alpha, \theta)$ including Exponential ($\alpha = 1$)	$\theta \sim \text{Gamma}(a, b)$	$a_n = a + n\alpha,$ $b_n = b + n\bar{y}$	Gamma-Gamma(a_n, b_n, α)
$y\|\theta \sim \prod_{i=1}^n \text{Uniform}(0, \theta)$	$\theta \sim \text{Pareto}(a, b)$	$a_n = a + n,$ $b_n = \max\{b, y\}$	$\begin{cases} \frac{a_n}{a_n+1}\text{Uniform}(0, b_n), & \tilde{y} \le b_n \\ \frac{1}{a_n+1}\text{Pareto}(a_n, b_n), & \tilde{y} > b_n \end{cases}$
$y\|\theta \sim \text{NegBin}(\theta, r)$ including Geometric ($r = 1$)	$\theta \sim \text{Beta}(a, b)$	$a_n = a + r,$ $b_n = b + y$	Negative-Binomial-Beta(a_n, b_n, r_p)

TABLE 3.1
(Continued.)

Sampling distribution	Conjugate prior	Posterior parameters	Predictive distribution
$y\|\theta \sim \prod_{i=1}^n \text{Pareto}(\theta, c)$	$\theta \sim \text{Gamma}(a, b)$	$a_n = a + n,$ $b_n = b + \sum_{i=1}^n \log\left(\frac{y_i}{c}\right)$	$\frac{\Gamma(a_n+1)}{\Gamma(a_n)} \frac{1}{b_n \tilde{y}} \left[1 + \frac{1}{b_n}\log\left(\frac{\tilde{y}}{c}\right)\right]^{-(a_n+1)}$
$y\|\theta \sim \prod_{i=1}^n \text{Pareto}(\alpha, \theta)$	$\theta \sim \text{Pareto}(a, b)$	$a_n = a - n\alpha,$ $b_n = b,$ truncated to $(b, u = \min\{y\})$	$\frac{a_n\alpha}{a_{n+1}\left(b^{-a_n} - u^{-a_n}\right)}$ $\times \begin{cases} [b^{-a_{n+1}} - \tilde{y}^{-a_{n+1}}]\tilde{y}^{-(\alpha+1)}, & \tilde{y} < u \\ b^{-a_{n+1}} - u^{-a_{n+1}}, & \tilde{y} \geq u \end{cases}$ $a_{n+1} = a_n - \alpha$

† Actually in the cases of binomial and negative-binomial likelihoods, the predictive distributions given are for the number of successes in m future Bernoulli trials and for the number of Bernoulli failures before r_p successes, respectively (where the success probability is θ in both cases).

‡ Here, as is conventional, we parameterise the normal distribution in terms of its mean and variance. In BUGS, however, it is parameterised in terms of the mean and precision (inverse-variance) — see Appendix C.

§ Note the t distribution is parameterised in terms of mean, inverse-scale-squared, and degrees of freedom, respectively — see Appendix C.1.

Letting $y = 1$ denote the event that I observe a head and θ denote the probability of a head, we have that $\theta \in (0.25, 0.5, 0.75)$. Note that our goal here is to make an inference about the posterior distribution of θ itself (an unobservable parameter), *not* about whether we will get a head on the next throw (an observable event). Given that I select the coin at random, a reasonable prior distribution for θ is to assume

$$p(\theta = 0.25) = p(\theta = 0.5) = p(\theta = 0.75) = 0.33.$$

The sampling distribution or likelihood of the data can be represented by a Bernoulli distribution,

$$p(y|\theta) = \theta^y (1 - \theta)^{(1-y)}.$$

The resulting posterior distribution for θ is shown in the following table.

θ	Prior $p(\theta)$	Likelihood $p(y = 1|\theta)$	Un-normalised posterior $p(y = 1|\theta)p(\theta)$	Normalised posterior $\frac{p(y=1|\theta)p(\theta)}{p(y=1)†}$
0.25	0.33	0.25	0.0825	0.167
0.50	0.33	0.50	0.1650	0.333
0.75	0.33	0.75	0.2475	0.500
Sum	1.00	1.50	0.495	1.000

† The normalising constant can be calculated as $p(y = 1) = \sum_i p(y = 1|\theta_i)p(\theta_i) = 0.495$.

So observing a head on a single toss of the coin means that there is now a 50% probability that the chance of heads is 0.75 and only a 16.7% probability that the chance of heads in 0.25.

The following code shows how such a model may be implemented in BUGS.

```
y <- 1
####################################
y                ~ dbern(theta.true)
theta.true      <- theta[coin]
coin             ~ dcat(p[])
for(i in 1:3) {
  p[i]            <- 1/3
  theta[i]        <- 0.25*i
  coin.prob[i] <- equals(coin, i)
}
```

Language note: coin has a categorical distribution taking on values 1,2,3, and equals() is used to identify the individual probabilities in that distribution. This is sometimes called the pick trick for choosing a random element of a vector — see also §5.4 and pick in the index for other examples. The loop-index i can be used in the calculations: the term theta.true is not strictly necessary, as the nested index y ~ dbern(theta[coin]) could be used instead.

node	mean	sd	MC error	2.5%	median	97.5%	start	sample
coin.prob[1]	0.1662	0.3723	0.001141	0.0	0.0	1.0	1	100000
coin.prob[2]	0.3342	0.4717	0.001435	0.0	0.0	1.0	1	100000
coin.prob[3]	0.4997	0.5	0.001491	0.0	0.0	1.0	1	100000

Suppose we want to predict the probability that the next toss is a head. Now, algebraically

$$\Pr(\tilde{Y} = 1|y) = \sum_i \Pr(\tilde{Y} = 1|\theta_i)p(\theta_i|y)$$

$$= (0.25 \times 0.167) + (0.50 \times 0.333) + (0.75 \times 0.500) = 7/12.$$

In BUGS a generic method is to predict a new observation by adding the line

```
Y.pred ~ dbern(theta.true)
```

but of course in this case, we could just monitor theta.true directly, since it represents the required probability.

node	mean	sd	MC error	2.5%	median	97.5%	start	sample
Y.pred	0.5832	0.493	0.001611	0.0	1.0	1.0	1	100000
theta.true	0.5834	0.186	5.687E-4	0.25	0.5	0.75	1	100000

The same results are obtained by each method, but the Monte Carlo error is far smaller with the "direct" method, since we avoid the additional error in sampling an actual predictive observation.

3.5 Combinations of conjugate analyses

In many situations, more than one source of data is required to learn about some quantity (also see §11.4). We might be able to obtain exact posteriors for the parameter underlying each dataset, using independent conjugate analyses, but Monte Carlo simulation may still be required to combine the evidence from the different datasets.

Example 3.5.1. *Heart transplants: learning from data*
In Example 2.6.1, the expected survival of patients with heart failure undergoing heart transplantation was estimated using Monte Carlo simulation based on a fixed operative mortality rate (80%) and a given prior for the post-transplant survival rate (exponential with mean 5 years). We extend this example here to learn about these parameters from data.

Suppose that 10 patients in a particular centre received a heart transplant, and 8 of these survived the operation. These patients were followed up for the rest

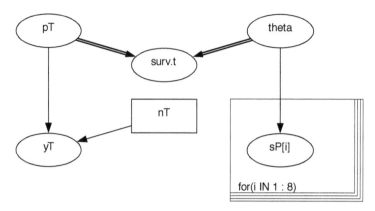

FIGURE 3.7
Graphical model for heart transplant survival. Evidence from the operative mortality yT and the long-term survival sP is combined to estimate the overall expected post-transplant survival surv.t.

of their lives, and survived for 2,3,4,4,6,7,10 and 12 years, respectively. Based on these data, we predict the expected lifetime for a similar patient about to undergo a transplant.

A binomial model with a conjugate uniform prior is used to estimate the probability of survival p_T during the operation. The post-transplant survival data are modelled as exponential with rate θ (mean $1/\theta$). A conjugate gamma prior (Table 3.1) is used for θ with parameters $a = b = 0.001$, which is vague relative to the data (see §5.2.6). Although the posteriors for p_T and θ are available in closed form as Beta(9,3) and Gamma(8.001,47.001) distributions, Monte Carlo integration is required to calculate the posterior of the *total* expected survival p_T/θ. This is then compared with the expected lifetime *without* transplant (assumed to be 2 years) to estimate the expected survival benefit from transplantation, labelled Is in the BUGS code. The graphical model (Figure 3.7) illustrates how the expected survival with transplant surv.t is inferred from two sources of data yT and sP.

```
yT        ~ dbin(pT, nT)
pT        ~ dunif(0, 1)
for (i in 1:8) {
  sP[i] ~ dexp(theta)
}
theta     ~ dgamma(0.001, 0.001)
surv.t <- pT/theta    # expected survival with transplant
Is     <- surv.t - 2
```

The data are supplied in a list.

```
list(yT=8, nT=10, sP=c(2,3,4,4,6,7,10,12))
```

An initial value is also provided for θ^*.

```
list(theta=1)
```

The posterior mean survival benefit is about 3 years, but with a wide posterior credible interval of about 0.2 to 9 years.

node	mean	sd	MC error	2.5%	median	97.5%	start	sample
Is	3.136	2.271	0.003046	0.2014	2.669	8.836	1001	500000
pT	0.7495	0.1202	1.702E-4	0.482	0.7639	0.9396	1001	500000
surv.t	5.136	2.271	0.003046	2.201	4.669	10.84	1001	500000

This analysis could easily be extended to include estimating the cost and cost effectiveness of transplantation, as in Example 2.6.1. See §11.4 for more complex examples of combining data from different sources in BUGS.

3.6 Bayesian and classical methods

There are three broad approaches to statistical inference.

1. The *Fisherian* approach is perhaps most prominent in current practice: based largely on the work of R. A. Fisher, the basic idea is to use the likelihood function as a basis for point and interval estimation, and *p*-values as a measure of the discrepancy of data with a claimed hypothesis.

2. The *Neyman–Pearson* philosophy of "inductive behaviour," originally proposed by Jerzy Neyman and Egon Pearson in the 1930s, is rooted in decision making and the error rates in choosing between null and alternative hypotheses H_0 and H_1, and procedures for estimation and testing are selected on the basis of their long-run properties.

3. Finally, the Bayesian approach uses the likelihood supplemented by a prior distribution to produce a posterior probability distribution for parameters of interest, which may be combined with a loss function if a formal decision is desired (Berger, 1985; Bernardo and Smith, 1994), though we do not cover decision theory in this book.

---

*See Section 4.3. This should not really be necessary, but without it WinBUGS (or current OpenBUGS) will automatically generate an extreme initial value from the flat gamma prior, giving numerical overflow and an error message. Ideally WinBUGS would sample from the tighter conjugate posterior, or use the prior mean, median, or mode as JAGS does.

For further comparison of inferential methods see, for example, Cox and Hinkley (1974). While the Fisherian and Neyman–Pearson approaches are generally considered as "classical" or "frequentist" methods and contrasted to Bayesian analysis, the fairly informal discussion below suggests there are perhaps more similarities between the Fisherian and Bayesian approaches to estimation than may at first be apparent. In contrast, there can be strong differences in approaches to model selection, as discussed in Chapter 8.

3.6.1 Likelihood-based inference

If we assume a set of independent and identically distributed observations $y = \{y_1, \ldots, y_n\}$ from a sampling model $p(y_i|\theta)$, $i = 1, \ldots, n$, with scalar θ, then the likelihood (as introduced in §3.1.2) is any function of θ that is proportional to $p(y|\theta) = \prod_i p(y_i|\theta)$. We shall denote such a function $L(\theta; y)$. The maximum likelihood estimate is the value $\hat{\theta}$ which maximises $L(\theta; y)$, or equivalently maximises the log-likelihood denoted $\ell(\theta; y)$.

Let

$$I(\theta) = -E_{Y|\theta}\left[\frac{d^2 \log p(Y|\theta)}{d\theta^2}\right] = E_{Y|\theta}\left[\left(\frac{d \log p(Y|\theta)}{d\theta}\right)^2\right]$$

be the "Fisher Information" contained in a single observation Y. Then under broad generality conditions the maximum likelihood estimator has an asymptotic normal distribution

$$\hat{\theta} \sim \mathrm{N}(\theta, (n\hat{I}(\hat{\theta}))^{-1}) \tag{3.11}$$

where $\hat{I}()$ is a sample-based estimate of $I()$. Thus the maximum likelihood estimator will converge to the true value of the parameter assuming the sampling model has been appropriately chosen. Similar results hold for multivariate θ: there are various procedures for dealing with nuisance parameters ψ, such as creating a "profile likelihood" $L(\theta, \hat{\psi}|\theta; y)$ for θ based on the conditional maximum likelihood estimates $\hat{\psi}|\theta$.

3.6.2 Exchangeability

"Exchangeability" is to Bayesian inference what "independently and identically distributed" is to classical inference. It is a formal expression of the idea that we find no systematic reason to distinguish individual variables: informally it is a *judgement* that they are "similar" but not identical. More formally, we judge that $Y_1, \ldots, Y_n$ are finitely exchangeable if the probability that we assign to any set of potential outcomes $p(y_1, \ldots, y_n)$ is unaffected by permutations of the labels attached to the variables, so that under any permutation $\pi(i)$, $i = 1, \ldots, n$, we would assume that $p(y_1, \ldots, y_n) = p(y_{\pi(1)}, \ldots, y_{\pi(n)})$.

For example, suppose Y_1, Y_2, Y_3 are the first three flips of a (possibly biased) coin, where $Y_1 = 1$ indicates a head, and $Y_1 = 0$ indicates a tail. We might judge $p(Y_1 = 1, Y_2 = 0, Y_3 = 1) = p(Y_2 = 1, Y_1 = 0, Y_3 = 1) = p(Y_1 = 1, Y_3 = 0, Y_2 = 1)$: i.e., the probability of getting 2 heads and a tail is unaffected by the particular flip on which the tail comes. This is a fairly strong assumption, but a natural judgement to make if we have no reason to think that one flip is systematically any different from another. Note that it does *not* mean we believe that $Y_1, \ldots, Y_n$ are independent: this would not allow us to learn about the chance of a head.

de Finetti (1931) proved a remarkable "representation theorem" — that if every finite sequence of an infinite sequence of binary variables $Y_1, \ldots, Y_n, \ldots$ is judged finitely exchangeable, then it implies that the joint density for any finite set can be written in the form

$$p(y_1, \ldots, y_n) = \int \prod_{i=1}^{n} p(y_i|\theta) p(\theta) d\theta$$

for some density $p(\theta)$ (assuming regularity conditions so that the density exists and is continuous).

It is easy to argue from "right to left" in this equation, since this is a standard expression for conditional and marginal probability. But the "left to right" identity is not at all obvious and has very powerful implications: when extended to a more general version, it says that exchangeable random quantities can be thought of as being *independently and identically distributed* and drawn from some common *parametric distribution* depending on an unknown parameter θ, which itself has a *prior distribution* $p(\theta)$. Thus, from a subjective judgement about the exchangeability of observable quantities, the whole apparatus of parametric models and Bayesian statistics is derived rather than assumed.

We will see in Chapter 10 how we can use these ideas to develop hierarchical models.

3.6.3 Long-run properties of Bayesian methods

The long-run properties of Bayesian methods provide an attractive link to more familiar procedures. Asymptotically (under broad regularity conditions), as the sample size increases, the influence of the prior distribution decreases and the posterior distribution tends to a form leading to numerically identical (although conceptually distinct) inferences as those obtained from a likelihood perspective. We emphasise that this is based on asymptotics in which $n \to \infty$, but the number of parameters p remains fixed. Very informally, for an exchangeable sequence we have $p(\theta|y) \propto \prod_i p(y_i|\theta) p(\theta)$, and so we can write

$$\log p(\theta|y) = \text{const} + \sum_i \log p(y_i|\theta) + \log p(\theta),$$

where the second term is $O(n)$ and will dominate the prior term, which remains fixed as the sample size increases. Hence expanding as a Taylor series around the maximum likelihood estimate $\hat{\theta}$ (so that the $(\theta - \hat{\theta})$ term disappears), we get

$$\log p(\theta|y) \approx \text{const} + \sum_i \log p(y_i|\hat{\theta}) + \frac{1}{2}(\theta - \hat{\theta})^2 \sum_i \frac{d^2}{d\theta^2} \log p(y_i|\theta)\Big|_{\hat{\theta}} + \dots,$$

where the quadratic term is $-n \times \hat{I}(\hat{\theta})$, and $\hat{I}(\hat{\theta})$ is a sample-based estimate of the Fisher Information $I(\theta) = -E[\frac{d^2}{d\theta^2} \log p(Y|\theta)]$. Hence, taking exponents of both sides gives

$$\theta \sim \text{N}(\hat{\theta}, (n\hat{I}(\hat{\theta}))^{-1}).$$

The posterior distribution will therefore give essentially the same asymptotic estimates and intervals as the maximum likelihood estimator (Equation 3.11). However, note that the posterior is a distribution for θ given $\hat{\theta}$, whereas (3.11) is the sampling distribution of $\hat{\theta}$ given θ.

3.6.4 Model-based vs procedural methods

Both Fisherian and Bayesian approaches are based on the assumption of a fully described parametric model. A distinction, however, can be drawn between Bayesian and *non-likelihood-based* frequentist methods. The latter may be termed "procedural," in that a statistical procedure that can be applied to data is invented, rather than being derived from a fully specified sampling model assumption, and then its properties explored in a range of possible circumstances. Such techniques include many classical nonparametric procedures such as the sign and Wilcoxon tests, generalised estimating equations, adaptive techniques such as M-estimation, survey weighting methods such as inverse probability weights, and so on. In some circumstances these procedures can be essentially reproduced within a Bayesian framework by assuming a suitably extended model, so that particular forms of tail behaviour can mimic M-estimation, and "nonparametric" methods can be obtained from a very flexible parametric model, as described in §11.8.

We can, however, explore what will happen in a Bayesian analysis if we make an erroneous assumption about the model. The asymptotic analysis shown above reveals that the posterior mean will tend to the maximum likelihood estimator $\hat{\theta}$, which will itself converge to the "true" parameter value θ, assuming that the true sampling distribution is $p(y|\theta)$ for some value θ. If this is not the case and the "true" sampling distribution is some other density $p_T(y)$, then the posterior mean will converge to the "closest" value θ_0 in the assumed family of distributions, where θ_0 minimises the Kullback–Leibler discrepancy $H(\theta)$ between $p(y|\theta)$ and $p_T(y)$, where

$$H(\theta) = \int \log \frac{p_T(y)}{p(y|\theta)} p_T(y) dy.$$

See Gelman et al. (2004), p. 585, for more discussion of convergence under the "wrong model."

The crucial point is that if the wrong model is assumed, then the confidence with which the Bayesian inferences are made may be inappropriate, since, due to the possibility of model error, the posterior distributions do not reflect the full uncertainty. This means that there is a strong responsibility to assure oneself of the adequacy of the chosen model, by both model checking and adopting a sufficiently broad family to ensure robustness to a range of different possible contingencies — see Chapter 8. Similar arguments apply in a non-Bayesian likelihood context.

3.6.5 The "likelihood principle"

This principle states that "all information about θ provided by data y is contained in the likelihood, that is, any function $\propto p(y|\theta)$:" i.e., if observations y and y' are dependent on the same parameter θ and have the same likelihood $L(\theta; y) \propto L(\theta; y')$, then the inferences about θ should be identical. This is clearly trivially true for Bayesian analysis, since the posterior only depends on the likelihood and prior.

This principle seems self-evident, until one considers frequentist Neyman–Pearson tests that, in order to conserve a fixed Type I error rate, force one to allow for how many times one intends to examine the data when deciding whether to reject a null hypothesis. In other words, to take into account what you would have done had you observed something different! For example, when making an inference about a proportion θ in a Neyman–Pearson (but not a Fisherian) framework, it will make a difference whether you decide in advance to carry out n trials and observe r successes, and hence adopt a binomial model, or decide to carry on until you have observed r successes, and happen to need to do n trials, which gives rise to a negative binomial model. There is no difference in a Bayesian analysis, although of course the prior may influence the results.

These issues become particularly important when conducting sequential or adaptive clinical trials, in which the data is periodically examined and, depending on the results observed, alterations may be made to the design or the trial stopped altogether. From a strict Bayesian perspective, no adjustment need be made to the conclusions as a result of these flexible designs, and indeed, there are frequentist methods for adjusting both estimates and intervals in a sequential trial, although they are seldom used. However, when designing a trial, funders and regulatory bodies may still demand control of Type 1 error and an idea of the power of a study to detect a certain effect, and both of these "operating characteristics" are affected by adaptive designs. It has therefore become standard practice within Bayesian adaptive trials to make no adjustment for the design when drawing inferences, but to allow for the design when making pre-trial assessments of the operating characteristics (Berry et al., 2010).

4

Introduction to Markov chain Monte Carlo methods

4.1 Bayesian computation

4.1.1 Single-parameter models

As discussed in Chapter 3, the posterior distribution contains all the information needed for Bayesian inference. In all of the examples encountered thus far there is a *single* unknown parameter, whose posterior distribution might be graphed to provide a complete picture of the current state of knowledge arising from the data and prior information. More generally, though, we wish to calculate numeric summaries of the posterior distribution via integration, e.g., $E[\theta|y] = \int_\theta \theta p(\theta|y) \, d\theta$. In the *conjugate* examples considered so far, the posterior distribution is available in closed form and so the required integrals are straightforward to evaluate. However, outside the conjugate family of models, the posterior is usually of non-standard form (although we can always write down its density function to within a constant of proportionality). As a consequence, at least some of the integrals required for summarising the distribution are difficult.

Various methods are available for evaluating such integrals. In cases where we can sample directly from the posterior, such as in conjugate problems, we could use Monte Carlo simulation (if we wished to venture beyond standard results). More generally, however, we could try to obtain an approximation to the posterior density that is analytically tractable, for example, assuming asymptotic normality of the posterior or more complex techniques such as Laplace's method (see, for example, Carlin and Louis (2008); Gelman et al. (2004) for further details). Alternatively, numerical integration methods can be used (Davis and Rabinowitz (1975); Press et al. (2002), Ch. 4). Standard techniques include Gaussian quadrature, or a form of (non-iterative) Monte Carlo integration, which differs from the form described in § 1.4. There we could obtain a direct sample from $p(\theta|y)$ — here we cannot, so we would integrate by sampling points uniformly from the region to be integrated over, averaging the values of the integrand at those points, finally multiplying by the size of the region.

Here, however, we focus exclusively on the class of iterative methods known as Markov chain Monte Carlo (MCMC) integration (Gelfand and Smith, 1990; Geman and Geman, 1984; Metropolis et al., 1953; Hastings, 1970). These are by far the most powerful and flexible class of algorithms available for Bayesian computation, though see § 4.6 for a brief discussion of situations where MCMC is not well suited. We first present an example in which the single parameter of interest has a non-standard posterior, to illustrate the ease with which complex integrals can be evaluated using MCMC in BUGS. Later, after discussing *multi-parameter* models, we will describe the types of MCMC algorithm used by BUGS for performing such computations.

Example 4.1.1. *Surgery (continued): non-conjugate inference*
Suppose we observe the number of deaths y in a given hospital for a high-risk operation. Let n denote the total number of such operations performed and suppose we wish to make inferences regarding the underlying *true* mortality rate, θ, say. The likelihood, up to a constant of proportionality, is given by

$$p(y|\theta) \propto \theta^y (1 - \theta)^{n-y}.$$

Note that θ must lie between 0 and 1, and suppose that to impose this constraint we choose a non-conjugate, normal prior for the logistic transform of θ:

$$\text{logit}\,\theta = \log\left(\frac{\theta}{1-\theta}\right) \sim \text{Normal}(\mu, \omega^2)$$

$$\Rightarrow p(\theta) = \frac{1}{\theta(1-\theta)} \times \frac{1}{\omega\sqrt{2\pi}} \exp\left\{-\frac{1}{2\omega^2}(\text{logit}\,\theta - \mu)^2\right\}.$$

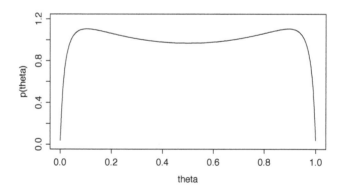

FIGURE 4.1
Prior density for θ in the case where $\text{logit}\,\theta \sim \text{Normal}(0, 2.71)$.

Figure 4.1 shows the prior density for θ with $\mu = 0$ and $\omega^2 = 2.71$, which correspond to a good approximation of the standard logistic density (Appendix C and §5.2.5), which would be uniform on the scale of θ. Multiplying prior and likelihood together gives the posterior

$$p(\theta|y) = A \times \theta^{y-1}(1-\theta)^{n-y-1}\exp\left\{-\frac{1}{2\omega^2}(\text{logit}\,\theta - \mu)^2\right\},$$

where A is the normalising constant required to make the density integrate to 1. A is analytically intractable, but even if we knew A, the posterior expectation

$$E[\theta|y] = A \times \int_0^1 \theta^y(1-\theta)^{n-y-1}\exp\left\{-\frac{1}{2\omega^2}(\text{logit}\,\theta - \mu)^2\right\}d\theta$$

would still be intractable.

In Example 3.3.1, where a conjugate Beta(. , .) prior was specified for θ, we were able to derive the posterior in closed form and perform Monte Carlo integration directly by specifying that closed form as a sampling distribution in BUGS. With a normal prior on logit θ, however, there is no closed-form posterior. In such cases BUGS can perform *Markov chain* Monte Carlo integration instead if we simply specify the likelihood and prior separately. Suppose $y = 10$ and $n = 100$:

```
y             <- 10
n             <- 100
#############################
y               ~ dbin(theta, n)
logit(theta) <- logit.theta
logit.theta   ~ dnorm(0, 0.368)   # precision = 1 / 2.71
```

The software knows how to derive the posterior distribution and subsequently sample from it. The resulting samples are used as in standard Monte Carlo integration to compute various posterior summaries, e.g.,

```
node  mean    sd       MC error 2.5%     median 97.5%  start sample
theta 0.1081 0.03029 3.387E-4 0.05725 0.1052 0.1744 1001   10000
```

Hence the posterior mean is 0.108 and an approximate 95% credible interval for θ is (0.0573, 0.174). Given the prior distribution and the observed data, we can be 95% sure that the "true" mortality rate lies between 0.0573 and 0.174. These results are very similar to those obtained in the conjugate case with θ assigned a fully uniform Beta(1, 1) prior, as opposed to the approximately uniform prior shown in Figure 4.1: mean = 0.108, interval = (0.0557, 0.175).

4.1.2   Multi-parameter models

More generally we are interested in models with more than one unknown parameter. As the number of parameters increases, however, it is increasingly difficult to identify a conjugate prior, to the extent that for all but the simplest of problems the *joint* posterior distribution is of non-standard form. In

addition, the integrals required for inference become high dimensional. For
example, suppose we have a joint posterior distribution for the vector of unknowns $\theta = \{\theta_1, \ldots, \theta_k\}$. We often want to base inference on the *marginal*
posterior of a subset of the parameters: the marginal posterior for θ_1, say, is

$$p(\theta_1|y) = \int_{\theta_2} \cdots \int_{\theta_k} p(\theta|y)\, d\theta_2 \ldots d\theta_k.$$

In such cases MCMC is often the *only* suitable method of integration.

Example 4.1.2. *A multi-parameter model*
Suppose we have observed data y_i, $i = 1, \ldots, n$, which we believe arise from
a heavy-tailed Student-t distribution with unknown mean μ, unknown inverse-
scale-squared r, and unknown degrees of freedom d (see Appendix C.1). Further
suppose that we specify independent Normal(γ, ω^2) and Gamma(α, β) priors for
μ and r, respectively, and an independent discrete-uniform prior for d on the set
$\{2, 3, \ldots, 30\}$. The *joint* posterior distribution is given by

$$p(\mu, r, d|y) \propto \left\{ \frac{\Gamma(\frac{d+1}{2})}{\Gamma(\frac{d}{2})} \sqrt{\frac{r}{d\pi}} \right\}^n \prod_{i=1}^{n} \left\{ 1 + \frac{r}{d}(y_i - \mu)^2 \right\}^{-(d+1)/2} \quad \text{[likelihood]}$$

$$\times \ \exp\left\{ -\frac{1}{2\omega^2}(\mu - \gamma)^2 \right\} \quad \text{[prior } \mu\text{]}$$

$$\times \ r^{\alpha-1} \exp(-\beta r) \quad \text{[prior } r\text{]}$$

$$\times \ 1/29 \quad \text{[prior } d\text{]},$$

which is certainly of non-standard form! Suppose we wish to make *marginal* inferences about the unknown degrees of freedom d. Then we need

$$p(d|y) = \int_{\mu} \int_{r} p(\mu, r, d|y)\, dr d\mu,$$

which is intractable. In BUGS we simply specify the likelihood and each prior as
follows:

```
for (i in 1:n) {y[i]    ~ dt(mu, r, d)}
mu                      ~ dnorm(gamma, inv.omega.squared)
r                       ~ dgamma(alpha, beta)
d                       ~ dcat(p[])
p[1]                    <- 0
for (i in 2:30) {p[i] <- 1/29}
```

The software then uses Markov chain Monte Carlo to generate samples from
the joint posterior distribution $p(\mu, r, d|y)$. These can be used to make arbitrary
inferences about the joint posterior or, by simply ignoring samples not pertaining
to the variable(s) of interest, to make marginal inferences about any subset of the
parameters. For example, we first generate a toy data set by simulating $n = 100$

values from a t-distribution with $\mu = 0$, $r = 1$, and $d = 4$. We then fit the above
BUGS model, with "vague" priors given by $\gamma = 0$, $\omega = 100$, and $\alpha = \beta = 10^{-3}$
(see Chapter 5 for discussion of why these might be suitable choices), and initial
values `list(mu = 0, r = 1, d = 10)`, to obtain

node	mean	sd	MC error	2.5%	median	97.5%	start	sample
d	12.82	7.584	0.1648	3.0	11.0	29.0	1	100000
mu	0.04393	0.09752	5.455E-4	-0.1467	0.0431	0.2368	1	100000
r	1.339	0.3203	0.005169	0.8774	1.282	2.123	1	100000

Hence, we can immediately infer that the degrees of freedom has a (marginal)
posterior median of 11 and a 95% credible interval of $[3, 29]$. Visual inspection
of the posterior (Figure 4.2) reveals that there is limited information in the data
regarding d but that the mode, 5, is close to the true value of 4.

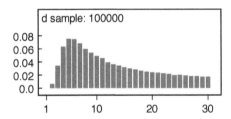

FIGURE 4.2
Approximate posterior distribution for number of degrees of freedom from analysis
of 100 observations from $t(0, 1, 4)$.

4.1.3 Monte Carlo integration for evaluating posterior integrals

As we have seen in §1.4 we can calculate arbitrary summaries of interest for a
given distribution by Monte Carlo integration. Hence, assuming we can obtain
a sample of realisations from the joint posterior $p(\theta|y)$, we have an entirely
general method for evaluating the integrals necessary for Bayesian inference.
So how might we obtain a sample from $p(\theta|y)$?

For all but the most tractable of posterior distributions, this cannot be
done directly. However, a number of algorithms exist for *indirect* sampling
from non-standard distributions. In general, these methods work by sampling
values from an approximate distribution and then correcting or adjusting the
values so that they better resemble a sample from the true distribution of
interest. The book by Ripley (1987) offers the interested reader a thorough

account of many such methods. Amongst the most widely used are importance sampling and rejection sampling. These are non-iterative algorithms, in the sense that the same approximation to the *target* distribution is used throughout. However, to use them for Bayesian computation necessitates finding a density that is a good approximation to the (log) joint posterior *and* that is easy to sample from directly. For many realistically complex Bayesian models, this is difficult or impossible to do (using generic methods that do not have to be tuned to specific applications).

The alternative is to use an *iterative* algorithm, in which a single realisation from the approximating distribution is drawn at each iteration, but the approximate distribution is *improved at each step*. Once the approximating distribution is sufficiently close to the target (i.e., the joint posterior), successive draws from this distribution can be considered to form a sample from the joint posterior of interest. Hence as the iterations proceed, the approximating distribution can be thought of as *converging* towards the posterior. Theorems exist which prove that if the approximating distribution is set up in a certain way (essentially so that the successive realisations form a Markov chain with appropriate transition probabilities — see below), then this convergence will occur almost surely as T (the number of iterations) $\to \infty$ (Tierney, 1994; Roberts and Rosenthal, 2004; Robert and Casella, 2004; Asmussen and Glynn, 2011). In practice, of course, only a finite number of iterations is possible, and as we shall see in §4.4, deciding at which point the approximating distribution is *close enough* to the target posterior is crucial when using these methods for Bayesian inference.

4.2 Markov chain Monte Carlo methods

As hinted above, one of the most reliable and general methods for choosing a suitable iterative approximating distribution for sampling from complex Bayesian posterior distributions is to use a Markov chain. Formally, a sequence of random variables $X^{(0)}, X^{(1)}, X^{(2)}, \ldots$ forms a Markov chain if, for all t, the distribution of the $t + 1^{th}$ variable in the sequence is given by

$$X^{(t+1)} \sim p_{\text{trans}}(x|X^{(t)} = x^{(t)}), \qquad (4.1)$$

that is, conditional on the value of $X^{(t)}$, the distribution of $X^{(t+1)}$ is independent of all other preceding values, $X^{(t-1)}, \ldots, X^{(0)}$. The right-hand side of (4.1) is called the *transition distribution* of the Markov chain and defines the *conditional* probability of moving to any particular new value given the current value of the chain. Subject to fairly general regularity conditions (including irreducibility and aperiodicity, see Cox and Miller (1965)), the *marginal* (or *un*conditional) distribution of $X^{(t+1)}$ will converge to a unique *stationary dis-*

tribution as $t \to \infty$. In simple terms, this means that although each variable in the chain depends directly on its predecessor, eventually (as t increases) we reach a point such that for practical purposes, all subsequent values are distributed *marginally* according to the same fixed distribution, which, crucially, is independent of the starting value $X^{(0)}$. In other words, the chain eventually forgets where it started and conforms to an underlying "equilibrium" distribution.

So how does this help us to generate realisations of θ from the joint posterior distribution $p(\theta|y)$ in a Bayesian analysis? Replacing the random variable X above by the random vector θ, the answer is to choose a transition distribution suitable for generating (from an arbitrary initial state $\theta^{(0)}$) a sequence of realisations $\theta^{(1)}, \theta^{(2)}, \theta^{(3)}, \ldots$ whose unique stationary distribution is the joint posterior of interest $p(\theta|y)$. The marginal distributions of the $\theta^{(t)}$s ($t = 1, 2, 3, \ldots$) play the role of the approximating distributions discussed earlier, with the approximation becoming successively closer to the target posterior as the Markov chain converges to its stationary distribution.

Many methods exist for designing and sampling from such transition distributions, and their suitability depends on the nature of the joint posterior distribution to be explored. As Bayesian models have become more and more sophisticated, so people have invented cleverer and cleverer algorithms for constructing efficient Markov chains to sample from required posterior distributions. Inevitably there is a trade-off between the generality of a particular method and its ability to sample efficiently from complex, high-dimensional densities through fine-tuning. Here we focus on the main algorithms used by the BUGS software, which, of necessity, are designed to be robust in a wide range of applications rather than optimised for specific cases. Some strategies and tricks for improving the efficiency of BUGS simulations in certain situations are discussed, for example, in §6.1, §10.5, §11.2.

Note that, in general, MCMC methods generate a *dependent* sample from the joint posterior of interest, since each realisation depends directly on its predecessor. We can still use this sample as the basis for Monte Carlo integration, however, as all of the results discussed in §1.4 still hold.

4.2.1 Gibbs sampling

The Gibbs sampler (Geman and Geman, 1984; Gelfand and Smith, 1990; Casella and George, 1992) is one of the most widely used algorithms for simulating Markov chains. It is a special case of the Metropolis–Hastings algorithm (Metropolis et al., 1953; Hastings, 1970) and generates a multi-dimensional Markov chain by splitting the vector of random variables θ into subvectors (often scalars) and sampling each subvector in turn, conditional on the *most recent* values of all other elements of θ. The algorithm proceeds as follows. Let the vector of unknowns θ consist of k sub-components, i.e., $\theta = (\theta_1, \theta_2, \ldots, \theta_k)$:

1. Choose arbitrary starting values $\theta_1^{(0)}, \theta_2^{(0)}, \ldots, \theta_k^{(0)}$ for each component,

where subscripts denote sub-components of θ and superscripts denote the iteration number (iteration zero being the initial state of the Markov chain).

2. Sample new values for each element of θ by cycling through the following steps:

- Sample a new value for θ_1, from the *full conditional distribution* of θ_1 given the most recent values of all other elements of θ and the data:

$$\theta_1^{(1)} \sim p(\theta_1 | \theta_2^{(0)}, \theta_3^{(0)}, \ldots, \theta_k^{(0)}, y).$$

- Sample a new value $\theta_2^{(1)}$ for the second component of θ, from its full conditional distribution $p(\theta_2 | \theta_1^{(1)}, \theta_3^{(0)}, \ldots, \theta_k^{(0)}, y)$. Note that as a new value for θ_1 has already been sampled, it is this "most recent" value that is conditioned upon, together with the starting values for all other elements of θ.

- . . .

- Sample $\theta_k^{(1)}$ from $p(\theta_k | \theta_1^{(1)}, \theta_2^{(1)}, \ldots, \theta_{k-1}^{(1)}, y)$.

This completes one iteration of the Gibbs sampler and generates a new realisation of the vector of unknowns, $\theta^{(1)}$.

3. Repeat stage 2 many times, always conditioning on the most recent values of other parameters, to obtain a sequence of dependent realisations of the vector of unknowns $\theta^{(1)}, \theta^{(2)}, \ldots, \theta^{(T)}$ (where T is typically of the order of many thousands).

Figure 4.3 graphically illustrates the algorithm in the case of a hypothetical two-parameter problem ($k = 2$). The beauty of Gibbs sampling is that simulation from a complex, high-dimensional joint posterior distribution is reduced to a sequence of algorithms for sampling from one- or low-dimensional distributions. As we shall see in the next subsection, these univariate or low-dimensional full conditional distributions can usually be simplified by exploiting the conditional independence structure of the model, and in many cases are available in closed form (see Example 4.2.2 below), in which case direct sampling is straightforward using a specialized, distribution-specific random number generator (Ripley, 1987).

4.2.2 Gibbs sampling and directed graphical models

Suppose the model of interest can be represented as a directed acyclic graph with stochastic nodes $\mathcal{G}$ and directed links $\mathcal{L}$. As discussed in § 2.1.2, the conditional independence assumptions expressed through the DAG structure allow us to write $p(\mathcal{G}) = \prod_{v \in \mathcal{G}} p(v | \text{pa}[v])$ (Lauritzen et al., 1990). That is, the joint distribution of all nodes is given by the product, over all nodes, of the

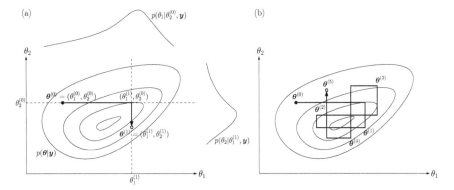

FIGURE 4.3
(a) First iteration of the Gibbs sampler for an illustrative two-parameter (bivariate) problem. The contours show the "height" of the true bivariate posterior distribution $p(\theta|y) = p(\theta_1, \theta_2|y)$. The starting point of the Gibbs sampler $\theta^{(0)} = (\theta_1^{(0)}, \theta_2^{(0)})$ is shown by the solid dot, and the pair of values $(\theta_1^{(1)}, \theta_2^{(1)})$ sampled in the first iteration is shown by the open circle. The univariate density projected onto the top horizontal axis shows the full conditional distribution $p(\theta_1|\theta_2^{(0)}, y)$, which is obtained by taking a horizontal "slice" through the joint posterior distribution at the value $\theta_2 = \theta_2^{(0)}$ (indicated by the horizontal dashed line). A new value for θ_1 $(\theta_1^{(1)})$ is generated from this full conditional, and then a "slice" parallel to the θ_2 axis is taken through the joint posterior at $\theta_1 = \theta_1^{(1)}$ (vertical dashed line). This gives the univariate full conditional $p(\theta_2|\theta_1^{(1)}, y)$, which is shown projected onto the right-hand vertical axis, and from which a new value for θ_2 $(\theta_2^{(1)})$ is sampled. (b) First five iterations of the Gibbs sampler shown in (a). Note that the sampler always moves parallel to the axes.

assumed distribution of each node conditional on its parents; in other words, the product of all distributional assumptions. The set of all unknowns θ and the set of all data y together form a partition of $\mathcal{G}$, and so

$$p(\theta, y) = \prod_{v \in \mathcal{G}} p(v|\text{pa}[v]). \tag{4.2}$$

From the definition of conditional probability, $p(\theta|y) = p(\theta, y)/p(y)$, which is proportional to $p(\theta, y)$ when considered as a function of θ. Hence the joint posterior can be obtained trivially for any DAG, up to a constant of proportionality. Similarly, *any* conditional distribution involving all nodes in the graph is also proportional to $p(\theta, y)$. Thus for any unobserved node (or set of nodes) θ_i, say, the full conditional distribution $p(\theta_i|\theta_{\backslash i}, y)$, where $\theta_{\backslash i}$ denotes "all elements of θ except θ_i," is proportional to $p(\theta, y)$ and can therefore be expressed as the right-hand side of (4.2). However, we are seeking to identify a distribution in θ_i and so any factor in (4.2) not involving θ_i can be ignored, since it forms part of the normalising constant. Hence we obtain

$$p(\theta_i|\theta_{\backslash i}, y) \propto p(\theta_i|\mathrm{pa}[\theta_i]) \times \prod_{v \in \mathrm{ch}[\theta_i]} p(v|\mathrm{pa}[v]), \qquad (4.3)$$

and so the full conditional is dependent only on $\mathrm{pa}[\theta_i]$, $\mathrm{ch}[\theta_i]$ and all *co-parents* of θ_i's children. Collectively these three sets of nodes form a neighbourhood in the graph around θ_i known as the *Markov blanket*; θ_i is *conditionally independent* of all other nodes in the graph, *given* the Markov blanket. The subsequent derivation of a closed form for the full conditional (where available) is exactly analogous to the derivation of a closed-form posterior in conjugate, single-parameter models. The first term on the right-hand side of (4.3) plays the role of the prior distribution and is referred to as the "prior component." The product in (4.3) plays the role of the likelihood and is known as the "likelihood component." Finally, as we are conditioning on the most recent values of all other nodes, it is as if those nodes have *known* values, and so there is effectively only one unknown, θ_i.

Example 4.2.1. *The Markov blanket*
Consider the directed acyclic graph shown in Figure 4.4. Suppose we wish to derive the full conditional distribution for node C. This is proportional to the product of distributions for C and all of its children, conditional on their parents, i.e.,

$$p(\mathsf{C}|\mathsf{A}, \mathsf{B}, \mathsf{D}, \ldots, \mathsf{I}) \propto p(\mathsf{C}|\mathsf{A}, \mathsf{B}) \times p(\mathsf{E}|\mathsf{C}, \mathsf{D}) \times p(\mathsf{F}|\mathsf{C}, \mathsf{D}),$$

since $\mathrm{ch}[\mathsf{C}] = \{\mathsf{E}, \mathsf{F}\}$. The Markov blanket for C is given by its parents, its children, and all co-parents of its children, in this case $\{\mathsf{A}, \mathsf{B}, \mathsf{D}, \mathsf{E}, \mathsf{F}\}$. *Given* the Markov blanket, a node is conditionally independent of all other nodes in the graph (G, H, and I here), and so $p(\mathsf{C}|\mathsf{A}, \mathsf{B}, \mathsf{D}, \ldots, \mathsf{I})$ can be rewritten as $p(\mathsf{C}|\mathsf{A}, \mathsf{B}, \mathsf{D}, \mathsf{E}, \mathsf{F})$.

Example 4.2.2. *Full conditional distributions*
Suppose we observe data $y_1, \ldots, y_n$, assumed to form a random sample from a normal distribution with unknown mean μ and unknown precision τ. Further suppose that we specify independent priors on μ and τ as follows:

$$\mu \sim \mathsf{Normal}(\gamma, \omega^2); \quad \tau \sim \mathsf{Gamma}(\alpha, \beta).$$

Note that these are the conjugate priors for the cases in which the precision and the mean, respectively, have known values (see Table 3.1). Note also, however, that the joint prior $p(\mu, \tau) = \mathsf{Normal}(\gamma, \omega^2) \times \mathsf{Gamma}(\alpha, \beta)$ does *not* lead to a closed-form joint posterior. Thus, in order to make inferences about μ and τ, we might run a Gibbs sampler. We first choose arbitrary starting values, $\mu^{(0)} = 5$ and $\tau^{(0)} = 10$, say, and then an equally arbitrary updating order, μ then τ, say.

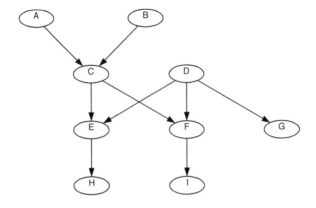

FIGURE 4.4
Directed acyclic graph for Example 4.2.1, showing parent–child relationships between nine nodes.

At iteration t of the Gibbs sampler we draw $\mu^{(t)} \sim p(\mu|\tau^{(t-1)}, y)$ and $\tau^{(t)} \sim p(\tau|\mu^{(t)}, y)$, where the full conditional sampling distributions can be derived from the DAG shown in Figure 4.5. The full conditional for μ is proportional to the prior for μ multiplied by the distribution of each child of μ conditional on that child's parents. From the graph, $\mathrm{ch}[\mu] = \{y_i, i = 1, \ldots, n\}$, and so

$$p(\mu|\tau^{(t-1)}, y) \propto \exp\left\{-\frac{1}{2\omega^2}(\mu - \gamma)^2\right\} \times \exp\left\{-\frac{\tau^{(t-1)}}{2} \sum_{i=1}^{n}(y_i - \mu)^2\right\}.$$

Conditioning on $\tau = \tau^{(t-1)}$ is essentially the same as assuming τ to be known. Hence this is exactly the same type of calculation as is required for deriving the single-parameter posterior (for μ) in the unknown mean, known precision/variance case — see §3.3.2. Therefore,

$$p(\mu|\tau^{(t-1)}, y) = \mathrm{Normal}\left(\frac{\tau^{(t-1)} \sum y_i + \omega^{-2}\gamma}{n\tau^{(t-1)} + \omega^{-2}}, \frac{1}{n\tau^{(t-1)} + \omega^{-2}}\right).$$

Similarly, $\mathrm{ch}[\tau] = \{y_i, i = 1, \ldots, n\}$, and so

$$p(\tau|\mu^{(t)}, y) \propto \tau^{\alpha-1}\exp\{-\beta\tau\} \times \tau^{\frac{n}{2}}\exp\left\{-\frac{\tau}{2} \sum_{i=1}^{n}(y_i - \mu^{(t)})^2\right\},$$

which is the same as in the known mean, unknown precision case presented in Table 3.1. Hence

$$p(\tau|\mu^{(t)}, y) = \mathrm{Gamma}\left(\alpha + \frac{n}{2}, \beta + \frac{1}{2}\sum_{i=1}^{n}(y_i - \mu^{(t)})^2\right).$$

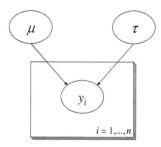

FIGURE 4.5
Directed acyclic graph depicting the assumption $y_i \sim$ Normal(μ, τ^{-1}), $i = 1, \ldots, n$. Note the use of rectangular "plates" to denote repetition, i.e., the "loop" over i.

4.2.3   Derivation of full conditional distributions in BUGS

Equation (4.3) demonstrates that only *local* knowledge of the graph is needed to derive the full conditional distribution of any node (or group of nodes). Indeed, in order to derive all of the full conditionals required for Gibbs sampling on any DAG, we simply need to know how each node is related to its parents and which nodes are its children. BUGS stores this information in the form of an object-oriented version of the specified graph, with "objects" representing nodes and "pointer variables" linking nodes together, in particular linking each node to its children. The software uses an expert system to visit each node in turn and classify the form of the full conditional by considering the "compatibility" of the prior component (the distribution of the node itself, conditional on its parents) with each term in the likelihood component (the distribution of each child conditional on its parents). If the full conditional is available in closed form then a specialized, distribution-specific algorithm will be used to derive and sample from that closed form. If a closed form is not available, however, the expert system chooses a more general algorithm from a range of suitable methods, based on any important features of the full conditional that might have been gleaned from the aforementioned compatibility considerations, e.g., whether the density is "log-concave," or has infinite support, say. Some of the more commonly used alternative sampling methods are discussed briefly in the following subsection; the reader is referred to Lunn et al. (2000) and Lunn et al. (2009b) for further details regarding the internal workings of BUGS.

4.2.4   Other MCMC methods

As mentioned above, if direct sampling from the full conditional is not possible, then the WinBUGS software implements a number of alternative, more general algorithms, including "slice" sampling (Neal, 2003), Metropolis sampling (Metropolis et al., 1953; Hastings, 1970), and various types of rejection method, such as "adaptive rejection sampling" (Gilks, 1992; Gilks and Wild, 1992) — see Lunn et al. (2000) or the WinBUGS manual for more details. Further algorithms are available in the OpenBUGS implementation of BUGS, for example, delayed rejection (Green and Mira, 2001) and hybrid/Hamiltonian methods (Duane et al., 1987; Neal, 2010) — please see the OpenBUGS documentation. Note that such methods are used only as a means of updating full conditionals within a Gibbs sampling scheme; use of the Metropolis algorithm, for example, thus leads to a type of sampling known as Metropolis-within-Gibbs.

Metropolis algorithms We do not wish to go into the details of various sampling methods in this text, but we feel it is instructive to provide an overview of Metropolis-based algorithms since they lead to a characteristic form of output that could be perceived as erroneous by the inexperienced user. Metropolis algorithms work by first sampling values from a *proposal* distribution, which approximates the relevant full conditional (or, more generally, *target* distribution) but is easy to sample from. An *acceptance probability* α is then calculated for the proposed value, and a Uniform$(0, 1)$ random number u is drawn to convert that probability into an accept/reject decision (reject if $u > \alpha$, accept otherwise). If the proposed value is to be accepted then the Markov chain moves to that value, otherwise it remains at its current value. This leads to a Markov chain that stays in the same place for a number of iterations, on a regular basis, particularly if the proposal distribution does not approximate the target distribution well (giving typically low values for α). If we plot a continuous line joining successive samples together, then the resulting *trace* plot — see §4.4.1 — can have the appearance of a cityscape (a metropolis perhaps), with the roofs of buildings corresponding to multiple successive rejections (see Figure 4.6, for example). Despite the chain containing multiple repeated values, it may be used, for inference, in exactly the same way as if all values had been sampled directly, as long as sufficient time is allowed for the chain to fully explore the target distribution.

In principle it is possible to use the Metropolis algorithm to jointly update the entire posterior distribution in one go, rather than using the alternating conditional updating scheme of the Gibbs sampler. However, this often requires problem-specific fine-tuning to obtain a multivariate proposal distribution that is a good approximation to the joint posterior of interest; hence, such an approach is not used in any implementation of BUGS, although note that various types of "block updating," where a group of variables is updated together according to their joint full conditional, are possible. Even when used within a Gibbs sampling scheme to sample from univariate full conditionals, the Metropolis algorithm requires some tuning to obtain efficient simulations. This tuning process has been fully automated in WinBUGS and is known as the *adapting phase* (see also §12.4.3, and the WinBUGS manual). Samples

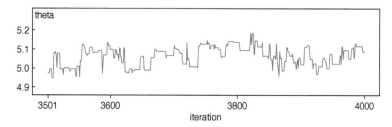

FIGURE 4.6

Trace plot, whereby successive samples for a single parameter are joined together by a continuous line (plotted against iteration number). The plot illustrates how output from Metropolis-based samplers contains multiple sequences of repeated values, which can give the plot the appearance of a cityscape.

generated during this phase, however, cannot be considered to arise from the posterior distribution, and so they must be discarded before performing any Monte Carlo integration. The "slice" sampler used in all BUGS implementations also fine-tunes itself during an initial adapting phase, and the same rule about discarding samples generated during this phase applies. These and other issues relating to the interpretation of output from different MCMC algorithms will be further discussed as they arise in context.

4.3   Initial values

Before any MCMC method can be started, we must first initialise the Markov chain, i.e., provide starting values for each unknown parameter in the statistical model. WinBUGS and OpenBUGS can often do this automatically, by sampling from the assumed ("prior") distribution of each parameter. However, such distributions may have very large variances, say, to reflect a lack of prior knowledge, and wildly inappropriate starting values may result. The disparity between such initial values and those values supported by the posterior often then causes the program to crash. Hence we have the facility to specify initial values manually instead. JAGS, on the other hand, when asked to generate initial values automatically, chooses a "central" value from the distribution, such as the mean or median, which avoids this problem. In WinBUGS initial values can be supplied as a list after the model description or in a separate file (see §12.4.2). In practice a mixture of these two strategies typically works well, with some values specified by the user and the remaining values generated automatically by the BUGS engine. All of the remaining examples in this book have been run with a mixture of manually specified and automatically generated initial values. Except in cases where the manually specified values have been given explicitly, it will not generally be possible for the reader to reproduce *exactly* any results quoted using their own code, although results should agree up to Monte Carlo error — see §4.5.

Example 4.3.1. *Initial values*

Consider the multi-parameter Student-t problem described in Example 4.1.2. There are three unknown parameters, each of which must be initialised. The prior for μ, the mean of the Student-t distribution, is Normal(γ, ω^2). With $\gamma = 0$ and $\omega = 100$ as in Example 4.1.2, typical values from this prior lie in the range $(-20000, 20000)$. Hence we might prefer to simply set $\mu^{(0)} = 0$ rather than sample from the prior and obtain 14,603, say, which is unlikely to be a good starting point! The prior for r is Gamma($0.001, 0.001$), which WinBUGS cannot even sample from (due to numerical inaccuracies caused by the small size of the first parameter), and so a user-specified initial value is essential, e.g., $r^{(0)} = 1$. We may also provide our own starting value for the degrees-of-freedom parameter, e.g., $d^{(0)} = 8$, but any value from the discrete-uniform prior on $\{2, 3, \ldots, 30\}$ would represent a reasonable starting point and so we might prefer to let Win-BUGS generate this automatically. The user-specified values for μ and r can be provided in the form of a list given after the model description:

```
list(mu = 0, r = 1)
```

In Example 2.1.2, we described the basic steps for running a model in WinBUGS. In this example, we would now need to load this list of initial values in Step 7, as follows, which was previously ignored.

- Highlight the word list by double-clicking on it, and click load inits in the Specification Tool.

Initial values for the remaining parameters (d in this case) are generated automatically using the *gen inits* command (see also §12.4.3); note that this must be used *after* the user-specified values have been loaded.

4.4 Convergence

As discussed in §4.2, if we simulate realisations from a Markov chain transition distribution, then under broad conditions, eventually the simulated values will be marginally distributed according to the Markov chain's unique stationary distribution. To fully exploit this fact we need to be able to detect when the marginal behaviour of the chain is sufficiently close to stationarity, so that we can harvest all subsequent realisations as a dependent sample from the stationary distribution; note that the initial, non-stationary portion of the chain is referred to as the "burn-in."

It is important to note that convergence here is to a distribution, not to a fixed value. In practice we diagnose convergence retrospectively, by guessing for how long to run the simulation (we can always keep going if our initial guess proves insufficient) and then trying to determine if some latter portion of the chain can be considered stationary. We typically have to run the simulation well beyond the point of convergence, as detecting stationarity (or a lack thereof) requires a substantial sample size, as does accurate inference. The actual post-convergence sample size required depends very much on the *efficiency* of the Markov chain. We will discuss efficiency more formally in §4.5, but we note here that, basically, the number of samples required, both for detecting (non-) stationarity and for drawing accurate inferences, increases with an increasing level of serial dependence in the chain (also known as serial- or auto-correlation).

So why might we have a significant level of autocorrelation in the chain? First note that all Markov chains, by definition, exhibit at least some serial correlation. Second, imagine the two-parameter Gibbs sampler depicted in Figure 4.3 and suppose there is a high degree of correlation in the target distribution, such that the contours form narrow ellipses along the line $y = x$, say. The Gibbs sampler is capable of exploring the entire joint posterior but can only move perpendicular to the axes. Hence the shape of the distribution prevents the sampler from taking anything other than small steps, and so successive values of each variable are close together. All Metropolis-based sampling algorithms also induce a level of autocorrelation in the sample, due to their tendency to remain in the same place for a number of iterations, as discussed in §4.2.4.

4.4.1 Detecting convergence/stationarity by eye

Our task is made easier by the fact that it is often straightforward to detect (lack of) convergence *informally* by eye. Figure 4.7 illustrates this via several *history* (or *trace*) plots, whereby a continuous line joining successive realisations of a specific variable is plotted against Gibbs iteration number. The various plots comprising the figure are discussed below.

Our model may contain many parameters but it is not feasible to visually examine more than one at a time. This does not preclude convergence diagnosis, however, since we can simply assume convergence at the point where all parameters have reached stationarity — note that it is quite normal for different parameters to converge at different rates. Many realistically complex models, however, have hundreds or even thousands of parameters, and performing an individual assessment for each one may be impractical. In such cases it is important to at least assess all parameters (and functions thereof) of interest. It is also prudent to examine at least a random selection of the remaining parameters.

As demonstrated by Figure 4.7(a), a Markov chain that has reached stationarity should look like a random scatter about a stable mean value. In addition, if plotted from the point of convergence onwards, a stationary Markov chain that contains sufficient information for reliable inferences, informally speaking, has the appearance of a "fat hairy caterpillar" (see Figure 4.7(b)). However, if the Markov chain has a "snake-like" appearance instead (Figure 4.7(c)), this does not necessarily mean that it hasn't converged, since there may simply be a high degree of autocorrelation in the sample. Hence the time scale over which the chain is plotted may be too short to clearly demonstrate stationarity, meaning that more samples are required. If we were to continue sampling indefinitely, repeatedly redrawing the entire history plot within the same area, it would eventually take on the required form, i.e., that of a fat caterpillar (see Figure 4.7(d)), since any "periodicity" would be "squeezed" to the point where it looked like random scatter — note that this also applies to regions of any Metropolis output where rejections have occurred. Markov chains generated using the Gibbs sampler often converge surprisingly quickly, within the first few iterations, say. However, it is important to note that the rate of convergence can be very slow, so slow in fact that a Markov chain may appear stationary even when it is not (the apparently "stable" mean may be drifting very slowly).

Of course, there is no reason why we cannot simulate two or more Markov chains and *pool* the resulting samples (after convergence) for making inferences. Such an approach provides us with a very effective means of checking convergence. If the multiple chains are given widely differing initial states, then we can be reasonably confident that stationarity has been reached when the chains come together and start behaving similarly. Figure 4.7(e) illustrates this by showing the same chain as in Figure 4.7(a) but with another chain, initialised at a different starting point, superimposed. This approach forms the basis of perhaps the most reliable *formal* convergence diagnostic, which is discussed in the following subsection. We note here, however, that running multiple chains can be inefficient, particularly if the rate of convergence is slow, since we effectively have to wait for convergence multiple times. This problem is exacerbated by the need to use widely differing initial states, which increase confidence in the diagnosis but often also substantially increase the number of iterations required to reach convergence.

4.4.2   Formal detection of convergence/stationarity

Numerous techniques for formally diagnosing convergence can be found in the literature. Many of these have been implemented within the CODA (Best et al., 1995; Plummer et al., 2006) and BOA (Smith, 2000) software packages for R and S-Plus, which are designed to take BUGS output as input. No method should be used blindly, however, as none can provide conclusive

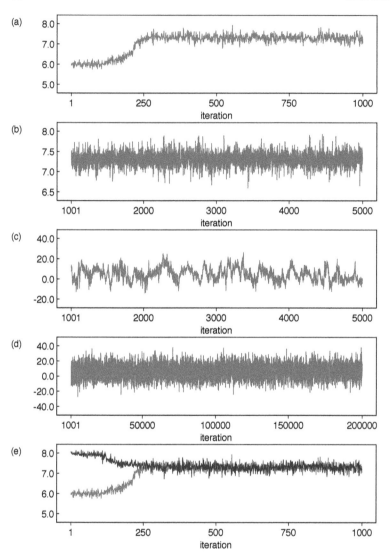

FIGURE 4.7

History/trace plots depicting five different scenarios. A continuous line joining
successive realisations of a given parameter is plotted against Gibbs iteration
number: (a) After convergence to its stationary distribution, a Markov chain
typically looks like a random scatter about some stable mean value (iteration
~250 onwards); (b) A converged chain that contains sufficient information
for accurate inferences looks like a "fat hairy caterpillar;" (c) A "snake"-like
chain may have converged but contains too much serial-/auto-correlation for
accurate inferences to be drawn — more samples required; (d) The same chain
as in (c) but extended to iteration 200,000 and plotted over the same width
— the chain now takes on the required form; (e) The same chain as in (a) but
with another chain, initialised at a different starting point, superimposed.

proof of convergence and all are fallible. Indeed they are invariably founded on a null hypothesis of convergence and as such are designed only to detect *non*-convergence. Different methods examine/highlight different stochastic features of the chain(s), and so it is prudent to always consider a range of methods. For any given method, it is usually possible to construct an example whereby that method will fail to detect non-convergence but others may succeed. The reader is referred to Cowles and Carlin (1996) and Mengersen et al. (1999) for detailed reviews of the various methods available. One method that seems to stand out as being particularly effective, in our experience, is that originally proposed by Gelman and Rubin (1992) and subsequently modified in Brooks and Gelman (1998). Here multiple chains starting at "overdispersed" initial values* are simulated and convergence is assessed by comparing within- and between-chain variability. This is the only convergence diagnostic currently implemented in WinBUGS (and OpenBUGS), but we reiterate the wide range of methods available in BOA and CODA.

Again, we examine only one variable at a time, although multivariate extensions of the approach do exist (Brooks and Gelman, 1998). Suppose we simulate M chains, each of length $2T$, with a view to assessing the degree of stationarity in the final T iterations. We take as a measure of posterior variability the width of the $100(1-\alpha)\%$ credible interval for the parameter of interest, e.g., $\alpha = 0.2$. From the final T iterations we calculate the empirical $100(1-\alpha)\%$ credible interval for each chain. We then calculate the average width of these intervals across the M chains and denote this by W. Finally, we compute the width B of the empirical $100(1-\alpha)\%$ credible interval based on all MT samples pooled together. The ratio $\hat{R} = B/W$ of pooled to average interval widths should be > 1 if the starting values are suitably overdispersed; it will also tend to 1 as convergence is approached, and so we can assume convergence for practical purposes if $\hat{R} < 1.05$, say.

Rather than calculating a single value of $\hat{R}$, we can examine the behaviour of $\hat{R}$ over iteration time by performing the above procedure repeatedly for an increasingly large fraction of the total iteration range, ending with all of the final T iterations contributing to the calculation as described above. Brooks and Gelman (1998) propose splitting the total iteration range $(1, \ldots, 2T)$ into Q batches of length a and calculating $\hat{R}(q)$, $B(q)$, and $W(q)$ based on the latter halves of iterations $1, \ldots, qa$, for $q = 1, \ldots, Q$. A plot of $\hat{R}(q)$, $B(q)$, and $W(q)$ against some appropriate function of q then allows us not only to assess the rate of convergence of $\hat{R}$ to 1, but also to check that both B and W have stabilised, which is crucial for a reliable diagnosis. WinBUGS chooses $a = \max(100, \lfloor 2T/100 \rfloor)^\dagger$ and plots against the starting iteration of each range, $\lfloor qa/2 \rfloor + 1$, so that the plot directly indicates the vicinity of the

---

*In this context "overdispersed" means that the initial values are more variable than they would have been if they had somehow been drawn from the target posterior.

$^\dagger \lfloor x \rfloor$ denotes the largest integer not greater than x.

point of convergence.

Example 4.4.1. *Brooks–Gelman–Rubin diagnostic*
Consider the two converging Markov chains shown in Figure 4.7(e). Clearly convergence occurs at around 250 iterations. Figure 4.8 shows the corresponding Brooks–Gelman–Rubin (BGR) diagnostics for the iteration ranges 51–100, 101–200, 151–300, 201–400, ..., 501–1000 ($2T = 1000$, $a = 100$, $Q = 2T/a = 10$). These are plotted against the starting iteration of each range ($\lfloor qa/2 \rfloor + 1 = 51$, 101, 151, 201, ..., 501) so that the approximate point of convergence can be read directly off the figure. The $\hat{R}$ line suggests convergence at around 200 iterations. However, note that the B and W lines do not stabilise until slightly later.

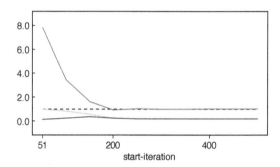

FIGURE 4.8
$\hat{R}(q)$, $B(q)$ and $W(q)$ for the two Markov chains shown in Figure 4.7(e), plotted against the starting iteration, $\lfloor qa/2 \rfloor + 1$, of each range for which the diagnostics are calculated. The upper line represents $\hat{R}$, which should converge to the value 1 (indicated by the horizontal dashed line). The upper and lower of the other two lines represent B and W, respectively. For plotting purposes, so that they can be clearly seen on the same scale as $\hat{R}$, these are normalised such that the maximum estimated interval width is equal to 1.

This strategy works because both the length of the chains used in the calculation and the start iteration are always increasing. Hence we will always eventually (with an increasing sample size) discard any "burn-in" iterations and also include a sufficient number of stationary samples to conclude convergence.

4.5 Efficiency and accuracy

For a given sample size, the accuracy of our inferences is dependent on the efficiency of our posterior sample, which decreases with an increasing level of autocorrelation. One way of increasing efficiency is to reparameterise the model so that the posterior correlation between parameters is reduced (see, for example, §6.1). However, identifying such a parameterisation is not always straightforward. Another way to improve efficiency is to perform a process known as *thinning* whereby only every νth value from the Gibbs sampler is actually retained for inference (the rest are still generated but are subsequently discarded). However, this only represents an efficiency gain in terms of storing and post-processing the sample: for the same computational cost of simulation, the full sample will always contain more information and hence lead to better accuracy.

4.5.1 Monte Carlo standard error of the posterior mean

The easiest way in which to improve accuracy is to simply increase the posterior sample size, but what sample size should we choose to achieve a specific level of accuracy? Indeed, how should we define accuracy? Suppose we are interested in estimating the posterior expectation of some (scalar-valued) function $g(\theta_i)$. Further suppose we have T posterior samples for θ_i and denote our Monte Carlo estimate, based on those T samples, by

$$\bar{g}_T = \frac{1}{T} \sum_{t=1}^{T} g(\theta_i^{(t)}).$$

As discussed in §1.4, if the samples $\theta_i^{(1)}, \ldots, \theta_i^{(T)}$ were independent, then the *Central Limit Theorem* would provide us with a (Monte Carlo) standard error for $\bar{g}_T$ of $\sqrt{Var[g(\theta_i)|y]/T}$, which of course could be estimated by $s/\sqrt{T}$ where s^2 is the sample variance:

$$s^2 = \frac{1}{T-1} \sum_{t=1}^{T} \left\{ g(\theta_i^{(t)}) - \bar{g}_T \right\}^2.$$

Central limit theorems also exist for *dependent* samples (see Jones (2004), for example). In particular, $\bar{g}_T$ tends, in distribution, to Normal$(E[g(\theta_i)|y], \rho/T)$ as $T \to \infty$, for some positive constant ρ. We cannot use the sample variance to estimate ρ as $\rho \neq Var[g(\theta_i)|y]$ when the sample is dependent. Instead we split the sample into Q batches of length a and assume that a is sufficiently large that the central limit theorem approximately holds for each batch:

$$\bar{g}_{a,q} = \frac{1}{a} \sum_{t=(q-1)a+1}^{qa} g(\theta_i^{(t)}) \sim_{\text{approx}} \text{Normal}\left(E[g(\theta_i)|y], \frac{\rho}{a}\right), \quad q = 1, \ldots, Q.$$

Thus the batch means, $\bar{g}_{a,q}$, $q = 1, \ldots, Q$, form a sample from some distribution with variance ρ/a. Hence

$$\rho \approx \hat{\rho} = \frac{a}{Q-1} \sum_{q=1}^{Q} (\bar{g}_{a,q} - \bar{g}_T)^2,$$

and so the standard error of $\bar{g}_T$ can now be approximated, by $\sqrt{\hat{\rho}/T}$. This "Monte Carlo standard error" (MCSE) tells us how accurately the mean of our Monte Carlo samples for g estimates the *true* posterior expectation of g. It should not be confused with the posterior standard deviation, which reflects the inherent uncertainty about g given the model and the data.

The above is known as the *batch means* method of calculating the MCSE. There do exist other approaches, in particular methods adapted from time series analysis (e.g., Geweke (1992)), but BUGS implements batch means due to its simplicity. The value chosen for a is given by $\lfloor \sqrt{T} \rfloor$ (Chien, 1988). In cases where multiple Markov chains have been run, we split each chain in the way described above and the number of batches becomes MQ, where M is the number of chains.

4.5.2 Accuracy of the whole posterior

If we were only interested in accurate inference about the posterior mean of $g(\theta)$, we could simply run the chain for long enough that $\bar{g}_T \pm \delta \times \text{MCSE}$ ($\delta = 2$ typically) agree to the desired number of significant figures. However, we would usually want to characterise the whole posterior distribution of $g(\theta)$ and present a credible interval or posterior probability. Raftery and Lewis (1992) describe a general method for determining whether a posterior tail probability is estimated to within a particular degree of accuracy with a specified probability. This is implemented in the CODA package for R and S-Plus. For example, about 4000 *independent* samples after convergence are sufficient to estimate the cumulative distribution function of the 2.5% quantile of a well-behaved posterior within ± 0.005 with probability 0.95. Reported 95% credible intervals would then have actual posterior probability between 0.94 and 0.96. More iterations would be necessary if the chains are autocorrelated.

Alternatively, if we compare the MCSE to the posterior standard deviation, this will tell us whether the inaccuracy about estimating the *posterior mean* of $g(\theta)$ is large in the context of the *overall* uncertainty about $g(\theta)$. To elucidate this, with *independent* samples the MCSE would be asymptotically $s/\sqrt{T}$, from the central limit theorem. This suggests that comparing the MCSE to the estimated posterior standard deviation s gives us an estimate of the *effective sample size* $T^* = (s/\text{MCSE})^2$ of an autocorrelated chain, which represents the amount of information about the posterior distribution it contains. Therefore, if $\text{MCSE} < ks$, then the effective sample size is $> 1/k^2$. Conventionally (see the WinBUGS manual), chains are run until the MCSE is less than $k = 5\%$ of the posterior standard deviation, giving $T^* = 400$. This is sufficient for many practical purposes — however, the $T^* = 4000$ suggested by Raftery and Lewis's diagnostic for estimating tail probabilities requires around $k = 0.015$, or 1.5%.

A cruder but pragmatic strategy is to run the chains for increasingly more iterations, repeatedly recalculating whatever posterior summary statistics are of interest, until they do not appear to change within the desired accuracy. Three or four significant figures are usually enough in our experience — excessive precision can impede clarity when presenting results, and given the neglected uncertainties in most real statistical analyses (such as selection effects or measurement error) more precision may be spurious.

Example 4.5.1. *Monte Carlo standard error*
Consider again the two Markov chains shown in Figure 4.7(e) and suppose we wish to accurately estimate the mean of the underlying posterior distribution. Shown below are summary statistics, including MCSE (labelled MC error), calculated for various iteration ranges after extending the simulation to 8250 iterations. Note that the "burn-in" values from iterations 1–250 have been discarded and that samples from the two chains have been pooled together.

node	mean	sd	MC error	2.5%	median	97.5%	start	sample
theta	7.333	0.1591	0.007993	7.008	7.339	7.637	251	500
theta	7.314	0.1597	0.006135	6.992	7.321	7.627	251	1000
theta	7.313	0.1604	0.004867	6.985	7.319	7.623	251	2000
theta	7.31	0.1627	0.003654	6.986	7.313	7.629	251	4000
theta	7.312	0.1629	0.002417	6.984	7.313	7.629	251	8000
theta	7.316	0.1625	0.001625	6.993	7.315	7.641	251	16000

The MCSE (fourth column) is less than 5% of the posterior standard deviation (third column) after collecting as few as 1000 samples. Note, however, that if we require MCSE $< 0.01s$, i.e., a standard error that is less than 1% of the posterior standard deviation, then we must collect at least 16,000 samples in this case. This is due to the fact that the estimated MCSE only decreases by a factor of $\sim\sqrt{2}$ as we double the sample size (since MCSE $\approx \sqrt{\hat{\rho}/T}$).

4.6   Beyond MCMC

There are a number of modelling scenarios for which standard MCMC, and Gibbs sampling in particular, are not well suited. One such setting is the analysis of time series data, due to the potentially long chains of serial dependence in the data, although the fact that such models can at least be specified in the BUGS language (and analysed, albeit somewhat inefficiently) is testament to the power of the language — see §11.2. Over the last decade or so there have been many proposals for alternative formulations of the update mechanism, to develop faster and more efficient techniques for improving mixing and convergence rates. Some techniques use approximations to the true likelihood; others, such as Lagrangian–Hamiltonian updates (Girolami and Calderhead, 2011) generate more efficient proposal distributions. As alluded to in §4.2.4 above, implementation of such methods in OpenBUGS has been, and continues to be, explored. The "Stan" software, under development at http://code.google.com/p/stan at the time of writing, implements generic graphical models using Hamiltonian Monte Carlo sampling, and we eagerly await its official release.

Beyond MCMC, there have been advances in *Sequential Monte Carlo* (*SMC*; Doucet et al. (2001); Del Moral et al. (2006)), whereby sets of particles are propagated through sequential *importance samplers* (e.g., Ripley (1987)), rather than constructing a Markov chain under MCMC. There are issues of particle degeneracy, however, whereby the subset of particles (samples) that are consistent with the observed data becomes too small. Hybrid schemes combining MCMC within SMC have been proposed to gain the benefits of both approaches (Andrieu et al., 2010).

Recently, there has been a growing interest in so-called likelihood-free approaches, for example, *approximate Bayesian computation* (*ABC*; Pritchard et al. (1999); Beaumont et al. (2002); Beaumont (2010)), where simulation of the process or model is computationally cheap comparative to evaluating the likelihood in an MCMC approach. There are, however, many unresolved issues, such as model selection (Robert et al., 2011), and the theoretical justification is weaker than for MCMC. Approximation methods that allow fast and precise inference for specific classes of problem have been developed, such as for latent Gaussian models (Rue et al., 2009), but these are not generally applicable outside this class of models.

Variational Bayesian methods are based on approximating an intractable posterior distribution $p(\theta|y)$ by a distribution $q(\theta)$ from a family with a known analytical form and have been applied to machine learning. See, for example, Bishop (2006) or Mackay (2003) for a detailed introduction, and the Infer.NET software (Minka et al., 2011).

5

Prior distributions

The prior distribution plays a defining role in Bayesian analysis. In view of the controversy surrounding its use it may be tempting to treat it almost as an embarrassment and to emphasise its lack of importance in particular applications, but we feel it is a vital ingredient and needs to be squarely addressed. In this chapter we introduce basic ideas by focusing on single parameters, and in subsequent chapters consider multi-parameter situations and hierarchical models. Our emphasis is on understanding what is being used and being aware of its (possibly unintentional) influence.

5.1 Different purposes of priors

A basic division can be made between so-called "non-informative" (also known as "reference" or "objective") and "informative" priors. The former are intended for use in situations where scientific objectivity is at a premium, for example, when presenting results to a regulator or in a scientific journal, and essentially means the Bayesian apparatus is being used as a convenient way of dealing with complex multi-dimensional models. The term "non-informative" is misleading, since all priors contain some information, so such priors are generally better referred to as "vague" or "diffuse." In contrast, the use of informative prior distributions explicitly acknowledges that the analysis is based on more than the immediate data in hand whose relevance to the parameters of interest is modelled through the likelihood, and also includes a considered judgement concerning plausible values of the parameters based on external information.

In fact the division between these two options is not so clear-cut — in particular, we would claim that any "objective" Bayesian analysis is a lot more "subjective" than it may wish to appear. First, any statistical model (Bayesian or otherwise) requires qualitative judgement in selecting its structure and distributional assumptions, regardless of whether informative prior distributions are adopted. Second, except in rather simple situations there may not be an agreed "objective" prior, and apparently innocuous assumptions can strongly influence conclusions in some circumstances.

In fact a combined strategy is often reasonable, distinguishing parameters of primary interest from those which specify secondary structure for the model. The former will generally be location parameters, such as regression coefficients, and in many cases a vague prior that is locally uniform over the region supported by the likelihood will be reasonable. Secondary aspects of a model include, say, the variability between random effects in a hierarchical model. Often there is limited evidence in the immediate data concerning such parameters and hence there can be considerable sensitivity to the prior distribution, in which case we recommend thinking carefully about reasonable values in advance and so specifying fairly informative priors — the inclusion of such external information is unlikely to bias the main estimates arising from a study, although it may have some influence on the precision of the estimates and this needs to be carefully explored through sensitivity analysis. It is preferable to construct a prior distribution on a scale on which one has has a good interpretation of magnitude, such as standard deviation, rather than one which may be convenient for mathematical purposes but is fairly incomprehensible, such as the logarithm of the precision. The crucial aspect is not necessarily to avoid an influential prior, but to be aware of the extent of the influence.

5.2 Vague, "objective," and "reference" priors

5.2.1 Introduction

The appropriate specification of priors that contain minimal information is an old problem in Bayesian statistics: the terms "objective" and "reference" are more recent and reflect the aim of producing a baseline analysis from which one might possibly measure the impact of adopting more informative priors. Here we illustrate how to implement standard suggestions with BUGS. Using the structure of graphical models, the issue becomes one of specifying appropriate distributions on "founder" nodes (those with no parents) in the graph.

We shall see that some of the classic proposals lead to "improper" priors that do not form distributions that integrate to 1: for example, a uniform distribution over the whole real line, no matter how small the ordinate, will still have an infinite integral. In many circumstances this is not a problem, as an improper prior can still lead to a proper posterior distribution. BUGS in general requires that a full probability model is defined and hence forces all prior distributions to be proper — the only exception to this is the `dflat()` distribution (Appendix C.1). However, many of the prior distributions used are "only just proper" and so caution is still required to ensure the prior is not having unintended influence.

5.2.2   Discrete uniform distributions

For discrete parameters it is natural to adopt a discrete uniform prior distribution as a reference assumption. We have already seen this applied to the degrees of freedom of a t-distribution in Example 4.1.2, and in §9.8 we will see how it can be used to perform a non-Bayesian bootstrap analysis within BUGS.

5.2.3   Continuous uniform distributions and Jeffreys prior

When it comes to continuous parameters, it is tempting to automatically adopt a uniform distribution on a suitable range. However, caution is required since a uniform distribution for θ does not generally imply a uniform distribution for functions of θ. For example, suppose a coin is known to be biased, but you claim to have "no idea" about the chance θ of it coming down heads and so you give θ a uniform distribution between 0 and 1. But what about the chance (θ^2) of it coming down heads in both of the next two throws? You have "no idea" about that either, but according to your initial uniform distribution on θ, $\psi = \theta^2$ has a density $p(\psi) = 1/(2\sqrt{\psi})$, which can be recognised to be a Beta(0.5, 1) distribution and is certainly not uniform.

Harold Jeffreys came up with a proposal for prior distributions which would be invariant to such transformations, in the sense that a "Jeffreys" prior for θ would be formally compatible with a Jeffreys prior for any 1–1 transformation $\psi = f(\theta)$. He proposed defining a "minimally informative" prior for θ as $p_J(\theta) \propto I(\theta)^{1/2}$ where $I(\theta) = -E[\frac{d^2}{d\theta^2} \log p(Y|\theta)]$ is the Fisher information for θ (§3.6.1). Since we can also express $I(\theta)$ as

$$I(\theta) = E_{Y|\theta}\left[\left(\frac{d\log p(Y|\theta)}{d\theta}\right)^2\right],$$

we have

$$I(\psi) = I(\theta)\left|\frac{d\theta}{d\psi}\right|^2.$$

Jeffreys' prior is therefore invariant to reparameterisation since

$$I(\psi)^{1/2} = I(\theta)^{1/2}\left|\frac{d\theta}{d\psi}\right|,$$

and the Jacobian terms cancel when transforming variables via the expression in §2.4. Hence, a Jeffreys prior for θ transforms to a Jeffreys prior for any 1–1 function $\psi(\theta)$.

As an informal justification, Fisher information measures the curvature of the log-likelihood, and high curvature occurs wherever small changes in parameter values are associated with large changes in the likelihood: Jeffreys' prior gives more weight to these parameter values and so ensures that the influence of the data and the prior essentially coincide. We shall see examples of Jeffreys priors in future sections.

Finally, we emphasise that if the specific form of vague prior is influential in the analysis, this strongly suggests you have insufficient data to draw a robust conclusion based on the data alone and that you should not be trying to be "non-informative" in the first place.

5.2.4 Location parameters

A location parameter θ is defined as a parameter for which $p(y|\theta)$ is a function of $y - \theta$, and so the distribution of $y - \theta$ is independent of θ. In this case Fisher's information is constant, and so the Jeffreys procedure leads to a uniform prior which will extend over the whole real line and hence be improper. In BUGS we could use dflat() to represent this distribution, but tend to use proper distributions with a large variance, such as dunif(-100,100) or dnorm(0,0.0001): we recommend the former with appropriately chosen limits, since explicit introduction of these limits reminds us to be wary of their potential influence. We shall see many examples of this use, for example, for regression coefficients, and it is always useful to check that the posterior distribution is well away from the prior limits.

5.2.5 Proportions

The appropriate prior distribution for the parameter θ of a Bernoulli or binomial distribution is one of the oldest problems in statistics, and here we illustrate a number of options. First, both Bayes (1763) and Laplace (1774) suggest using a uniform prior, which is equivalent to Beta(1, 1). A major attraction of this assumption, also known as the Principle of Insufficient Reason, is that it leads to a discrete uniform distribution for the predicted number y of successes in n future trials, so that $p(y) = 1/(n+1)$, $y = 0, 1, ..., n$,[*] which seems rather a reasonable consequence of "not knowing" the chance of success. On the $\phi = \text{logit}(\theta)$ scale, this corresponds to a standard logistic distribution, represented as dlogis(0,1) in BUGS (see code below).

Second, an (improper) uniform prior on ϕ is formally equivalent to the (improper) Beta(0, 0) distribution on the θ scale, i.e., $p(\theta) \propto \theta^{-1}(1 - \theta)^{-1}$: the code below illustrates the effect of bounding the range for ϕ and hence making these distributions proper. Third, the Jeffreys principle leads to a Beta(0.5, 0.5) distribution, so that $p_J(\theta) = \pi^{-1}\theta^{\frac{1}{2}}(1 - \theta)^{\frac{1}{2}}$. Since it is common to use normal prior distributions when working on a logit scale, it is of interest to consider what normal distributions on ϕ lead to a "near-uniform"

---

[*]See Table 3.1 — the posterior predictive distribution for a binomial observation and beta prior is a beta-binomial distribution. With no observed data, $n = y = 0$ in Table 3.1, this posterior predictive distribution becomes the *prior predictive* distribution, which reduces to the discrete uniform for $a = b = 1$.

distribution on θ. Here we consider two possibilities: assuming a prior variance of 2 for ϕ can be shown to give a density for θ that is "flat" at $\theta = 0.5$, while a normal with variance 2.71 gives a close approximation to a standard logistic distribution, as we saw in Example 4.1.1.

```
theta[1]          ~ dunif(0,1)      # uniform on theta
phi[1]            ~ dlogis(0,1)

phi[2]            ~ dunif(-5,5)      # uniform on logit(theta)
logit(theta[2]) <- phi[2]

theta[3]          ~ dbeta(0.5,0.5) # Jeffreys on theta
phi[3]            <- logit(theta[3])

phi[4]            ~ dnorm(0,0.5)    # var=2, flat at theta = 0.5
logit(theta[4]) <- phi[4]

phi[5]            ~ dnorm(0,0.368) # var=2.71, approx. logistic
logit(theta[5]) <- phi[5]
```

We see from Figure 5.1 that the first three options produce apparently very different distributions for θ, although in fact they differ at most by a single implicit success and failure (§5.3.1). The normal prior on the logit scale with variance 2 seems to penalise extreme values of θ, while that with variance 2.71 seems somewhat more reasonable. We conclude that, in situations with very limited information, priors on the logit scale could reasonably be restricted to have variance of around 2.7.

Example 5.2.1. *Surgery (continued): prior sensitivity*
What is the sensitivity to the above prior distributions for the mortality rate in our "Surgery" example (Example 3.3.2)? Suppose in one case we observe 0/10 deaths (Figure 5.2, left panel) and in another, 10/100 deaths (Figure 5.2, right panel). For 0/10 deaths, priors 2 and 3 pull the estimate towards 0, but the sensitivity is much reduced with the greater number of observations.

5.2.6   Counts and rates

For a Poisson distribution with mean θ, the Fisher information is $I(\theta) = 1/\theta$ and so the Jeffreys prior is the improper $p_J(\theta) \propto \theta^{-\frac{1}{2}}$, which can be approximated in BUGS by a dgamma(0.5, 0.00001) distribution. The same prior is appropriate if θ is a rate parameter per unit time, so that $Y \sim \text{Poisson}(\theta t)$.

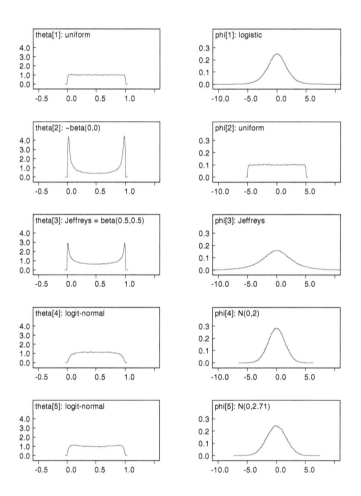

FIGURE 5.1

Empirical distributions (based on 100,000 samples) corresponding to various different priors for a proportion parameter.

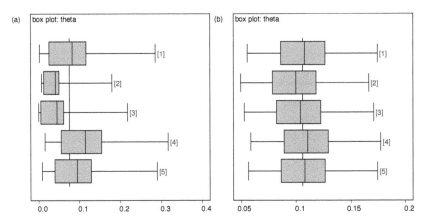

FIGURE 5.2

Box plots comparing posterior distributions arising from the five priors discussed above for mortality rate: (a) 0/10 deaths observed; (b) 10/100 deaths observed.

5.2.7 Scale parameters

Suppose σ is a scale parameter, in the sense that $p(y|\sigma) = \sigma^{-1} f(y/\sigma)$ for some function f, so that the distribution of Y/σ does not depend on σ. Then it can be shown that the Jeffreys prior is $p_J(\sigma) \propto \sigma^{-1}$, which in turn means that $p_J(\sigma^k) \propto \sigma^{-k}$, for any choice of power k. Thus for the normal distribution, parameterised in BUGS in terms of the precision $\tau = 1/\sigma^2$, we would have $p_J(\tau) \propto \tau^{-1}$. This prior could be approximated in BUGS by, say, a `dgamma(0.001,0.001)`, which also can be considered an "inverse-gamma distribution" on the variance σ^2. Alternatively, we note that the Jeffreys prior is equivalent to $p_J(\log \sigma^k) \propto$ const, i.e., an improper uniform prior. Hence it may be preferable to give $\log \sigma^k$ a uniform prior on a suitable range, for example, `log.tau ~ dunif(-10, 10)` for the logarithm of a normal precision. We would usually want the bounds for the uniform distribution to have negligible influence on the conclusions.

We note some potential conflict in our advice on priors for scale parameters: a uniform prior on $\log \sigma$ follows Jeffreys' rule but a uniform on σ is placing a prior on an interpretable scale. There usually would be negligible difference between the two — if there is a noticeable difference, then there is clearly little information in the likelihood about σ and we would recommend a weakly informative prior on the σ scale.

Note that the advice here applies only to scale parameters governing the variance or precision of *observable* quantities. The choice of prior for the variance of *random effects* in a hierarchical model is more problematic — we discuss this in §10.2.3.

5.2.8   Distributions on the positive integers

Jeffreys (1939) [p. 238] suggested that a suitable prior for a parameter N, where $N = 0, 1, 2, ...$, is $p(N) \propto 1/N$, analogously to a scale parameter.

Example 5.2.2. *Coin tossing: estimating number of tosses*
Suppose we are told that a fair coin has come up heads $y = 10$ times. How many times has the coin been tossed? Denoting this unknown quantity by N we can write down the likelihood as

$$p(y|N) = \text{Binomial}(0.5, N) \propto \frac{N!}{(N-y)!}0.5^N.$$

As N is integer-valued we must specify a *discrete* prior distribution.

Suppose we take Jeffreys' suggestion and assign a prior $p(N) \propto 1/N$, which is improper but could be curtailed at a very high value. Then the posterior distribution is

$$p(N|y) \propto \frac{N!}{(N-y)!}0.5^N/N \propto \frac{(N-1)!}{(N-y)!}0.5^N, \quad N \geq y,$$

which we can recognise as the kernel of a negative binomial distribution with mean $2y = 20$. This has an intuitive attraction, since if instead we had fixed $y = 10$ in advance and flipped a coin until we had y heads, then the sampling distribution for the random quantity N would be just this negative binomial. However, it is notable that we were *not* told that this was the design — we have no idea whether the final flip was a head or not.

Alternatively, we may wish to assign a uniform prior over integer values from 1 to 100, i.e., $\Pr(N = n) = 1/100$, $n = 1, ..., 100$. Then the posterior for N is proportional to the likelihood, and its expectation, for example, is given by

$$E[N|y] = \sum_{n=1}^{100} n \Pr(N = n|y) = A \sum_{n=1}^{100} \frac{n \times n!}{(n-y)!}0.5^n, \tag{5.1}$$

where A is the posterior normalising constant. The right-hand side of (5.1) cannot be simplified analytically and so is cumbersome to evaluate (although this is quite straightforward with a little programming). In BUGS we simply specify the likelihood and the prior as shown below.

```
y <- 10
y   ~ dbin(0.5, N)
N   ~ dcat(p[])
for (i in 1:100) {p[i] <- 1/100}
```

BUGS can use the resulting samples to summarise the posterior graphically as well as numerically. Numeric summaries, such as the one shown below, allow us to make formal inferences; for example, we can be 95% certain that the coin has been tossed between 13 and 32 times. Graphical summaries, on the other hand,

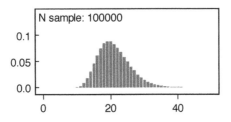

FIGURE 5.3
Approximate posterior distribution for number of (unbiased) coin tosses leading to ten heads.

might reveal interesting features of the posterior. Figure 5.3 shows the posterior density for N. Note that the mode is 20, which is the intuitive answer, as well as being the MLE and the posterior mean using the Jeffreys prior. Note also that although the uniform prior supports values in $\{1, ..., 9\}$, which are impossible in light of the observed data (10 heads), the posterior probability for these values is, appropriately, zero.

node	mean	sd	MC error	2.5%	median	97.5%	start	sample
N	21.01	4.702	0.01445	13.0	20.0	32.0	1	100000

In Example 5.5.2 we consider a further example of a prior over the positive integers which reveals the care that can be required.

5.2.9 More complex situations

Jeffreys' principle does not extend easily to multi-parameter situations, and additional context-specific considerations generally need to be applied, such as assuming prior independence between location and scale parameters and using the Jeffreys prior for each, or specifying an ordering of parameters into groups of decreasing interest.

5.3 Representation of informative priors

Informative prior distributions can be based on pure judgement, a mixture of data and judgement, or data alone. Of course, even the selection of relevant data involves a substantial degree of judgement, and so the specification of an informative prior distribution is never an automatic procedure.

We summarise some basic techniques below, emphasising the mapping of relevant data and judgement onto appropriate parametric forms, ideally representing "implicit" data.

5.3.1   Elicitation of pure judgement

Elicitation of subjective probability distributions is not a straightforward task due to a number of potential biases that have been identified. O'Hagan et al. (2006) provide some *"Guidance for best practice,"* emphasising that probability assessments are constructed by the questioning technique, rather than being "pre-formed quantifications of pre-analysed belief" (p. 217). They say it is best to interview subjects face-to-face, with feedback and continual checking for biases, conducting sensitivity analysis to the consequence of the analysis, and avoiding verbal descriptions of uncertainty. They recommend eliciting intervals with moderate rather than high probability content, say by focusing on 33% and 67% quantiles: indeed one can simply ask for an interval and afterwards elicit a 'confidence' in that assessment (Kynn, 2005). They suggest using multiple experts and reporting a simple average, but it is also important to acknowledge imperfections in the process, and that even genuine "expertise" cannot guarantee a suitable subject. See also Kadane and Wolfson (1998) for elicitation techniques for specific models.

In principle any parametric distribution can be elicited and used in BUGS. However, it can be advantageous to use conjugate forms since, as we have seen in Chapter 3, the prior distribution can then be interpreted as representing "implicit data," in the sense of a prior estimate of the parameter and an "effective prior sample size." It might even then be possible to include the prior information as "data" and use standard classical methods (and software) for statistical analysis.

Below we provide a brief summary of situations: in each case the "implicit data" might be directly elicited, or measures of central tendency and spread requested and an appropriate distribution fitted. A simple moment-based method is to ask directly for the mean and standard deviation, or elicit an approximate 67% interval (i.e., the parameter is assessed to be twice as likely to be inside the interval as outside it) and then treat the interval as representing the mean $\pm$ 1 standard deviation, and solve for the parameters of the prior distribution. In any case it is good practice to iterate between alternative representations of the prior distribution, say as a drawn distribution, percentiles, moments, and interpretation as "implicit data," in order to check the subject is happy with the implications of their assessments.

- Binomial proportion θ. Suppose our prior information is equivalent to having observed y events in a sample size of n, and we wanted to derive a corresponding Beta(a, b) prior for θ. Combining an improper Beta$(0,0)$ "pre-prior" with these implicit data gives a conjugate "posterior" of Beta$(y, n - y)$, which we can interpret as our elicited prior. The mean of this elicited prior is $a/(a+b) = y/n$, the intuitive point estimate for θ, and the implicit sample size is $a+b=n$. Using a uniform "pre-prior" instead of the Beta(0,0) gives $a = y+1$ and $b = n-y+1$.

Alternatively, a moment-based method might proceed by eliciting a prior standard deviation as opposed to a prior sample size, and by then solving the mean and variance formulae (Appendix C.3) for a and b: $a = mb/(1-m)$, $b = m(1-m)^2/v + m - 1$, for an elicited mean $m = \hat{\theta}$ and variance v.

- Poisson rate θ: if we assume θ has a Gamma(a,b) distribution we can again elicit a prior estimate $\hat{\theta} = a/b$ and an effective sample size of b, assuming a Gamma(0,0) pre-prior (see Table 3.1, Poisson-gamma conjugacy), or we can use a moment-based method instead.

- Normal mean μ: a normal distribution can be obtained be eliciting a mean γ and standard deviation ω directly or via an interval. By conditioning on a sampling variance σ^2, we can calculate an effective prior sample size $n_0 = \sigma^2/\omega^2$ which can be fed back to the subject.

- Normal variance σ^2: $\tau = \sigma^{-2}$ may be assumed to have a Gamma(a,b) distribution, where a/b is set to an estimate of the precision, and $2a$ is the effective number of prior observations, assuming a Gamma(0,0) pre-prior (see Table 3.1, normal y with unknown variance σ^2).

- Regression coefficients: In many circumstances regression coefficients will be unconstrained parameters in standard generalised linear models, say log-odds ratios in logistic regression, log-rate-ratios in Poisson regression, log-hazard ratios in Cox regression, or ordinary coefficients in standard linear models. In each case it is generally appropriate to assume a normal distribution. Kynn (2005) described the elicitation of regression coefficients in GLMs by asking an expert for expected responses for different values of a predictor. Lower and upper estimates, with an associated degree of confidence, were also elicited, and the answers used to derive piecewise-linear priors.

Example 5.3.1. *Power calculations*

A randomised trial is planned with n patients in each of two arms. The response within each treatment arm is assumed to have between-patient standard deviation σ, and the estimated treatment effect Y is assumed to have a Normal$(\theta, 2\sigma^2/n)$ distribution. A trial designed to have two-sided Type I error α and Type II error β in detecting a true difference of θ in mean response between the groups will require a sample size per group of

$$n = \frac{2\sigma^2}{\theta^2}(z_{1-\beta} + z_{1-\alpha/2})^2,$$

where $\Pr(Z < z_p) = p$ for a standard normal variable $Z \sim \text{Normal}(0, 1)$. Alternatively, for fixed n, the power of the study is

$$\text{Power} = \Phi\left(\sqrt{\frac{n\theta^2}{2\sigma^2}} - z_{1-\alpha/2}\right).$$

If we assume $\theta = 5, \sigma = 10, \alpha = 0.05, \beta = 0.10$, so that the power of the trial is 90%, then we obtain $z_{1-\beta} = 1.28$, $z_{1-\alpha/2} = 1.96$, and $n = 84$.

Suppose we wish to acknowledge uncertainty about the alternative hypothesis θ and the standard deviation σ. First, we assume past evidence suggests θ is likely to lie anywhere between 3 and 7, which we choose to interpret as a 67% interval (± 1 standard deviation), and so $\theta \sim \text{Normal}(5, 2^2)$. Second, we assess our estimate of $\sigma = 10$ as being based on around 40 observations, from which we assume a $\text{Gamma}(a, b)$ prior distribution for $\tau = 1/\sigma^2$ with mean $a/b = 1/10^2$ and effective sample size $2a = 40$, from which we derive $\tau \sim \text{Gamma}(20, 2000)$.

```
tau      ~ dgamma(20, 2000)
sigma <- 1/sqrt(tau)
theta  ~ dnorm(5, 0.25)
n        <- 2*pow((1.28 + 1.96)*sigma/theta, 2)   # n for 90% power
power <- phi(sqrt(84/2)*theta/sigma - 1.96)       # power for n = 84
p70    <- step(power - 0.7)                        # Pr(power > 70%)
```

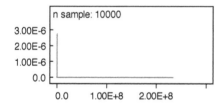 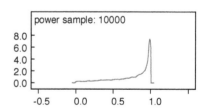

FIGURE 5.4
Empirical distributions based on 10,000 simulations for: n, the number of subjects required in each group to achieve 90% power, and power, the power achieved with 84 subjects in each group.

node	mean	sd	MC error	2.5%	median	97.5%	start	sample
n	38740.0	2.533E+6	25170.0	24.73	87.93	1487.0	1	10000
p70	0.7012	0.4577	0.004538	0.0	1.0	1.0	1	10000
power	0.7739	0.2605	0.002506	0.1151	0.8863	1.0	1	10000

Note that the median values for n (88) and power (0.89) are close to the values derived by assuming fixed θ and σ (84 and 0.90, respectively), but also note the huge uncertainty. It is quite plausible, under the considered prior for θ and σ, that to achieve 90% power the trial may need to include nearly 3000 subjects. Then again, we might get away with as few as 50! A trial involving 84 subjects in each group could be seriously underpowered, with 12% power being quite plausible. Indeed, there is a 30% chance that the power will be less than 70%.

5.3.2 Discounting previous data

Suppose we have available some historical data and we could obtain a prior distribution for the parameter θ based on an empirical estimate $\hat{\theta}_H$, say, by matching the prior mean and standard deviation to $\hat{\theta}_H$ and its estimated standard error. If we were to use this prior directly then we would essentially be pooling the data in a form of meta-analysis (see §11.4), in which case it would be preferable (and essentially equivalent) to use a reference prior and include the historical data directly in the model.

If we are reluctant to do this, it must be because we do not want to give the historical data full weight, perhaps because we do not consider it to have the same relevance and rigour as the data directly being analysed. We may there-fore wish to *discount* the historical data using one of the methods outlined below.

- *Power prior:* this uses a prior mean based on the historical estimate $\hat{\theta}_H$, but discounts the "effective prior sample size" by a factor κ between 0 and 1: for example, a fitted Beta(a, b) would become a Beta$(\kappa a, \kappa b)$, a Gamma(a, b) would become a Gamma$(\kappa a, \kappa b)$, a Normal(γ, ω^2) would become a Normal$(\gamma, \omega^2/\kappa)$ (Ibrahim and Chen, 2000).

- *Bias modelling:* This explicitly considers that the historical data may be biased, in the sense that the estimate $\hat{\theta}_H$ is estimating a slightly different quantity from the θ of current interest. We assume that $\theta = \theta_H + \delta$, where δ is the bias whose distribution needs to be assessed. We further assume $\delta \sim [\mu_\delta, \sigma_\delta^2]$, where $[,]$ indicates a mean and variance but otherwise unspecified distribution. Then if we assume the historical data give rise to a prior distribution $\theta_H \sim [\gamma_H, \omega_H^2]$, we obtain a prior distribution for θ of

$$\theta \sim [\gamma_H + \mu_\delta, \omega_H^2 + \sigma_\delta^2].$$

Thus the prior mean is shifted and the prior variance is increased.

The power prior only deals with variability — the discount factor κ essen-tially represents the "weight" on a historical observation, which is an attractive concept to communicate but somewhat arbitrary to assess. In contrast, the bias modelling approach allows biases to be added, and the parameters can be defined in terms of the size of potential biases.

Example 5.3.2. *Power calculations (continued)*
We consider the power example (Example 5.3.1) but with both prior distributions
discounted. We assume each historical observation informing the prior distribution
for σ is only worth half a current observation, so that the prior for σ is only based
on 10 rather than 20 observations. This discounts the parameters in the gamma
distribution for τ by a factor of 2. For the treatment effect, we assume that the
historical experiment could have been more favourable than the current one, so
that the historical treatment effect had a bias with mean -1 and SD 2, and so
would be expected to be between -5 and 3. Thus an appropriate prior distribution
is $\theta \sim \text{Normal}(5 - 1, 2^2 + 2^2)$ or $\text{Normal}(4, 8)$ — this has been constrained to
be > 0 using the I(,) construct (see Appendix A.2.2 and §9.6). This leads to
the code:

```
# tau      ~ dgamma(20, 2000)
  tau      ~ dgamma(10, 1000)       # discounted by 2
# theta    ~ dnorm(5, 0.25)
  theta    ~ dnorm(4, 0.125)I(0,)   # 4 added to var and shifted
                                    # by -1, constrained to be >0
```

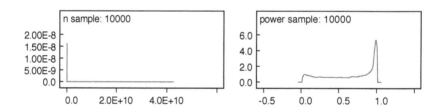

FIGURE 5.5
Empirical distributions based on 10,000 simulations for: n, the number of subjects
required in each group to achieve 90% power, and power, the power achieved with
84 subjects in each group. Discounted priors for tau and theta used.

node	mean	sd	MC error	2.5%	median	97.5%	start	sample
n	4.542E+6	4.263E+8	4.26E+6	20.96	125.6	14270.0	1	10000
p70	0.5398	0.4984	0.005085	0.0	1.0	1.0	1	10000
power	0.6536	0.3315	0.003406	0.04353	0.7549	1.0	1	10000

This has raised the median sample size to 126, but with huge uncertainty. There
is a 46% probability that the power is less than 70% if the sample size stays at
84.

5.4 Mixture of prior distributions

Suppose we want to express doubt about which of two or more prior distributions is appropriate for the data in hand. For example, we might suspect that *either* a drug will produce a similar effect to other related compounds, *or* if it doesn't behave like these compounds we are unsure about its likely effect.

For two possible prior distributions $p_1(\theta)$ and $p_2(\theta)$ for a parameter θ, the overall prior distribution is then a *mixture*

$$p(\theta) = qp_1(\theta) + (1-q)p_2(\theta),$$

where q is the assessed probability that p_1 is "correct." If we now observe data y, it turns out that the posterior for θ is

$$p(\theta|y) = q'p_1(\theta|y) + (1-q')p_2(\theta|y)$$

where

$$p_i(\theta|y) \propto p(y|\theta)p_i(\theta),$$
$$q' = \frac{qp_1(y)}{qp_1(y) + (1-q)p_2(y)},$$

where $p_i(y) = \int p(y|\theta)p_i(\theta)\, d\theta$ is the predictive probability of the data y assuming $p_i(\theta)$. The posterior is a mixture of the respective posterior distributions under each prior assumption, with the mixture weights adapted to support the prior that provides the best prediction for the observed data.

This structure is easy to implement in BUGS for any form of prior assumptions. We first illustrate its use with a simple example and then deal with some of the potential complexities of this formulation. In the example, `pick` is a variable taking the value j when the prior assumption j is selected in the simulation.

Example 5.4.1. *A biased coin?*
Suppose a coin is either unbiased or biased, in which case the chance of a "head" is unknown and is given a uniform prior distribution. We assess a prior probability of 0.9 that it is unbiased, and then observe 15 heads out of 20 tosses — what is the chance that the coin is biased?

```
r <- 15; n <- 20            # data
#########################################
r            ~ dbin(p, n)   # likelihood
p            <- theta[pick]
pick         ~ dcat(q[])    # 2 if biased, 1 otherwise
q[1]         <- 0.9
```

```
q[2]       <- 0.1
theta[1] <- 0.5          # if unbiased
theta[2]  ~ dunif(0, 1)  # if biased
biased    <- pick - 1    # 1 if biased, 0 otherwise
```

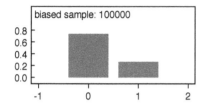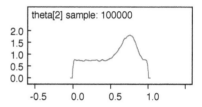

FIGURE 5.6

Biased coin: empirical distributions based on 100,000 simulations.

```
node      mean   sd      MC error  2.5%    median  97.5%  start sample
biased    0.2619 0.4397  0.002027  0.0     0.0     1.0    1     100000
theta[2]  0.5594 0.272   9.727E-4  0.03284 0.6247  0.9664 1     100000
```

So the probability that the coin is biased has increased from 0.1 to 0.26 on the basis of the evidence provided. The rather strange shape of the posterior distribution for theta[2] is explained below.

If the alternative prior assumptions for theta in Example 5.4.1 were from the same parametric family, e.g., beta, then we could formulate this as p ~ dbeta(a[pick], b[pick]), say, with specified values of a[1], a[2], b[1], and b[2]. However, the more general formulation shown in the example allows prior assumptions of arbitrary structure.

It is important to note that when pick=1, theta[1] is sampled from its *posterior* distribution, but theta[2] is sampled from its *prior* as pick=1 has essentially "cut" the connection between the data and theta[2]. At another MCMC iteration, we may have pick=2 and so the opposite will occur, and this means that the posterior for each theta[j] recorded by BUGS is a mixture of "true" (model specific) posterior and its prior. This explains the shape of the posterior for theta[2] in the example above. If we are interested in the posterior distribution under each prior assumption individually, then we could do a separate run under each prior assumption, or only use those values for theta[j] simulated when pick=j: this "post-processing" would have to be performed outside BUGS.

We are essentially dealing with alternative model formulations, and our q's above correspond to posterior probabilities of models. There are well-known difficulties with these quantities both in theory, due to their potential dependence on the within-model prior distributions, and in particular when calculating within MCMC: see §8.7. In principle we can use the structure above to handle a list of arbitrary alternative models, but in practice considerable care is needed if the sampler is not to go "off course" when sampling from the prior distribution at each iteration when that model is not being "picked." It is possible to define "pseudo-priors" for these circumstances, where pick also dictates the prior to be assumed for theta[j] when pick $\neq j$ — see §8.7 and Carlin and Chib (1995).

5.5   Sensitivity analysis

Given that there is no such thing as the *true* prior, sensitivity analysis to alternative prior assumptions is vital and should be an integral part of Bayesian analysis. The phrase "community of priors" (Spiegelhalter et al., 2004) has been used in the clinical trials literature to express the idea that different priors may reflect different perspectives: in particular, the concept of a "sceptical prior" has been shown to be valuable. Sceptical priors will typically be centred on a "null" value for the relevant parameter with the spread reflecting plausible but small effects. We illustrate the use of sceptical and other prior distributions in the following example, where the evidence for an efficacious intervention following myocardial infarction is considered under a range of priors for the treatment effect, namely, "vague," "sceptical," "enthusiastic," "clinical," and "just significant."

Example 5.5.1. *GREAT trial*
Pocock and Spiegelhalter (1992) examine the effect of anistreplase on recovery from myocardial infarction. 311 patients were randomised to receive either anistreplase or placebo (conventional treatment); the number of deaths in each group is given in the table below.

		Treatment		total
		anistreplase	placebo	
Event	death	13	23	36
	no death	150	125	275
total		163	148	311

Let r_j, n_j, and π_j denote the number of deaths, total number of patients, and underlying mortality rate, respectively, in group $j \in \{1,2\}$ ($1 =$ anistreplase; $2 =$ placebo). Inference is required on the log-odds ratio ($\log(\text{OR})$) for mortality in the anistreplase group compared to placebo, that is,

$$\delta = \log \left\{ \frac{\pi_1/(1-\pi_1)}{\pi_2/(1-\pi_2)} \right\} = \text{logit}\,\pi_1 - \text{logit}\,\pi_2. \tag{5.2}$$

A classical maximum likelihood estimator and approximate variance are given by

$$\hat{\delta} = \log\left\{\frac{r_1/(n_1 - r_1)}{r_2/(n_2 - r_2)}\right\}, \quad V(\hat{\delta}) \approx s^2 = \frac{1}{r_1} + \frac{1}{r_2} + \frac{1}{n_1 - r_1} + \frac{1}{n_2 - r_2}.$$

For the above data these give $\hat{\delta} = -0.753$ with $s = 0.368$. An approximate Bayesian analysis might proceed via the assumption $\hat{\delta} \sim \text{Normal}(\delta, s^2)$ with a locally uniform prior on δ, e.g., $\delta \sim \text{Uniform}(-10, 10)$. A more appropriate likelihood is a binomial assumption for each observed number of deaths: $r_j \sim \text{Binomial}(\pi_j, n_j)$, $j = 1, 2$. In this case we could be "vague" by specifying Jeffreys priors for the mortality rates, $\pi_j \sim \text{Beta}(0.5, 0.5)$, $j = 1, 2$, and then deriving the posterior for δ via (5.2). Alternatively we might parameterise the model directly in terms of δ:

$$\text{logit}\,\pi_1 = \alpha + \delta/2, \quad \text{logit}\,\pi_2 = \alpha - \delta/2,$$

which facilitates the specification of informative priors for δ. Here α is a nuisance parameter and is assigned a vague normal prior: $\alpha \sim \text{Normal}(0, 100^2)$. Our first informative prior for δ is a "clinical" prior based on expert opinion: a senior cardiologist, informed by one unpublished and two published trials, expressed belief that *"an expectation of 15–20% reduction in mortality is highly plausible, while the extremes of no benefit and a 40% relative reduction are both unlikely."* This is translated into a normal prior with a 95% interval of -0.51 to 0 (0.6 to 1.0 on the OR scale): $\delta \sim \text{Normal}(-0.26, 0.13^2)$. We also consider a "sceptical" prior, which is designed to represent a reasonable expression of doubt, perhaps to avoid early stopping of trials due to fortuitously positive results. For example, a hypothetical sceptic might find treatment effects more extreme than a 50% reduction or 100% increase in mortality largely implausible, giving a 95% prior interval (assuming normality) of -0.69 to 0.69 (0.5 to 2 on the OR scale): $\delta \sim \text{Normal}(0, 0.35^2)$.

As a counterbalance to the sceptical prior we might specify an "enthusiastic" or "optimistic" prior, as a basis for conservatism in the face of early negative results, say. Such a prior could be centred around some appropriate beneficial treatment effect with a small prior probability (e.g., 5%) assigned to negative treatment benefits. We do not construct such a prior in this example, however, since the clinical prior described above also happens to be "enthusiastic" in this sense. Another prior of interest is the "just significant" prior. Assuming that the treatment effect is significant under a vague prior, it is instructive to ask how sceptical we would have to be for that significance to vanish. Hence we assume $\delta \sim \text{Normal}(0, \sigma_\delta^2)$ and we search for the largest value of σ_δ such that the 95% posterior credible interval (just) includes zero. BUGS code for performing such a search is presented below along with code to implement the clinical, sceptical, and vague priors discussed above. (Note that a preliminary search had been run to identify the approximate value of σ_δ as somewhere between 0.8 and 1, though closed form approximations exist for this "just signficant" prior (Matthews, 2001; Spiegelhalter et al., 2004)).

```
model {
  for (i in 1:nsearch) {                    # search for "just
    pr.sd[i]          <- start + i*step # significant" prior
    pr.mean[i]        <- 0
  }
  pr.mean[nsearch+1] <- -0.26
  pr.sd[nsearch+1]   <- 0.13              # clinical prior
  pr.mean[nsearch+2] <- 0
  pr.sd[nsearch+2]   <- 0.35              # sceptical prior

  # replicate data for each prior and specify likelihood...
  for (i in 1:(nsearch+3)) {
    for (j in 1:2) {
      r.rep[i,j]      <- r[j]
      n.rep[i,j]      <- n[j]
      r.rep[i,j]       ~ dbin(pi[i,j], n.rep[i,j])
    }
  }
  delta.mle           <- -0.753
  delta.mle            ~ dnorm(delta[nsearch+4], 7.40)

  # define priors and link to log-odds...
  for (i in 1:(nsearch+2)) {
    logit(pi[i,1])    <- alpha[i] + delta[i]/2
    logit(pi[i,2])    <- alpha[i] - delta[i]/2
    alpha[i]           ~ dnorm(0, 0.0001)
    delta[i]           ~ dnorm(pr.mean[i], pr.prec[i])
    pr.prec[i]        <- 1/pow(pr.sd[i], 2)
  }
  pi[nsearch+3,1]      ~ dbeta(0.5, 0.5)
  pi[nsearch+3,2]      ~ dbeta(0.5, 0.5) # Jeffreys prior
  delta[nsearch+3]    <- logit(pi[nsearch+3,1])
                         - logit(pi[nsearch+3,2])
  delta[nsearch+4]     ~ dunif(-10, 10) # locally uniform prior
}

list(r = c(13, 23), n = c(163, 148),
  start = 0.8, step = 0.005, nsearch = 40)
```

The derived value of σ_δ is ~ 0.925, corresponding to the 25th element of delta[] above. Selected posterior and prior distributions are summarised below. We note the essentially identical conclusions of the classical maximum likelihood approach and the two analyses with vague priors. The results suggest we should conclude that anistreplase is a superior treatment to placebo if we are either (a priori) completely ignorant of possible treatment effect sizes, or we trust the senior cardiologist's expert opinion, or perhaps if we are otherwise enthusiastic about the new treatment's efficacy. If, on the other hand, we wish to claim prior indifference as to the sign of the treatment effect but we believe "large" treatment effects to be implausible, we should be more cautious. The "just significant" prior has a 95% interval of $(\exp(-1.96 \times 0.925), \exp(1.96 \times 0.925)) = (0.16, 6.1)$ on the OR scale, corresponding to reductions/increases in mortality as extreme as 84%/610%. These seem quite extreme, implying that only a small degree of scepticism is required to render the analysis "non-significant." We might conclude that the GREAT trial alone does not provide "credible" evidence for superiority, and larger-scale trials are required to quantify the treatment effect precisely.

node	mean	sd	MC error	2.5%	median	97.5%	start	sample
delta[25]	-0.6635	0.3423	5.075E-4	-1.343	-0.6609	3.598E-4	1001	500000
delta[41]	-0.317	0.1223	1.741E-4	-0.5562	-0.317	-0.07745	1001	500000
delta[42]	-0.3664	0.2509	3.497E-4	-0.8608	-0.366	0.1245	1001	500000
delta[43]	-0.7523	0.367	5.342E-4	-1.487	-0.7479	-0.04719	1001	500000
delta[44]	-0.7534	0.3673	5.432E-4	-1.475	-0.7529	-0.0334	1001	500000

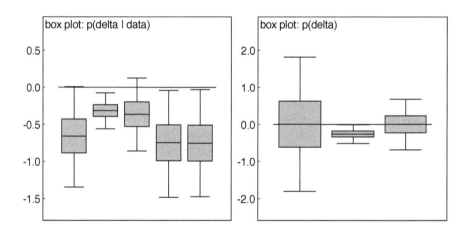

FIGURE 5.7

Left-hand side: Posterior distributions for δ from analysis of GREAT trial data. From left to right: corresponding to "just significant," "clinical," "sceptical," "Jeffreys" and "locally uniform" priors. Right-hand side: Prior distributions for analysis of GREAT trial data. From left to right: "just significant," "clinical" and "sceptical."

A primary purpose of trying a range of reasonable prior distributions is to find unintended sensitivity to apparently innocuous "non-informative" assumptions. This is reflected in the following example.

Example 5.5.2. *Trams: a classic problem from Jeffreys (1939)*

Suppose you enter a town of unknown size whose trams you know are numbered consecutively from 1 to N. You first see tram number $y = 100$. How large might N be?

We first note that the sampling distribution is uniform between 1 and N, so that $p(y|N) = \frac{1}{N}$, $y = 1, 2, \ldots, N$. Therefore the likelihood function for N is $\propto 1/N$, $N \geq y$, so that y maximises the likelihood function and so is the maximum likelihood estimator. The maximum likelihood estimate is therefore 100, which does not appear very reasonable.

Suppose we take a Bayesian approach and consider the prior distributions on the positive integers explored earlier (Example 5.2.2) — we will first examine the consequences using WinBUGS and then algebraically. We first consider a prior that is uniform on the integers up to an arbitrary upper bound M, say 5000. Y is assumed drawn from a categorical distribution: the following code shows how to set a uniform prior for N over the integers 1 to 5000 (as in Example 5.2.2) and how to use the step function to create a uniform sampling distribution between 1 and N.

```
Y <- 100
#######################
Y                ~ dcat(p[])
# sampling distribution is uniform over first N integers
# use step function to change p[j] to 0 for j>N
for (j in 1:M) {
  p[j]       <- step(N - j + 0.01)/N
}
N                ~ dcat(p.unif[])
for (j in 1:M) {
  p.unif[j] <- 1/M
}
```

node	mean	sd	MC error	2.5%	median	97.5%	start	sample
N	1274.0	1295.0	10.86	109.0	722.0	4579.0	1001	10000

The posterior mean is 1274 and the median is 722, reflecting a highly skewed distribution. But is this a sensible conclusion? For an improper uniform prior over the whole of the integers, the posterior distribution is

$$p(N|y) \propto p(y|N)p(N) \propto 1/N, \ N \geq y.$$

This series diverges and so this produces an improper posterior distribution. Although our bounded prior is proper and so our posterior distribution is formally proper, this "almost improper" character is likely to lead to extreme sensitivity to prior assumptions. For example, a second run with $M = 15{,}000$ results in a posterior mean of 3041 and median 1258. In fact we could show algebraically that the posterior mean increases as $M/\log(M)$; thus we can make it as big as we want by increasing M (proof as exercise).

We now consider Jeffreys' suggestion of a prior $p(N) \propto 1/N$, which is improper but can be constructed as follows if an upper bound, say 5000, is set.

```
N                      ~ dcat(p.jeffreys[])
for (j in 1:5000) {
  reciprocal[j] <- 1/j
  p.jeffreys[j] <- reciprocal[j]/sum.recip
}
sum.recip        <- sum(reciprocal[])
```

The results show a posterior mean of 409 and median 197, which seems more reasonable — Jeffreys approximated the probability that there are more than 200 trams as $1/2$.

```
node mean   sd     MC error 2.5%  median 97.5%  start sample
N    408.7 600.4 4.99      102.0 197.0  2372.0 1001  10000
```

Suppose we now change the arbitrary upper bound to $M = 15,000$. Then the posterior mean becomes 520 and median 200. The median, but not the mean, is therefore robust to the prior. We could show that the conclusion about the median is robust to the arbitrary choice of upper bound M by proving that as M goes to infinity the posterior median tends to a fixed quantity (proof as exercise).

Finally, if a sensitivity analysis shows that the prior assumptions make a difference, then this finding should be welcomed. It means that the Bayesian approach has been worthwhile taking, and you will have to think properly about the prior and justify it. It will generally mean that, at a minimum, a weakly informative prior will need to be adopted.

6

Regression models

As in classical regression, Bayesian regression models are formulated by specifying a sampling distribution for the data (which we also loosely term the likelihood) and then a form of relationship between the assumed distribution of the response variable and any explanatory variables. The only difference is that we also specify prior distributions for the regression coefficients and any other unknown (nuisance) parameters. As we will see in this chapter, there are several advantages to a Bayesian approach, however, such as it being relatively straightforward to include parameter restrictions, use non-linear models, "robustify" against outliers, make predictions and inferences about functions of regression parameters, and handle missing data.

6.1 Linear regression with normal errors

Suppose our response variable is denoted y_i, $i = 1, ..., n$, and we have p covariates $x_{1i}, ..., x_{pi}$. We specify

$$y_i \sim \text{Normal}(\mu_i, \sigma^2), \quad \mu_i = \beta_0 + \sum_{k=1}^{p} \beta_k x_{ki},$$

along with prior distributions for $\beta_0, \beta_1, ..., \beta_p$ and σ. For example,

$$\beta_k \sim \text{Normal}(0, 100^2), \quad \log \sigma \sim \text{Uniform}(-100, 100)$$

or the alternative priors discussed in §5.2.4 and §5.2.7. Again we emphasise that if the specific choice of vague prior is influential, this suggests that a robust conclusion cannot be drawn from the data alone and more informative priors based on background information should be considered.

In Bayesian regression analysis it is generally advisable to consider "centering" any covariates, that is, subtracting the empirical mean from each value, as illustrated in the following example. This has the effect of reducing the posterior correlation between each coefficient $(\beta_1, ..., \beta_p)$ and the intercept term β_0, because the intercept is essentially relocated to the "centre" of the data. As discussed in §4.4, high levels of posterior correlation are problematic for Gibbs sampling.

Example 6.1.1. *Growth curve*
Gelfand et al. (1990) examine growth data from 30 young rats whose weights
were measured weekly for five weeks. In this example we fit a linear regression to
the 9th rat's data. The response variable y_i, $i = 1, ..., 5$, is the weight, in grams,
on day x_i.

```
model {
  for (i in 1:5) {
    y[i]          ~ dnorm(mu[i], tau)
    mu[i]        <- alpha + beta*(x[i] - mean(x[]))
  }
  # Jeffreys priors
  alpha          ~ dflat()
  beta           ~ dflat()
  tau           <- 1/sigma2
  log(sigma2) <- 2*log.sigma
  log.sigma      ~ dflat()
}

list(y = c(177,236,285,350,376), x = c(8,15,22,29,36))
```

We specify improper uniform priors for all parameters, and so the posterior mode
will be equal to the maximum likelihood estimates: $\hat{\alpha} = 284.8$, $\hat{\beta} = 7.31$, $\hat{\sigma}^2 = 71.3$ — note the posterior of σ^2 is extremely skewed. Figure 6.1 shows the posterior
distribution of the model fit, produced through the `Inference->Compare` dialog
box in WinBUGS.

node	mean	sd	MC error	2.5%	median	97.5%	start	sample
alpha	284.8	7.89	0.078	269.9	284.8	300.1	4001	10000
beta	7.316	0.7814	0.008582	5.82	7.316	8.819	4001	10000
sigma2	316.3	743.6	26.14	37.24	145.6	1586.0	4001	10000

Linear models where all or some of the covariates are categorical are sometimes called *analysis of variance* or *analysis of covariance* models, respectively, since the interest is often in comparing the variation of the outcome
within and between categories. In BUGS these are treated just like any other
linear regression — as a linear model with coefficients for each explanatory
variable.

Example 6.1.2. *New York crime*
Press (1971) presents data on the effects of increasing police manpower in New
York City. The response variable is the (seasonally adjusted) change in the number

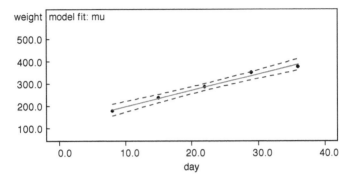

FIGURE 6.1
Model fit from Bayesian linear regression of rat 9's data in Example 6.1.1. The posterior median, 2.5% and 97.5% percentiles for each mu[i] are joined together by straight lines: the solid line joins the medians, whereas the 95% credible intervals are joined by dashed lines. The observed weights are shown by dots.

of thefts in 23 precincts of New York City from a 27-week base period in 1966 to a 58-week experimental period in 1966–1967. The percentage increase in manpower in each precinct (MAN[]) is also recorded, as is the district (DIST[]) to which each precinct belongs (1 = Downtown, 2 = Midtown, 3 = Uptown). The DIST covariate is a categorical variable and requires a slightly different approach to covariates that represent quantities. It doesn't make sense to include a term like beta*DIST[i] in the model because we can't realistically assume that the effect of going from downtown to midtown is the same as going from midtown to uptown, or even that they have the same sign. Instead, we can create and incorporate into the model two new covariates, one equal to one for midtown precincts, and zero otherwise (D2[]), and another equal to one for uptown precincts, and zero otherwise (D3[]). The model code for this *multiple regression* (where more than one covariate is included) is then

```
for (i in 1:23) {
   y[i]                 ~ dnorm(mu[i], tau)
   D2[i]                <- equals(DIST[i], 2)
   D3[i]                <- equals(DIST[i], 3)
   mu[i]                <- beta0 + beta[1]*MAN[i]
                        + beta[2]*D2[i] + beta[3]*D3[i]
}
beta0                   ~ dnorm(0, 0.0001)
for (j in 1:3) {
   beta[j]              ~ dnorm(0, 0.0001)
}
tau                     <- 1/pow(sigma, 2)
# uniform prior on an interpretable scale
sigma                   ~ dunif(0, 100)
```

Posterior summaries for the model parameters are shown below.

node	mean	sd	MC error	2.5%	median	97.5%	start	sample
beta[1]	-0.2378	0.1188	0.001181	-0.4759	-0.2374	-0.006211	5001	10000
beta[2]	0.558	3.027	0.03123	-5.493	0.5386	6.595	5001	10000
beta[3]	-4.03	3.13	0.03011	-10.14	-4.014	2.239	5001	10000
beta0	2.573	1.992	0.01811	-1.291	2.581	6.533	5001	10000
sigma	5.837	1.037	0.01537	4.243	5.686	8.233	5001	10000

Another way to implement the same model is to make use of BUGS' nested indexing feature. In this case we can make use of the DIST covariate directly via the following modification (note that there is no need to calculate D2 and D3 in this case):

```
mu[i] <- beta0 + beta[1]*MAN[i] + gamma[DIST[i]]
```

with

```
gamma[1] <- 0
gamma[2]  ~ dnorm(0, 0.0001)
gamma[3]  ~ dnorm(0, 0.0001)
```

where gamma[1] is fixed because only two district contrasts are identifiable — we could instead remove the intercept term, beta0, and estimate gamma[1] in its place. To provide initial values for a vector such as gamma, which contains both unknown parameters and constants (or logical nodes), we simply specify NA for any elements that are constant/logical, e.g., gamma = c(NA,0,0).

Typically in multiple regression problems, such as in Example 6.1.2 above, we are aiming for a parsimonious model. With this in mind we might wonder whether including a particular covariate in the model is worthwhile. Intuitively it may seem reasonable to require covariates appearing in the final model to have coefficients with high posterior probabilities of being non-zero. Informally we could say that a covariate effect is "significant" (at the 95% level) if the 95% posterior credible interval for the associated coefficient does not include zero. Credible intervals will vary as we include/remove different covariates in/from the model, and so, adopting this strategy, we are faced with the usual problems of forwards and backwards selection.

We will look at model criticism and comparison in more detail in Chapter 8. The Bayesian framework can actually accommodate situations in which the choice of covariates to be included in a given model is a model parameter itself — these methods are reviewed briefly in §8.8.2.

Example 6.1.3. *New York crime (continued)*
The credible intervals obtained in Example 6.1.2 above suggest that the effect of police manpower is "significant," whereas the district effects are not. Hence we might consider removing the D2 and D3 variables from the regression equation:

```
mu[i] <- beta0 + beta[1]*MAN[i]
```

Posterior summaries for beta0, beta[1], and sigma are given below, and the resulting model fit is shown on the left-hand side of Figure 6.2. Note how the coefficients have changed values considerably, and that the effect of police manpower is no longer conventionally "significant," although it is close.

node	mean	sd	MC error	2.5%	median	97.5%	start	sample
beta[1]	-0.1761	0.1096	0.001085	-0.3921	-0.1765	0.04073	5001	10000
beta0	1.97	1.362	0.01394	-0.7489	1.963	4.632	5001	10000
sigma	5.873	0.9813	0.01054	4.297	5.74	8.136	5001	10000

6.2   Linear regression with non-normal errors

In classical linear modelling, the errors are usually assumed to be normally distributed, for example, the "least squares" estimators for linear regression are equivalent to maximum likelihood estimators under this assumption. However, we are not restricted to normality, and BUGS makes it easy to use any appropriate distribution. If we suspect outlying observations, for example, we can provide some robustness against their effects by assuming the data arise from a heavy-tailed t-distribution. Thus the outliers can be accommodated within the tails without necessarily forcing the location of the posterior to be moved significantly. The following example illustrates.

Example 6.2.1. *New York crime (continued): robust regression*
Note from the model fit shown on the left-hand-side of Figure 6.2 that the rightmost point is rather influential — without this point in place, positive and negative regression lines might seem equally plausible. The point corresponds to the 20th precinct (between the Hudson River and Central Park on the southwest side of Central Park). During the study, police manpower in the 20th precinct was *experimentally* increased by 40%, but no experimental changes were made in other precincts. Hence we might have cause to suspect that the corresponding observation could be an outlier. We robustify our analysis against the potential effects of such outliers with the following simple modification:

```
y[i] ~ dt(mu[i], tau, dof)
```

where dt(x,y,z) denotes a Student-t distribution with mean x, precision parameter y and degrees of freedom z (see Appendix C.1 — note the variance is $\frac{z}{y(z-2)}$ for $z > 2$). In principle, we could estimate the degrees of freedom, as in Example 4.1.2, by assigning an appropriate prior, but this can be problematic unless there are many observations. Instead, here, we set dof <- 4 to give a very heavy tailed distribution for the residuals. The model fit is shown on the right-hand side of Figure 6.2 and posterior summaries are given in the table below:

node	mean	sd	MC error	2.5%	median	97.5%	start	sample
beta0	1.699	1.632	0.02879	-1.191	1.59	5.222	5001	10000
beta[1]	-0.1244	0.1449	0.002557	-0.358	-0.1455	0.2097	5001	10000
sigma	4.883	1.035	0.01803	3.266	4.756	7.214	5001	10000

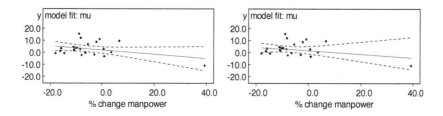

FIGURE 6.2
Model fits for New York crime data in Examples 6.1.3 and 6.2.1 with manpower effect alone in the regression equation: left-hand side, normal residuals; right-hand side, t_4 distributed residuals.

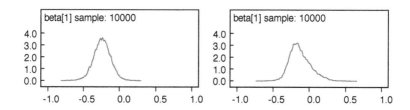

FIGURE 6.3
Posterior density estimates for manpower effect in regression analysis of New York crime data in Examples 6.1.3 and 6.2.1: left-hand side, normal residuals; right-hand side, t_4 distributed residuals.

Note that sigma is no longer the residual standard deviation; this is given by sd <- sigma*sqrt(dof/(dof-2)), which is sigma*1.414 in this case. Hence the posterior median residual standard deviation is 6.73, a little higher than before. Just looking at the posterior median model fit we might think that assuming a t distribution for the residuals has had a negligible effect. However, note that the posterior summaries show the effect of manpower has been attenuated, with the posterior median for beta[1] reduced in size from -0.177 (with normal errors) to

−0.146. This is consistent with less of the overall variability apparent in the data being explained by the model, as indicated by the increased residual standard deviation. Also note that use of the t distribution may affect our substantive inferences, since we can no longer consider the effect of manpower "significant" — considerably more posterior probability now lies to the right of zero. This is reflected by the (now positive) slope of the upper end of the credible interval for the model fit, and by the posterior density estimates for beta[1] shown in Figure 6.3.

6.3 Non-linear regression with normal errors

MCMC methods can easily accommodate non-linear regressions. The only additional effort required in fitting such models might be ensuring that the parameters always have meaningful values by imposing appropriate constraints. This is illustrated in the following example. Note that we can easily extend non-linear models to non-normal errors, as in the previous subsection.

Example 6.3.1. *Dugongs*
Carlin and Gelfand (1991) consider data on length (y_i) and age (x_i) measurements for $i = 1, ..., n = 27$ dugongs (sea cows) captured off the coast of Queensland. The data are shown in Figure 6.5. A frequently used nonlinear growth curve with no inflection point and an asymptote as $x \to \infty$ is the Von Bertalanffy growth model, given by

$$y_i \sim \mathsf{Normal}(\mu_i, \sigma^2), \quad \mu_i = L_\infty - (L_\infty - L_0) e^{-K x_i}, \quad i = 1, ..., n,$$

where $L_\infty > L_0 > 0$ and $K > 0$. L_∞ represents the maximum expected length achievable, and L_0 is the length at time 0. We can impose such constraints in various ways, e.g., $L_0, K \sim \mathsf{Uniform}(0, 100)$, $L_\infty = L_0 + \beta, \beta \sim \mathsf{Uniform}(0, 100)$. We illustrate the use of this particular prior in the code below (model 1). We also illustrate the use of truncated normal priors for L_∞ and L_0 (model 2). We can use the I(,) syntax to represent truncated distributions, as discussed in §9.6 and Appendix A.2.2, as there are no unknown parameters in the prior distribution. In addition, we present two further constrained priors based on the fact that the von Bertalanffy model can be rewritten as $\mu_i = \alpha - \beta \gamma^{x_i}$, with $\alpha = L_\infty > 0$, $\beta = L_\infty - L_0 > 0$, and $0 < \gamma = e^{-K} < 1$. In model 3 we assume $\alpha, \beta \sim \mathsf{Uniform}(0, 100)$ and $\gamma \sim \mathsf{Uniform}(0, 1)$. Model 4 is the same as model 3 except that we use approximately the same prior for γ as in model 1, by assuming $\gamma \sim \mathsf{Gamma}(0.001, 0.001)I(0, 1)$ (since $p(K) \propto 1$ is equivalent to $e^{-K} \sim \mathsf{Gamma}(0, 0)I(0, 1)$). Four copies of the data are supplied and we compare posterior distributions for α, β, γ, and σ^2 between the four priors.

```
model {
  for(j in 1:N) {
    for (i in 1:4) {
      y[i,j]                      ~ dnorm(mu[i,j], tau[i])
    }
    mu[1,j] <- Linf[1] - (Linf[1] - L0[1])*exp(-K[1]*x[1,j])
    mu[2,j] <- Linf[2] - (Linf[2] - L0[2])*exp(-K[2]*x[2,j])
    mu[3,j] <- alpha[3] - beta[3]*pow(gamma[3], x[3,j])
    mu[4,j] <- alpha[4] - beta[4]*pow(gamma[4], x[4,j])
  }
  L0[1]                     ~ dunif(0, 100)
  L0[2]                     ~ dnorm(0, 0.0001)I(0, Linf[2])
  Linf[1]                   <- L0[1] + beta[1]
  Linf[2]                   ~ dnorm(0, 0.0001)I(L0[2], )
  K[1]                      ~ dunif(0, 100)
  K[2]                      ~ dunif(0, 100)
  for (i in 1:2) {alpha[i] <- Linf[i]}
  for (i in 3:4) {alpha[i]  ~ dunif(0, 100)}
  beta[1]                   ~ dunif(0, 100)
  beta[2]                   <- Linf[2] - L0[2]
  for (i in 3:4) {beta[i]   ~ dunif(0, 100)}
  for (i in 1:2) {gamma[i] <- exp(-K[i])}
  gamma[3]                  ~ dunif(0, 1)
  gamma[4]                  ~ dgamma(0.001, 0.001)I(0, 1)
  for (i in 1:4) {
    tau[i]                  <- 1/sigma2[i]
    log(sigma2[i])          <- 2*log.sigma[i]
    log.sigma[i]            ~ dunif(-10, 10)
  }
}
```

node	mean	sd	MC error	2.5%	median	97.5%	start	sample
alpha[1]	2.65	0.07281	0.001407	2.527	2.644	2.809	10001	50000
alpha[2]	2.651	0.07263	0.001245	2.529	2.644	2.814	10001	50000
alpha[3]	2.656	0.07748	0.001929	2.532	2.647	2.829	10001	50000
alpha[4]	2.654	0.07424	0.001737	2.528	2.647	2.819	10001	50000
beta[1]	0.9751	0.07746	0.001512	0.8275	0.9736	1.129	10001	50000
beta[2]	0.9747	0.07807	6.402E-4	0.8263	0.9733	1.129	10001	50000
beta[3]	0.9759	0.07796	0.001022	0.828	0.9744	1.135	10001	50000
beta[4]	0.9759	0.07727	9.129E-4	0.8288	0.9742	1.132	10001	50000
gamma[1]	0.8607	0.03351	7.849E-4	0.7833	0.8646	0.9146	10001	50000
gamma[2]	0.8613	0.03373	5.879E-4	0.7845	0.8651	0.9161	10001	50000
gamma[3]	0.8632	0.03293	7.05E-4	0.7892	0.8665	0.9189	10001	50000
gamma[4]	0.8623	0.03386	6.953E-4	0.7839	0.8662	0.917	10001	50000
sigma2[1]	0.009987	0.003213	2.702E-5	0.005568	0.009403	0.01791	10001	50000
sigma2[2]	0.009961	0.003191	2.136E-5	0.005532	0.009387	0.01774	10001	50000
sigma2[3]	0.009973	0.003169	2.552E-5	0.005552	0.009384	0.01777	10001	50000
sigma2[4]	0.009975	0.003194	2.505E-5	0.005582	0.009389	0.01786	10001	50000

The results are virtually identical for all four prior distributions, even though the priors themselves differ considerably, as illustrated in Figure 6.4. The model fit is shown in Figure 6.5.

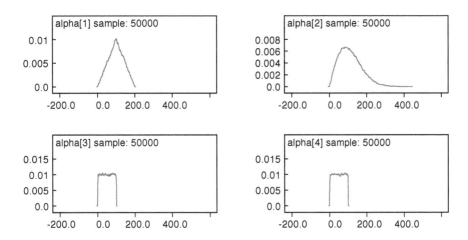

FIGURE 6.4
Prior distributions for α used in dugongs analyses — Example 6.3.1.

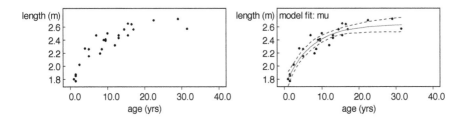

FIGURE 6.5
Left-hand side: dugong lengths (m) plotted against age in years. Right-hand side: same data with accompanying model fit.

6.4   Multivariate responses

Suppose we have observed n measurements on each of a number of individuals. Let i $(= 1, ..., N)$ index individuals and j $(= 1, ..., n)$ index measurements, and let Y_{ij} denote the jth observation made on individual i. Suppose also that measurements are made on the whole real line, so that a normality assumption might be appropriate. To account for the fact that observations made on the same individual may be correlated, we can assume that they follow a multivariate normal (MVN) distribution with unknown mean vector μ and variance-covariance matrix Σ. That is,

$$Y_i = (Y_{i1}, Y_{i2}, ..., Y_{in})' \sim \text{MVN}_n(\mu, \Sigma), \quad i = 1, ..., N.$$

If we have also observed covariates, such as the age at which each measurement was taken, specification of appropriate forms for the elements of μ leads to a multivariate regression model:

$$\mu_j = \beta_0 + \sum_{k=1}^{p} \beta_k x_{kj}, \quad j = 1, ..., n,$$

where x_{kj} denotes the jth value of covariate k. Typically we would specify vague normal priors for the coefficients, i.e. $\beta_. \sim \text{Normal}(0, 100^2)$, and an inverse-Wishart prior (see Appendix C.4) for the covariance Σ, via $\Sigma^{-1} \sim W(R, \rho)$. Here the right-hand side denotes a Wishart distribution with "scale matrix" R and degrees of freedom ρ. The Wishart distribution is the multivariate analogue of the gamma distribution and arises in classical statistics as the distribution of the sum-of-squares-and-products matrix in multivariate normal sampling. It is the conjugate prior for the precision matrix of a multivariate normal distribution. The least informative, proper Wishart prior is given by setting $\rho = p$, where p is the dimension of the distribution. The prior mean is ρR^{-1} and so a good choice for R is $\rho \Sigma_0$, where Σ_0 is some prior guess for the covariance.

Example 6.4.1. *Jaws*
Elston and Grizzle (1962) present repeated measurements of jawbone height on 20 boys. Each boy's jawbone was measured at ages 8, 8.5, 9, and 9.5 years, and interest focuses on describing the average growth curve of the jawbone. BUGS code for a multivariate regression model is given below.

```
model {
   for (i in 1:20) {Y[i, 1:4] ~ dmnorm(mu[], Sigma.inv[,])}
   for (j in 1:4)  {mu[j]     <- alpha + beta*x[j]}
   alpha                       ~ dnorm(0, 0.0001)
   beta                        ~ dnorm(0, 0.0001)
```

```
    Sigma.inv[1:4, 1:4]          ~ dwish(R[,], 4)
    Sigma[1:4, 1:4]              <- inverse(Sigma.inv[,])
}

list(Y = structure(
        .Data = c(47.8, 48.8, 49.0, 49.7,
                  46.4, 47.3, 47.7, 48.4,

                  . . . . . . . . . . . . . . .
                  46.3, 47.6, 51.3, 51.8),
        .Dim =  c(20, 4)),
    x = c(8.0, 8.5, 9.0, 9.5),
    R = structure(
        .Data = c(4, 0, 0, 0,
                  0, 4, 0, 0,
                  0, 0, 4, 0,
                  0, 0, 0, 4),
        .Dim =  c(4, 4)))
```

Array quantities in BUGS, such as Sigma.inv, must have their dimensions (1:4, 1:4 in this case) specified when they are defined, but not when they are used in the definitions of other nodes. See Appendix A.5. Note the use of the structure() syntax to specify data in matrix format — the data for R is supplied as a vector formed by concatenating successive rows of the matrix — see §12.4.2. The value of R is set equal to $\rho\Sigma_0$ where $\Sigma_0 = I$ is chosen by guessing the order of magnitude of variation between responses. Peeking at the data in order to set the prior is generally inappropriate as it is, strictly speaking, using the data twice in the analysis. However, assessing the order of magnitude of the variability is reasonable.

Also note that the multivariate normal distribution (dmnorm) in BUGS follows its univariate counterpart in being parameterised in terms of precision (Σ^{-1}). The matrix-valued inverse() function then allows inference on Σ. The model fit is shown in Figure 6.6 below. As an alternative for these data, we could have used a hierarchical "random coefficients" model — see Chapter 10.

Multivariate linear regressions are easily extended to nonlinear regressions, as in the univariate case. In addition, we can also specify a multivariate t-distribution (mvt) for the errors, to accommodate any outlying individuals. One area in which we do not have much freedom, however, is with the Wishart prior. Covariance matrices must always be positive-definite in order for them to make sense. The Wishart distribution is the only standard distribution that imposes this constraint naturally. If we wish to use an alternative form of prior then we must take responsibility for imposing the constraint ourselves, through appropriate parameterisation, say — the software will almost certainly crash if the constraint is not satisfied. Alternative priors for covariance matrices are further discussed in §10.2.3 and in Gelman et al. (2004),

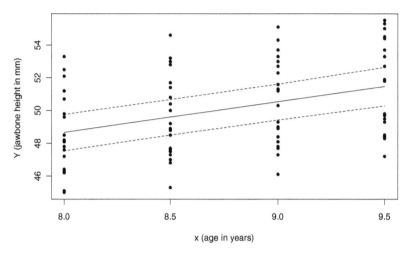

FIGURE 6.6

Model fit for jawbone data. The posterior median fit and 95% credible interval are indicated by the solid and dashed lines, respectively.

p. 483. Multivariate normal distributions with two specific structural forms for the covariance matrix are implemented as separate distributions in the BUGS language — see §11.3.6 for details.

6.5 Generalised linear regression models

Specification of Bayesian generalised linear models (GLMs) follows straightforwardly from the above discussion of linear models. No closed-form solution is available, but as we have seen for nonlinear models, it is still straightforward to obtain posterior samples using MCMC. The main differences with GLMs are that the sampling distribution of the data is typically non-normal and that we use a "link function" to transform parameters of that distribution onto a scale where a linear model can be used appropriately. More formally we assume that the data, y_i, $i = 1, ..., n$, arise from a specific distribution in the exponential family (McCullagh and Nelder, 1989), with

$$E[y_i] = \mu_i = g^{-1}(\eta_i), \quad \eta_i = \beta_0 + \sum_{k=1}^{p} \beta_k x_{ki},$$

for covariates x_{ki}, $k = 1, ..., p$. The exponential family of distributions includes distributions such as normal, Poisson, and binomial. Appropriate link functions $g(.)$ for these would generally be the identity function, $\log(.)$ and $\mathrm{logit}(.)$, respectively. For the binomial distribution, alternatives to $\mathrm{logit}(p) = \log(p/(1-p))$ are the *probit*, $\Phi^{-1}(p)$, and *complementary log-log*, $\log(-\log(1-p))$.

In the binomial case, the data can be expressed as $y_i = r_i/\nu_i$ for consistency with the relation $\mathrm{logit}(E[y_i]) = \eta_i$, where r_i is the number of "successes" out of ν_i "trials." Such scaling is not necessary in BUGS, however, as illustrated in the following example, along with the use of alternative link functions. Link functions in BUGS are slightly special in that they may appear on the left-hand side of a logical relationship, as shown for binary and count data below.

Note that we are not restricted to the exponential family of models. Hierarchical regression models, which include random effects, are discussed in §10.3, and many of the specialised models in Chapter 11 involve regression terms.

Example 6.5.1. *Binary data: Beetles*

Dobson (1983) analyses binary dose–response data from a bioassay experiment in which the numbers of beetles killed after 5-hour exposure to carbon disulphide at $N = 8$ different concentrations are recorded. Denoting the numbers of beetles killed at, and exposed to, dose x_i, $i = 1, ..., 8$, by y_i and n_i, respectively, we fit the following logistic regression model.

$$y_i \sim \mathrm{Binomial}(p_i, n_i), \quad \mathrm{logit}(p_i) = \alpha + \beta(x_i - \bar{x}),$$

with vague Normal$(0, 100^2)$ priors for α and β. Note that, again, the covariate (dose) is centred, by subtracting $\bar{x} = N^{-1} \sum x_i$. This is because serious MCMC convergence issues arise when the x_is are used directly, due to very high posterior correlation between α and β — see §4.4; centering essentially relocates the y-axis to $x = \bar{x}$, which, in this case, vastly reduces the dependence of the intercept α on β. The likelihood is specified via

```
for (i in 1:8) {
    y[i]            ~ dbin(p[i], n[i])
    logit(p[i]) <- alpha + beta*(x[i] - mean(x[]))
    }
```

If we want to assess the model fit visually then we will need to either transform the data onto the same scale as the linear model or transform the model fit onto the same scale as the observations. We insert the following code in the loop above so that we can examine both:

```
phat[i] <- y[i]/n[i]
yhat[i] <- n[i]*p[i]
```

Instead of a logistic regression, we might prefer to perform a complementary log-log or probit regression by replacing the logit link function with cloglog or probit, respectively. In the latter case, specifying the relationship as

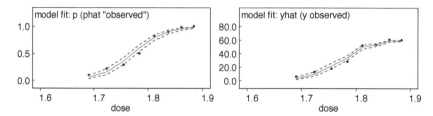

FIGURE 6.7
Model fits from logistic regression of "Beetles data." Left-hand side: 95% credible intervals for p[] and "observed" phat[] plotted against dose. Right-hand side: 95% credible intervals for yhat[] and observed y[] plotted against dose.

```
p[i] <- phi(alpha + beta*(x[i] - mean(x[])))
```

where phi denotes the cumulative distribution function of the standard normal distribution, is somewhat slower but can be more robust to numerical problems.

Example 6.5.2. *Count data: Salmonella*
Breslow (1984) analyses mutagenicity assay data (shown below) on salmonella in which three plates have each been processed at various doses of quinoline and the number of revertant colonies of TA98 salmonella subsequently measured.

Dose	0	10	33	100	333	1000
Plate 1	15	16	16	27	33	20
Plate 2	21	18	26	41	38	27
Plate 3	29	21	33	69	41	42

Denoting the dose by x_i, $i = 1, ..., 6$, and the number of colonies observed on plate j at dose x_i by y_{ij}, we fit the following GLM suggested by theory:

$$y_{ij} \sim \text{Poisson}(\mu_i), \quad \log \mu_i = \alpha + \beta \log(x_i + 10) + \gamma x_i,$$

with independent Normal$(0, 100^2)$ priors for α, β and γ.

```
for (i in 1:6) {
  for (j in 1:3) {
    y[i,j]      ~ dpois(mu[i])
  }
  log(mu[i]) <- alpha + beta*log(x[i] + 10) + gamma*x[i]
}
alpha          ~ dnorm(0, 0.0001)
beta           ~ dnorm(0, 0.0001)
gamma          ~ dnorm(0, 0.0001)
```

The model fit is shown on the left side of Figure 6.8. We also show 95% predictive intervals for the response variable at each dose, which are calculated by adding the following code to the model. These reflect uncertainty in α, β, and γ, as do the credible intervals for the model fit, but they also reflect sampling variation from the Poisson distribution.

```
for (i in 1:6) {y.pred[i] ~ dpois(mu[i])}
```

The predictive intervals indicate that the model may be deficient, since it cannot predict the level of variability apparent in the observed data. In particular, the largest observed response, at dose 100, is not realistically accommodated by the fitted model. One solution is to specify a hierarchical model instead — see Example 10.3.1. Another approach is to assume a negative binomial distribution (Appendix C.5) to explicitly model *over-dispersion* in the response variable. The negative binomial is more flexible than the Poisson but includes the Poisson distribution as a limiting case. If we make the following assumption for the responses:

```
y[i,j] ~ dnegbin(p[i], r)
```

then the mean is given by `mu[i] <- r*(1-p[i])/p[i]`. Hence we can rearrange to obtain `p[i] <- r/(mu[i] + r)` and model `log(mu[i])` as above. (Note that the Poisson distribution arises in the limit as $r \to \infty$.) We specify a discrete uniform prior for r via the following code:

```
r ~ dcat(pi[])
for (i in 1:max) {pi[i] <- 1/max}
```

with max $= 1000$. The resulting model fit and prediction interval are shown on the right side of Figure 6.8, and posterior summaries for both models are given in the table below. Note that posterior medians for the common parameters are strikingly similar, but the posterior uncertainty is increased substantially with the negative binomial model. Also note that the negative binomial model better accommodates the observed data. We examine model comparison and criticism more formally for this example in Chapter 8.

node	mean	sd	MC error	2.5%	median	97.5%	start	sample
dpois:								
alpha	2.182	0.2169	0.0109	1.767	2.178	2.629	1001	20000
beta	0.3169	0.05666	0.002886	0.1993	0.3186	0.4254	1001	20000
gamma	-0.001006	2.452E-4	1.06E-5	-0.001483	-0.001009	-5.044E-4	1001	20000
dnegbin:								
alpha	2.183	0.3206	0.01339	1.581	2.176	2.843	4001	100000
beta	0.3166	0.08655	0.00366	0.1374	0.3188	0.4774	4001	100000
gamma	-9.956E-4	3.794E-4	1.38E-5	-0.001721	-0.001009	-2.066E-4	4001	100000
r	72.24	145.3	0.8461	8.0	27.0	617.0	4001	100000

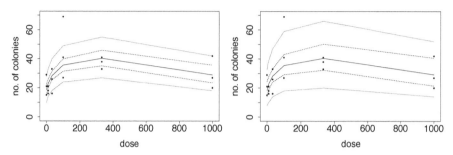

FIGURE 6.8
Posterior median model fits (——), 95% credible intervals (— —) and 95% prediction
intervals (.....) from regression analysis of salmonella data: left-hand side, Poisson
regression; right-hand side, negative-binomial regression.

6.6   Inference on functions of parameters

In a Bayesian context, MCMC makes inference easy for arbitrary functions of
parameters, such as coefficients in a regression model. For example, in a logistic
regression model, such as Example 6.5.1, the model is specified in terms of
the *log odds ratio* β, but the *odds ratio* $\exp(\beta)$ is usually more interpretable.
We simply evaluate the function of interest at every MCMC iteration, and the
resulting set of values represents a sample from the posterior distribution for
that function. In classical inference, the delta method is commonly used to
estimate standard errors or confidence intervals for functions of parameters,
though this can be inaccurate for very nonlinear functions. Bootstrapping
is a more accurate classical alternative with a similar computational cost to
MCMC.

Example 6.6.1. *Beetles (continued): ED95*
In Example 6.5.1, suppose we wish to estimate the *ED95*, that is, the dose that
will provide 95% of maximum efficacy:

$$\text{logit}\, 0.95 = \alpha + \beta(\textbf{\textit{ED95}} - \bar{x}) \quad \Rightarrow \quad \textbf{\textit{ED95}} = (\text{logit}\, 0.95 - \alpha)/\beta + \bar{x}$$

We simply add the following code into the logistic regression model and monitor
the ED95 variable:

```
ED95 <- (logit(0.95) - alpha)/beta + mean(x[])
```

The posterior mean and standard deviation are 1.857 and 0.00776, respectively
(500,000 iterations were necessary to achieve this level of precision).
   Using a classical logistic regression fitted in R, the estimated *ED95* is 1.858. The
delta method, based on the (multivariate normal) asymptotic distribution of the

maximum likelihood estimators of α and β, provides a standard error of 0.00773, which is reasonably accurate in this example. Note that classical standard errors obtained in this way will be underestimates of the true SE, due to the Cramer–Rao inequality, though will converge asymptotically to the true SE.

Obtaining a confidence interval via the delta method would rely on a normal approximation for some transformation of ED95, introducing further inaccuracies. A classical alternative would be to use bootstrap methods to obtain a set of values from the sampling distribution of the ED95 estimator. This could be achieved by resampling from the data and refitting the model, or by simulating from the asymptotic distribution of the estimators of α, β and computing the resulting ED95. The sample quantiles could then be used as the confidence interval. The latter method in this case gives a 95% interval of (1.844, 1.875), which closely matches the Bayesian 95% credible interval of (1.843, 1.874).

6.7 Further reading

Gelman and Hill (2007) give a detailed introduction to regression modelling, focusing on practical issues around building, fitting, criticising, and presenting models including linear and generalised linear regressions, and hierarchical or multilevel regression models (as we discuss in Chapter 10). Issues include predictive model checking (as in our Chapter 8), missing data (as in our §9.1), and causal inference. Many of the models discussed are Bayesian, and BUGS and R code is provided. Ntzoufras (2009) and Congdon (2003, 2006) give numerous examples of regression models in WinBUGS, and methods for selection of predictors. Ntzoufras (2009) gives a particularly detailed consideration of models for analysis of variance and covariance.

7

Categorical data

Regression models for binary and count data were introduced in Chapter 6. This chapter describes Bayesian models for more general forms of categorical or discrete data, starting with 2×2 tables which classify two binary variables, followed by multinomial models for single or multiple categorical outcomes and models for ordered categorical data. Regression techniques are introduced for relating multinomial and ordinal data to predictors.

As in any other Bayesian application, needing to specify prior distributions may be both an advantage and a challenge. While inferences are sometimes sensitive to the choice of prior, it can allow realistic information to be introduced and can stabilise estimates from data with small counts. The BUGS apparatus also allows models to be specified with arbitrary constraints on their parameters.

7.1   2 × 2 tables

Suppose N individuals are classified, according to two binary variables, in the following 2×2 table.

	Success	Failure	
Group 1	y_{11}	y_{12}	n_1
Group 2	y_{21}	y_{22}	n_2
Total	m_1	m_2	N

This type of data arises in three general situations.

One margin fixed N individuals are classified deterministically as n_1 in Group 1 and n_2 in Group 2. Each individual has a single random outcome, deemed "success" with probability p_1 for Group 1 and p_2 for Group 2. The total number of successes in Group i, y_{i1}, is then distributed as Binomial(n_i, p_i), for $i = 1, 2$. We are interested in how the success rate differs between groups, and we might make inferences about the *relative risk* p_2/p_1 or the *odds ratio* $(p_2/(1 - p_2))/(p_1/(1 - p_1))$. We already saw an example in §5.5, and we consider this situation further in §7.1.1.

Both margins fixed N individuals are classified as n_1 in Group 1 and n_2 in Group 2. The total number of successes and failures is *fixed* at m_1 and m_2, respectively. These are then allocated randomly to groups. A commonly cited example is the "lady tasting tea" experiment. A colleague of the statistician Ronald Fisher claimed to be able to tell whether the milk or the tea infusion had been poured into a cup first. N cups of tea with milk are prepared, n_1 where the milk is poured first and n_2 where the tea is poured first. The taster is told how many had the milk poured first, then she tries to guess which. m_1 is the number of cups guessed as "milk first," so that in this example, $m_1 = n_1$. This is mathematically more difficult and is discussed briefly in §7.1.3.

No margins fixed Two random binary outcomes are measured on N individuals, resulting in four possible combined outcomes with probabilities p_{11}, p_{12}, p_{21}, and p_{22}. For example, we might ask someone whether they smoke tobacco, drink alcohol, neither, or both. This is a 2×2 *contingency table*, governed by a multinomial model. The BUGS implementation of multinomial models is described in §7.2, and contingency table analysis is discussed briefly in §7.2.5.

7.1.1   Tables with one margin fixed

Here we concentrate on the case with one margin fixed, and analyse some fictitious data from the tea-tasting experiment introduced above. We deliberately choose an example with small counts, for which the choice of prior will be important and Bayesian inferences are more likely to differ from classical results. Example 5.5.1 discussed how informative priors can be placed on the (log-)odds ratio — here we place informative priors directly on the two outcome probabilities.

Example 7.1.1. *Lady tasting tea*
Suppose the tea-tasting experiment resulted in the following guesses.

		Guess Milk first	Tea first	
Actual	Milk first	3	1	4
	Tea first	1	3	4
	Total	4	4	8

For the purpose of this example, we suppose that the taster was not told beforehand how many cups had their milk poured first, so that the column totals are not fixed. The model is then two independent binomials, as in the BUGS code below. p_1, p_2 are the probabilities that she guesses that the milk was poured first, given that the milk or tea, respectively, were actually poured first. A classical analysis would normally test the null hypothesis that $p_1 = p_2$, but we estimate the posterior probability that $p_1 > p_2$, in other words that she has *some* ability to identify the pouring order. However, for many situations a more meaningful hypothesis may be that the difference between p_1 and p_2 is *practically* significant, so that $p_1 - p_2 > \epsilon$ for some value of ϵ.

```
for (i in 1:2) {
  y[i] ~ dbin(p[i], n[i])
  p[i] ~ dunif(0, 1)
}
```

The data are simply supplied as:

```
list(n=c(4,4), y=c(3,1))
```

We compare the conventional independent uniform priors with various alternatives.

"Reference" Independent Beta(0.5, 0.5) priors from Jeffreys' principle, or uniform priors for $\text{logit}(p_i)$, equivalent to Beta(0,0), as discussed in §5.2.5.

One parameter The probability of correct classification doesn't depend on whether the milk or the tea is poured first, so that $p_1 = 1 - p_2$. Then p_1 is given a uniform, Jeffreys, or logit uniform prior. In this case the alternative hypothesis is $p_1 > 0.5$.

One parameter, sceptical Again assuming $p_1 = 1 - p_2$, and following Lindley (1984), we could be sceptical and place substantial prior mass on the single point $p_1 = 0.5$ representing no discriminating skill. We give 50% prior probability to this point, assume zero prior probability to the situation $p_1 < 0.5$, where she consistently selects the reverse of the true pouring order, and place a uniform prior on the remaining region $p_1 > 0.5$ where she has some discriminating ability. As discussed in §8.7, studies of remarkable or supernatural abilities are one of the few occasions where a point null hypothesis is strictly realistic! This prior is implemented in BUGS using the pick formulation, introduced in §5.4.

```
p[1]       <- theta[pick]
pick        ~ dcat(q[])
q[1]       <- 0.5
q[2]       <- 0.5
theta[1] <- 0.5
theta[2]   ~ dunif(0.5, 1)
```

Dependent As discussed by Howard (1998), in many real 2×2 table situations, if p_1 is expected to be large, then so is p_2. Suppose we are told the lady is inclined to guess "milk first" for cups with the milk actually poured first, in other words p_1 is large. If we were sceptical about her tasting skill, we might believe that this is because she is more likely to guess "milk first"

TABLE 7.1
Posterior probabilities that the tea-taster has some discriminating
ability, for various priors.

| | Independent | One parameter | Dependent | |
			$\rho = 0.75$	$\rho = 0.875$
Uniform	0.89	0.91	0.81	0.75
Beta(0.5,0.5)	0.92	0.92	0.82	0.75
Beta(0,0)	0.95	0.94	0.83	0.76
Sceptical		0.65		

in all circumstances, and not because she can detect the pouring order.
This is equivalent to assuming a prior correlation. While this prior may
be less realistic in the tea-tasting experiment, it lets us illustrate a trick
described by Michael and Schucany (2002) for specifying identical marginal
priors for p_1 and p_2 while inducing a prior correlation between the two. If
$p_1 \sim \text{Beta}(\alpha, \beta)$ and $x|p_1 \sim \text{Binomial}(p_1, n)$, then, treating x as data, the
posterior of $p_1|x$ is $\text{Beta}(\alpha + x, \beta + n - x)$. However, if we define a new
random variable p_2 whose distribution is this posterior *integrated* over the
distribution of x, then the marginal distribution of p_2 will also be $\text{Beta}(\alpha, \beta)$,
and the correlation between p_1 and p_2 is $\rho = n/(\alpha + \beta + n)$. A similar trick
is available for the gamma distribution.

Therefore the following code specifies a joint prior for p_1 and p_2, where each
is marginally Uniform(0,1), and the correlation is $6/(6 + \alpha + \beta) = 0.75$ with
$\alpha = \beta = 1$ for the uniform ($\text{Beta}(1,1)$) distribution. With this correlation,
if $p_1 = 0.5$, then p_2 has a 95% chance of being between 0.1 and 0.92 —
not a very strong assumption.

```
alpha  <- 1;  beta <- 1;
p[1]      ~ dbeta(alpha, beta)
n.corr <- 6 # for rho=0.75, or n.corr <- 14 for rho=0.875
x         ~ dbin(p[1], n.corr)
a.post <- alpha + x
b.post <- alpha + n.corr - x
p[2]      ~ dbeta(a.post, b.post)
```

The posterior probabilities that $p_1 > p_2$ are obtained in each case as the
posterior mean of

```
post <- step(p[1] - p[2])
```

except for the one-parameter sceptical prior, where the (equivalent) probability
of $p_1 > 0.5$ is obtained as the posterior mean of `pick−1`. These are listed in
Table 7.1.

There is not much difference between the posterior probabilities under the vague uniform, Jeffreys, and logit uniform priors, or whether the success probability is assumed to be independent of the pouring order. However, the conclusions about the lady's tasting skill are more reserved under our more "subjective" priors. Under the sceptical prior, which assigned a probability of 0.5 to any tasting skill, the taster's six out of eight successful classifications only convert this to a posterior probability of 0.65. Under the dependent priors, as the prior correlation ρ between p_1 and p_2 increases, the posterior probability of discriminating ability becomes smaller, and the choice of marginal prior has even less impact.

Note that under certain priors, the Bayesian results agree with classical significance levels. The posterior probability of 0.92 under independent Jeffreys' priors is approximately the same (to 2 s.f.) as $1 - p$, where p is the one-sided p-value from the standard χ^2 test without continuity correction. The corresponding $1 - p$ from Fisher's exact test, however, is equal to the more conservative posterior probability of 0.76 under independent priors of $p_1 \sim \text{Beta}(0, 1)$, $p_2 \sim \text{Beta}(1, 0)$, which slightly favour $p_2 > p_1$ (Altham, 1969).

For some very sceptical priors, even if the taster had guessed all eight cups of tea correctly, then the posterior probability would still not be convincing; for example, it is 0.93 with a prior probability of 0.8 on $p_1 = 0.5$. A greater number of trials would then be required for stronger evidence! See Example 8.7.1 for a situation where even greater scepticism is appropriate.

7.1.2 Case-control studies

Case-control studies typically produce data as a 2×2 table, but with the *outcome totals* fixed, rather than the predictor totals. A fixed number of individuals with and without a certain outcome are collected and examined to see whether they were exposed to a particular predictor. These allow the estimation of the odds ratio of the outcome in terms of the exposure, but not the relative risk. They are typically used, however, for rare diseases where the odds ratio approximates the relative risk. Suppose Group 2 represents exposure. The number of exposed cases and controls are modelled as independent binomial, with odds p_{21}/p_{11} and p_{22}/p_{12}, respectively. The odds ratio $(p_{21}p_{12})/(p_{11}p_{22})$ between cases and controls for the probability of exposure equals the odds ratio between exposed and unexposed for the probability of the outcome.

Case-control studies are often analysed conditionally on fixed exposure totals as well as fixed outcome totals. Bayesian analyses of case-control studies, both conditional and unconditional, are discussed in the **Endo** example provided with WinBUGS and OpenBUGS.

7.1.3   Tables with both margins fixed

In the tea-tasting experiment as originally described by Fisher, the lady is told the margins of the table and thus always guesses the correct total number with milk or tea poured first. The number of cups y_{11}, for which milk was poured first and the lady also identifies them correctly, is no longer binomial but instead follows the *non-central hypergeometric* distribution. This is parameterised by the table margins and the odds ratio — see Appendix C.5 for the exact definition. The "null" distribution of y_{11}, when the odds ratio is 1 ($p_1 = p_2$), is the standard hypergeometric. This is the basis of Fisher's exact test, which is commonly used to test the null hypothesis instead of the (asymptotic) chi-squared test when the numbers in the table are small, whether or not the column totals are fixed.

In order to estimate the posterior distribution of the odds ratio in a Bayesian context, a non-central hypergeometric likelihood is required. This is provided by OpenBUGS and JAGS (with different parameterisations; see Appendix C.5) but not WinBUGS. In the tea-tasting example above, using a uniform prior for $p_1 = 1 - p_2$, the posterior probability that $p_1 > p_2$ is 0.90, similar to the probability of 0.91 obtained without the column totals fixed (labelled "One parameter" in Table 7.1).

In the tea-tasting example, the row totals equal the column totals. For more general tables with fixed margins, the margin totals imply complicated constraints on the internal cells (Wakefield, 2004) and thus on plausible prior values for the cell probabilities, assuming the prior is chosen after seeing the column totals but before seeing the data.

7.2   Multinomial models

Suppose we have data $y_1, \ldots, y_n$ which are arrays of counts, $y_i = (y_{i1}, y_{i2}, \ldots, y_{iR})$. Each y_i represents a set of M_i independent draws from R categories with probabilities $q = (q_1, \ldots, q_R)$ so that $\sum_r y_{ir} = M_i$ for all i, and $\sum_r q_r = 1$. For example, with $R = 3$ and $M_i = M = 9$ for all i: $y_1 = (1, 4, 4)$, $y_2 = (3, 3, 3)$, and so on. This is a *multinomial* model $y_i \sim \text{Multinomial}(q, M_i)$, with likelihood $\propto \prod_{i=1}^n \prod_{r=1}^R q_r^{y_{ir}}$:

```
y[i,1:3] ~ dmulti(q[1:3], M[i])
```

7.2.1   Conjugate analysis

The conjugate prior for the vector of multinomial probabilities q is the Dirichlet$(\alpha_1, \ldots, \alpha_R)$ distribution with $p(q) \propto \prod_r q_r^{\alpha_r - 1}$, a generalisation of the beta distribution for the probability of a binary outcome (see Appendix C.4):

```
q[1:3] ~ ddirch(alpha[1:3])
```

The parameter α_r is proportional to the expected probability q_r of the rth outcome, and the prior precision of q increases with the scale of the α_r. For example, $\alpha_r = 1$ for $r = 1, \ldots, R$ is a generalisation of the Uniform(0,1) or Beta(1,1) distribution. Given one y_i, the resulting posterior is

$$q \sim \text{Dirichlet}(\alpha_1 + y_{i1}, \ldots, \alpha_R + y_{iR}).$$

Note the greater posterior precision for larger sample sizes, and a greater influence of the prior for larger α_r.

Example 7.2.1. *Asthma: state transitions in a clinical trial*
Briggs, Ades, and Price (2003) examine transitions between five clinical states in a randomised trial of treatments (seretide and fluticasone) for asthma. The states are labelled STW: Successfully treated week, UTW: Unsuccessfully treated week, Hex: Hospital-managed exacerbation, Pex: Primary care-managed exacerbation, TF: Treatment failure. The number of occasions a patient occupied state a in one week, followed by state b the following week, is counted for all states a, b for 12 weeks. For patients randomised to the seretide arm, these are

| | To | | | | | |
| | STW | UTW | Hex | Pex | TF | |
From						Total
STW	210	60	0	1	1	272
UTW	88	641	0	4	13	746
Hex	0	0	0	0	0	0
Pex	1	0	0	0	1	2
TF	0	0	0	0	81	81

The aim is to estimate the transition probabilities between the states, which will inform a cost-effectiveness analysis. Although no patients entered the Hex state within the short 12 week follow-up, hospital admission is expensive and potentially important to long-term cost effectiveness.

We fit a discrete-time Markov model, equivalent to four independent multinomial models with probability vector q_i governing the state in the following week conditionally on the current state, each with a uniform Dirichlet prior on q_i. The fifth state, treatment failure, is absorbing; in other words, patients cannot move out of it, so $q_{5r} = 0$, $r = 1, \ldots, 4$, $q_{55} = 1$.

```
for (i in 1:4) {
  count[i, 1:5]          ~ dmulti(q[i, 1:5], M[i])
  q[i, 1:5]              ~ ddirch(alpha[])
}
for (r in 1:5) {alpha[r] <- 1}
```

One thousand samples from the conjugate posterior distribution produce posterior means of

From	To STW	UTW	Hex	Pex	TF
STW	0.76	0.22	0.004	0.007	0.007
UTW	0.12	0.86	0.001	0.007	0.02
Hex	0.20	0.20	0.20	0.20	0.20
Pex	0.28	0.14	0.14	0.14	0.30

The influence of the prior allows transitions not observed in the data to have a small but non-zero posterior probability. This represents the belief that asthma patients will occasionally need to be admitted to hospital after an exacerbation. This would not have been possible if the transition probabilities had been estimated simply by dividing the count data above by the appropriate denominator.

7.2.2   Non-conjugate analysis — parameter constraints

Conjugate Bayesian analyses for multinomial data, as above, can be performed in standard statistical software. The BUGS language and MCMC framework, however, enable more complex models in which the cell probabilities are functions of other parameters of interest, where no closed-form posterior distribution is available for those parameters.

Example 7.2.2. *Population genetics: self-fertilising plants*
Given the following frequencies of genotypes from a set of maternal plants (Holsinger, 2001–2010).

Maternal genotype	Offspring genotype AA	AB	BB
AA	427	95	
AB	108	161	71
BB		64	74

we wish to estimate the rate σ at which plants self-fertilise, and the frequency p of allele A in outcross pollen. The probabilities of each offspring genotype, conditionally on each maternal genotype, are functions of these parameters ($q = 1 - p$):

Maternal genotype	Offspring genotype		
	AA	AB	BB
AA	$(1-\sigma)p + \sigma$	$(1-\sigma)q$	0
AB	$(1-\sigma)p/2 + \sigma/4$	$1/2$	$(1-\sigma)q/2 + \sigma/4$
BB	0	$(1-\sigma)p$	$(1-\sigma)q + \sigma$

This can be written immediately in BUGS:

```
NAA[1:3]  ~ dmulti(XAA[1:3], KAA)
NAB[1:3]  ~ dmulti(XAB[1:3], KAB)
NBB[1:3]  ~ dmulti(XBB[1:3], KBB)
XAA[1]   <- (1 - sigma)*p + sigma
XAA[2]   <- (1 - sigma)*q
XAA[3]   <- 0
XAB[1]   <- (1 - sigma)*p/2 + sigma/4
XAB[2]   <- 0.5
XAB[3]   <- (1 - sigma)*q/2 + sigma/4
XBB[1]   <- 0
XBB[2]   <- (1 - sigma)*p
XBB[3]   <- (1 - sigma)*q + sigma
KAA      <- sum(NAA[])
KAB      <- sum(NAB[])
KBB      <- sum(NBB[])
p         ~ dunif(0, 1)
sigma     ~ dunif(0, 1)
q        <- 1 - p
```

obtaining posterior distributions for both the self-fertilisation rate and the allele A frequency, each acknowledging the uncertainty about the other (Figure 7.1).

```
node   mean    sd       MC error  2.5%    median  97.5%   start  sample
p       0.7049 0.02394  2.455E-4  0.6562  0.7053  0.7497  1001   10000
sigma  0.37    0.04158  4.236E-4  0.287   0.3703  0.4509  1001   10000
```

7.2.3 Categorical data with covariates

The multinomial logistic model allows responses $y_i \sim \text{Multinomial}(q_i, N)$ to be modelled in terms of a vector of covariates $x_i = (x_{1i}, \ldots, x_{pi})$. The *log odds ratio* for category r relative to category 1 is defined as

$$\eta_{ir} = \log\left(\frac{q_{ir}}{q_{i1}}\right) = \alpha_r + \sum_{k=1}^{p} \beta_{kr} x_{ki}, \quad r = 2, \ldots, R; \quad i = 1, \ldots, N$$

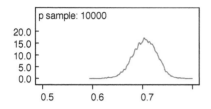
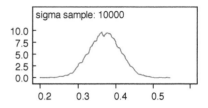

FIGURE 7.1
Posterior density estimates for allele frequency p and self-fertilisation rate `sigma` from population genetics example.

Conceptually, this is equivalent to $R-1$ binomial logistic regressions comparing category $r > 1$ with category 1. The category probabilities are then

$$q_{ir} = \frac{\phi_{ir}}{\sum_{s=1}^{R} \phi_{is}} \qquad \text{where } \phi_{ir} = e^{\eta_{ir}} = e^{\alpha_r + \sum_k \beta_{kr} x_{ki}}$$

with the constraint that $\phi_{i1} = 1$ (i.e., $\eta_{i1} = 0$).

Example 7.2.3. *Asthma (continued): including a treatment effect*
To compare the state transition probabilities between two treatment groups in the asthma trial example 7.2.1, we could fit independent multinomial models to the counts observed under each treatment. However, since the data are sparse, it would probably be more efficient to fit a multinomial logistic model to the dataset as a whole, with treatment as a covariate. We illustrate this for the transition from the first state (successfully treated week). The transition counts for the two treatments are:

| | To | | | | | |
	STW	UTW	Hex	Pex	TF	Total
Seretide	210	60	0	1	1	272
Fluticasone	66	32	0	0	2	100

The probability `q[i,2]` that a patient is unsuccessfully treated in the following week is allowed to vary between treatments `i` (where seretide is treatment 1 and fluticasone is treatment 2) but due to the small counts, all transition probabilities to other states are constrained to be the same between treatment groups. Diffuse normal priors are assumed for the baseline log odds `a[j]` for transition to state `j`, and for the log odds ratio `b.treat[2]` for treatment with fluticasone. Odds are relative to remaining in the baseline state (STW). These priors on the log-odds scale are not equivalent to the uniform Dirichlet priors used on the probability scale in Example 7.2.1, and sensitivity analysis to the prior variance may be advisable.

```
for (i in 1:2) {
    count[i, 1:5]          ~ dmulti(q[i, 1:5], M[i])
    for (r in 1:5) {
        q[i,r]             <- phi[i,r]/sum(phi[i,])
        log(phi[i,r])      <- a[r] + b.treat[r]*treat[i]
    }
}
for (r in 2:5) {a[r]       ~ dnorm(0, 0.00001)}
a[1]                       <- 0    #
b.treat[1]                 <- 0    # giving phi[i,1] = 1
b.treat[2]                 ~ dnorm(0, 0.00001)
or.treat                   <- exp(b.treat[2])
# no treatment effect on transitions other than to UTW
for (r in 3:5) {b.treat[r] <- 0}
treat[1]                   <- 0
treat[2]                   <- 1
```

After an adaptive phase of 4000 iterations, 6000 posterior samples result in a posterior mean of 1.72 (95% credible interval 1.00 to 2.80) for the odds ratio or.treat for treatment with fluticasone. An estimate of the odds ratio could also have been obtained "by hand" from the above data, as $(32/66)/(60/210) = 1.70$, but the Bayesian formulation also allows us to simultaneously obtain posterior distributions for this effect and for the other transition probabilities. These are substantively the same as those obtained from the multinomial model without covariates in Example 7.2.1.

This model could easily be extended to estimate the transition probabilities from the remaining states. Constraints can be applied by setting up the appropriate logical nodes — for example, assuming that the risk of hospital admission, or the effect of treatment on this risk, is independent of the current state.

7.2.4 Multinomial and Poisson regression equivalence

An alternative way of fitting a multinomial logistic regression in BUGS is to assume

$$y_{ir} \sim \text{Poisson}(\mu_{ir}), \quad \log(\mu_{ir}) = \lambda_i + \alpha_r + \sum_k \beta_{kr} x_{ki}.$$

With an improper uniform prior on λ_i, integrating over λ_i produces the same likelihood for the α_r and β_{kr} as the multinomial model (see Appendix C.6). This can be more efficient, though perhaps at the cost of clarity. The model in Example 7.2.3 could be expressed in this way as:

```
for (i in 1:2) {
```

```
for (r in 1:5) {
  count[i,r]      ~ dpois(mu[i,r])
  log(mu[i,r]) <- lambda[i] + a[r] + b.treat[r]*treat[i]
}
lambda[i]        ~ dflat()
}
```

7.2.5   Contingency tables

A common application of the multinomial/Poisson equivalence is to the analysis of *contingency tables*. These classify individuals according to two categorical outcomes, generalising the 2×2 table to any number of rows r and columns c. Suppose we observe y_{ij} individuals in row i and column j of the table, in other words, with level i of the first category and level j of the second. Assuming independent outcomes, the model for the y_{ij} is multinomial with corresponding probabilities p_{ij}: $i = 1, \ldots, r$; $j = 1, \ldots, c$. However, the Poisson log-linear formulation is more common: $y_{ij} \sim \text{Poisson}(\mu_{ij})$, where the log mean of y_{ij} is

$$\log(\mu_{ij}) = \alpha_i + \beta_j + \gamma_{ij}$$

under a saturated model, in which every cell has its own parameter. These are conceptually the same as the Poisson regression models illustrated in Example 6.5.2. Typically we would assess the hypothesis that the two factors are independent, so that $p_{ij} = p_i q_j$, or $\gamma_{ij} = 0$ in the Poisson formulation.

These generalise easily to *multiway* contingency tables with more than two categorical variables. Specific constraints on the probabilities may be employed to examine particular hypotheses and compared using the methods discussed in Chapter 8. Arbitrary constraints on related parameters are straightforward to implement using BUGS, as illustrated in Example 7.2.2. Many examples of models for contingency tables and their implementation in BUGS are given by Congdon (2005).

7.3   Ordinal regression

Ordinal data are discrete data with a natural ordering. These are commonly found in surveys, where respondents give preferences on a ranked scale such as the *Likert* (1. strongly disagree, 2. disagree, 3. neither agree nor disagree, 4. agree, 5. strongly agree). While ordinal data themselves are discrete, to aid modelling we can usually assume they represent an underlying continuous scale. Indeed, such data sometimes result from grouping an originally continuous response into intervals.

Suppose the data $y_1, \ldots, y_n$ are independent outcomes on an ordinal scale $1, 2, \ldots, R$. The R categories represent intervals $[a_0, a_1), [a_1, a_2), \ldots, [a_{R-1}, a_R]$ of a latent continuous variable Z_i which can take any real value, so that $a_0 = -\infty$ and $a_R = \infty$. The distribution of y_i is fully specified by the cumulative probabilities $q_{ir} = \Pr(Z_i \geq a_r)$ that the response is in category r or higher, for $r = 1, \ldots, R$. To model how the response varies with covariates x_i, we place a linear model on these probabilities on a suitable link-transformed scale. For example, using a logit link,

$$\mathrm{logit}(q_{ir}) = \mu_i - a_r, \qquad \mu_i = \alpha + \sum_k \beta_k x_{ki}$$

Implicitly, the latent variable Z_i has a logistic distribution with mean μ_i. Alternatively, a probit link function would imply a latent normally distributed variable (see Example 9.1.4 for an example of latent probit regression models).

The ordered logistic model is a *proportional odds* model. This means that the odds ratio of a higher score compared to a lower score, $\Pr(y_i \geq r)/\Pr(y_i < r)$, does not depend on which category r is chosen to define a "high" score. Equivalently for the latent variable, $\Pr(Z_i \geq a)/\Pr(Z_i < a)$ is independent of the cut-point a which distinguishes higher from lower scores.

The cut-points a_r defining the categories are considered to be unknown and must be given prior distributions which respect the ordering constraint $a_1 < a_2 < \ldots < a_R$.

Example 7.3.1. *Kidney transplants: ordered logistic regression*
Kidneys for transplantation are commonly obtained from donors who are brain dead but whose hearts are still beating. Because of the shortage of organs available this way, kidneys are increasingly also being obtained from donors after cardiac death. However, there are concerns about the extent of damage to these donor organs during the "agonal phase" from withdrawal of life-supporting treatment to cardiorespiratory arrest. Reid et al. (2011) investigated the impact of the duration of the agonal phase on organ damage. Each of 190 donor kidneys is given a score representing the presence of up to five indicators of kidney damage (acidaemia, lactic acidosis, hypotension, hypoxia, or oliguria), so that the score is 0 if none are present, up to 5 if all are present.

Figure 7.2 suggests the score increases with agonal phase duration. The question is how to quantify this increase. An ordered logistic regression is fitted to the agonal scores, shifted to take values from 1 to 6. The $N = 190$ scores Score[i] are considered as independent categorical outcomes. The logit of the cumulative probability Q[i,r] of scores of r or more is given a linear model with log(agonal phase duration in minutes) lAPD[i] as a predictor, and an unknown cut-point c[r] for each category r. A prior ordering is imposed on the cut-points by defining positive prior distributions on reasonably large ranges for the differences dc[r] between the cut-points.*

```
for (i in 1:N) {
  Score[i]                  ~ dcat(p[i,])
  # define in terms of cumulative probabilities
  p[i,1]                    <- 1 - Q[i,1]
  for (r in 2:5) {
    p[i,r]                  <- Q[i,r-1] - Q[i,r]
  }
  p[i,6]                    <- Q[i,5]
  for (r in 1:5) {
    logit(Q[i,r])           <- b.apd*lAPD[i] - c[r]
  }
}
for (i in 1:5) {dc[i] ~ dunif(0, 20)}
c[1]                        <- dc[1]
for (i in 2:5) {
  c[i]                      <- c[i-1] + dc[i]
}
b.apd                       ~ dnorm(0, 0.001)
or.apd                      <- exp(b.apd)
```

The odds ratio or.apd is interpreted as the increase in odds of a higher score corresponding to one unit increase in log(agonal phase minutes). After a burn-in of 500 and further 10,000 iterations, the posterior median odds ratio is 1.46 (95% CI 1.23, 1.74).

A more extensive analysis of these data (Reid et al., 2011) took account of the correlation between kidneys from the same donor using a hierarchical model (Chapter 10) and accounted for missing data on one or more of the indicators of damage using Bayesian multiple imputation (as in Example 9.1.4). The analysis presented above was based on a single imputation.

7.4 Further reading

Congdon (2005) gives a wide-ranging and detailed description of many Bayesian models for categorical data, including WinBUGS code for all examples. Some of the many topics covered in greater detail than in our book

---

*In JAGS and OpenBUGS, the elements of a vector can be sorted using c[1:5] <- sort(x[]).

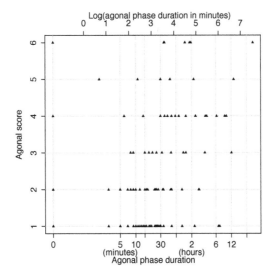

FIGURE 7.2

Agonal scores from donor kidneys compared with agonal phase duration.

include robust regression, flexible models including generalized additive and nonparametric models, dynamic linear models and contingency tables, and latent class and hierarchical extensions to many of these models.

8

Model checking and comparison

8.1 Introduction

The conclusions of a Bayesian analysis are conditional on the appropriateness of an assumed probability model, so we need to be satisfied that our assumptions are a reasonable approximation to reality, even though we do not generally believe any model is actually "true." Many aspects of an assumed model might be questioned: observations that don't fit, the distributional assumptions, qualitative structure, link functions, which covariates to include, and so on.

We can distinguish three elements that can be applied in an iterative manner (O'Hagan, 2003).

1. *Criticism:* exploratory checking of a single model, which may suggest...

2. *Extensions:* embedding a model in a list of alternatives, which leads to...

3. *Comparison:* assessing candidate models in terms of their evidential support and influence on conclusions of interest.

Classical statistical model assessment ideas, such as residuals and predictive ability on new data, generally depend on assumptions such as a linear model structure. Here we adapt these ideas to be generically applicable in arbitrary models, noting that the Bayesian approach means that parameters have distributions and so, for example, residuals and deviances will be quantities with posterior distributions. In addition, within a Bayesian framework there is a responsibility to check for unintended sensitivity to the prior and for conflict between prior and data. Fortunately, MCMC methods provide a flexible framework for implementing these ideas.

In this chapter we first focus on the *deviance* as a general measure of model adequacy. We go on to explore model criticism using residuals and methods based on generating replicate data and (possibly) parameters. We then consider embedding models in extended models, followed by deviance-based and traditional Bayesian measures for overall model comparison. Methods for accounting for uncertainty about model choice are then described, and we conclude by discussing the detection of conflict between prior and data. Assessment of hierarchical models is described in §10.7 and §10.8.

We emphasise that the techniques described in this chapter are more informal than the inferential methods covered elsewhere — model criticism and comparison inevitably involve a degree of judgement and cannot be reduced to a set of formal rules.

8.2   Deviance

We define the deviance as

$$D(\theta) = -2\log p(y|\theta) \tag{8.1}$$

which is explicitly a function of θ and so has a posterior distribution. This quantity is created automatically as a node by WinBUGS and OpenBUGS, named `deviance`. This can be monitored and plotted like any other node — note that since the full sampling distribution $p(y|\theta)$ is used, including the normalising constant, the absolute size of the deviance is generally difficult to interpret.

We note that the deviance depends on the specific formulation of a distribution, which becomes relevant when alternative parameterisations exist. For example, a Student's t_4 distribution can be expressed as `y ~ dt(mu,tau,4)` in terms of its mean μ and precision parameter τ, with corresponding density

$$p(y|\mu,\tau) = \frac{\Gamma(\frac{5}{2})\sqrt{\tau}}{\sqrt{4\pi}} \frac{1}{[1 + (y-\mu)^2\tau/4]^{\frac{5}{2}}}$$

(Appendix C.1) and hence the deviance is

$$D(\mu,\tau) = \log\pi - 2\log\Gamma(\frac{5}{2}) - \log\tau + \log(4) + 5\log[1 + (y-\mu)^2\tau/4]$$

Alternatively, we can express the t distribution indirectly as a scale mixture of normal densities. This follows from the standard result that if $Y \sim N(\mu,\sigma^2)$ and $\lambda \sim \chi_4^2$ are independent random quantities, then $\frac{(Y-\mu)}{\sigma}/\sqrt{\lambda/4} \sim t_4$, and so if $\tau = 4/(\lambda\sigma^2)$, then $(Y-\mu)\sqrt{\tau} \sim t_4$ (Appendix C.1). Then the deviance is simply $-2 \times$ log of the normal density (as a function of the mean and precision).

$$D(\mu,\sigma^{-2}) = \log 2\pi - \log\sigma^{-2} + (y-\mu)^2\sigma^{-2}$$

which may be expressed as

$$D(\mu,\tau,\lambda) = \log 2\pi - \log(\lambda\tau/4) + (y-\mu)^2\lambda\tau/4$$

showing the additional dependence on the random λ when using this representation.

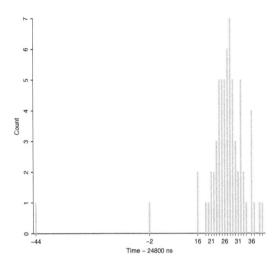

FIGURE 8.1
Newcomb's 66 measurements of the time taken for light to travel 7442 metres.

Example 8.2.1. *Newcomb's speed of light data*
Stigler (1977) presents 66 measurements made of the time taken for light to travel 7442 metres (recorded as deviations from 24,800 nanoseconds) made by Simon Newcomb in 1882. Sixty-four of these form a fairly symmetric distribution between around 17 and 40, while there are two gross outliers of -2 and -44. These are shown in Figure 8.1 and were initially analysed by Carlin and Louis (2008), who pointed out the rather clear non-normality.

A heavy-tailed t_4 distribution can be fitted to the data in two different ways. First as

```
y[i]    ~ dt(mu, tau ,4)
mu      ~ dunif(-100, 100)
tau     ~ dgamma(0.001, 0.001)
```

with the following results:

node	mean	sd	MC error	2.5%	median	97.5%	start	sample
deviance	436.4	2.02	0.02387	434.4	435.8	441.8	1001	10000
mu	27.48	0.6604	0.009961	26.2	27.47	28.78	1001	10000
tau	0.04919	0.01196	1.829E-4	0.02968	0.04776	0.07635	1001	10000

Or we can fit an identical model using the scale mixture of normals

```
y[i]              ~ dnorm(mu, invsigma2[i])
invsigma2[i] <- tau*lambda[i]/4
lambda[i]         ~ dchisqr(4)
```

with the same priors on μ and τ. If we monitor deviance for this representation, we shall be calculating at each iteration

$$D = \sum_i \left[\log 2\pi - \log \tau \lambda_i/4 + (y_i - \mu)^2 \tau \lambda_i/4) \right]$$

and obtain the following results:

node	mean	sd	MC error	2.5%	median	97.5%	start	sample
deviance	408.6	7.969	0.104	393.9	408.3	425.2	1001	10000
mu	27.49	0.6586	0.00904	26.2	27.49	28.78	1001	10000
tau	0.04911	0.01186	1.719E-4	0.02955	0.04774	0.07576	1001	10000

The parameter estimates are the same (up to Monte Carlo error) but the deviance is smaller in the representation in terms of a random scale parameter.

We shall examine this further when we consider model criticism by embedding in more complex models in §8.5.

8.3 Residuals

Residuals measure the deviation between observations and estimated expected values and should ideally be assessed on data that has not been used to fit the model. Classical residuals are generally calculated, however, from the fitted data and used to identify potential inadequacies in the model by, for example, plotting against covariates or fitted values, checking for auto-correlations, distributional shape, and so on. This analysis is generally carried out informally, and different forms of residual all have their Bayesian analogues.

8.3.1 Standardised Pearson residuals

A Pearson residual is defined as

$$r_i(\theta) = \frac{y_i - E(y_i|\theta)}{\sqrt{Var(y_i|\theta)}} \tag{8.2}$$

which is a function of θ and therefore has a posterior distribution. If it is considered as a function of random y_i for fixed θ, it has mean 0 and variance

1, and so we might broadly expect values between -2 and 2. For discrete sampling distributions the residuals can be "binned up" by, say, creating new variables made from adding y's corresponding to covariates within a specified range (Gelman et al., 2004). If we want to create a single-valued residual rather than a random quantity, alternative possibilities include using a single draw θ^t, plugging in the posterior means of θ, or using the posterior mean residual.

We note that the residuals are not independent since their posterior distributions all depend on θ, and they are best used informally. Nevertheless it seems natural to examine a summary measure such as $X^2 = \Sigma_i r_i^2$ as an overall measure of residual variation (McCullagh and Nelder, 1989).

Example 8.3.1. *Bristol surgery mortality*
The following data represent the mortality rates from 12 English hospitals carrying out heart surgery on children under 1 year old between 1991 and 1995 (Marshall and Spiegelhalter, 2007).

TABLE 8.1
Numbers of open-heart operations and deaths for children under 1 year of age carried out in 12 hospitals in England between 1991 and 1995, as recorded by Hospital Episode Statistics. The "tenth" data represents similar mortality rates but based on approximately one tenth of the sample size.

	Hospital	Full data		Tenth data	
		Operations n_i	Deaths y_i	Operations	Deaths
1	Bristol	143	41	14	4
2	Leicester	187	25	19	3
3	Leeds	323	24	32	2
4	Oxford	122	23	12	2
5	Guys	164	25	16	3
6	Liverpool	405	42	41	4
7	Southampton	239	24	24	2
8	Great Ormond St	482	53	48	5
9	Newcastle	195	26	20	3
10	Harefield	177	25	18	3
11	Birmingham	581	58	58	6
12	Brompton	301	31	30	3

Suppose we fit a binomial model under the assumption that all 12 hospitals had the same underlying risk θ, which we give a uniform prior. Then, since $E[Y_i|\theta] = n_i\theta$, $Var[Y_i|\theta] = n_i\theta(1-\theta)$ under this model, we can calculate the standardised residuals $r_i = (y_i - n_i\theta)/\sqrt{n_i\theta(1-\theta)}$ as well as the sum of the squared standardised residuals.

```
for (i in 1:12) {
  y[i]      ~ dbin(theta, n[i])
  res[i]   <- (y[i] - n[i]*theta)/sqrt(n[i]*theta*(1-theta))
  res2[i]  <- res[i]*res[i]
}
theta       ~ dunif(0, 1)
X2.obs      <- sum(res2[])    # sum of squared stand. resids
```

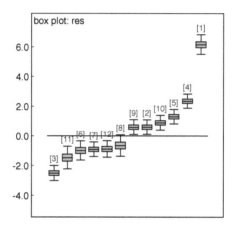

FIGURE 8.2
Standardised residuals for child heart surgery mortality rates in Table 8.1.

These are plotted in Figure 8.2 using the `Inference->Compare->box plot` facility in WinBUGS. To rank the boxes in increasing order of their mean, right-click on the plot, select `Properties...` then `Special...` and check the box labelled `rank`. The residual for Bristol is a clear outlier. `X2.obs` has a posterior mean of 59.1: for known θ we might expect this to be around 12, which informally suggests a severe lack of fit of a constant risk model. See §8.4.5 for more on goodness-of-fit tests, and Example 10.1.1, where we consider a hierarchical model for these data.

8.3.2 Multivariate residuals

In multivariate settings such as

$$Y_i = (Y_{i1}, Y_{i2}, ..., Y_{in})' \sim \text{MVN}_n(\mu, \Sigma),$$

for some response variable Y_{ij} $(i = 1, ..., N, j = 1, ..., n)$, we can also calculate standardised Pearson residuals, via, for example,

$$r_{ij}(\mu, \Sigma) = \frac{Y_{ij} - \mu_j}{\sqrt{\Sigma_{jj}}}.$$

This may highlight outlying observations but does not address the model's performance in terms of its ability to describe the correlation among observations from the same "unit" (e.g., individual). We might instead look at the Mahalanobis distance (or its square):

$$M_i = \sqrt{(Y_i - E(Y_i|\theta))' V(Y_i|\theta)^{-1} (Y_i - E(Y_i|\theta))},$$

where θ denotes the set of all model parameters. In the following example we show how to calculate M_i and M_i^2 in the case of a multivariate normal assumption for the response. Under multivariate normality, the distribution of M_i^2 (considered as a function of random Y_i with $\theta = \{\mu, \Sigma\}$ fixed) is known to be χ_n^2.

Example 8.3.2. *Jaws (continued): model checking*
We return to the "Jaws" example of §6.4. For each boy constituting the data set, the Mahalanobis distance from the fitted model (and its square) can be calculated by inserting the following code.

```
for (i in 1:20) {
   for (j in 1:4) {
      res[i, j]  <- Y[i, j] - mu[j]
      temp[i, j] <- inprod(Sigma.inv[j, 1:4], res[i, 1:4])
   }
   M.squared[i] <- inprod(res[i, 1:4], temp[i, 1:4])
   M[i]         <- sqrt(M.squared[i])
}
```

Box plots for both quantities are shown in Figure 8.3. If the data Y_i, $i = 1, ..., N$, were truly realisations from MVN(μ, Σ) for some fixed μ and Σ, then each M_i^2 (for random Y_i) would be $\sim \chi_4^2$. Hence we might expect the majority of individuals, in this case, to have squared distances less than the 95th percentile of a χ_4^2 distribution ≈ 9.5. In light of this we might be concerned as to whether individual 12's data are well accounted for by the model. Note, however, that we would actually expect 1 boy out of 20 to give a squared distance > 9.5.

8.3.3 Observed *p*-values for distributional shape

If the cumulative distribution function $F(Y|\theta)$ is available in closed form, then the quantity $P_i(\theta) = F(y_i|\theta)$ would have a uniform sampling distribu-

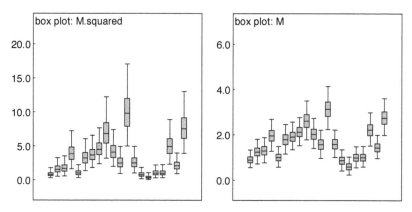

FIGURE 8.3
Squared Mahalanobis distance (left) and Mahalanobis distance (right) for each boy in "Jaws" Example 8.3.2.

tion for fixed θ — these are essentially the one-sided p-values for each observation.* Therefore a very informal assessment of the fit of a model could be achieved by taking the posterior means of the P_is, ranking them, plotting against $1, 2, \ldots, N$, and examining the plot for linearity. This plot is poor at visual detection of outliers since P_is near 0 and 1 do not show up well, but it can be effective in detecting inappropriate distributional shape.

Example 8.3.3. *Dugongs (continued): residuals*
Recall Example 6.3.1 about nonlinear regression for the length of dugongs. If we assume that the observation Y_i in Model 3 is normally distributed around the regression line μ_i, we have that $F(y_i|\mu_i, \sigma) = \Phi(r_i)$, where Φ is the standard normal cumulative distribution function, and $r_i = (y_i - \mu_i)/\sigma$ is the Pearson residual.

```
y[i]       ~ dnorm(mu[i], inv.sigma2)
mu[i]      <- alpha - beta*pow(gamma, x[i])
res[i]     <- (y[i] - mu[i])/sigma
p.res[i]   <- phi(res[i])
```

The posterior distributions of the standardised residuals, res, and their p-values, p.res, are shown in Figure 8.4, with the ordered mean p-values showing approximate linearity and so supporting the normal assumption.

---

*OpenBUGS and JAGS have functions to automatically compute the cumulative density of any stochastic node; see Appendix B.4.

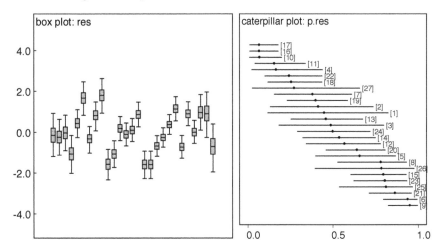

FIGURE 8.4
Box plot (left) and caterpillar plot (right) showing standardised residuals and
p-values, respectively, for dugongs, Example 8.3.3.

8.3.4 Deviance residuals and tests of fit

Residuals can be based directly on a standardised or "saturated" version of
the deviance (McCullagh and Nelder (1989), p. 398), defined as

$$D_S(\theta) = -2\log p(y|\theta) + 2\log p(y|\hat{\theta}_S(y)) \qquad (8.3)$$

where $\hat{\theta}_S(y)$ are appropriate "saturated" estimates: typically, when $E[Y] = \theta$,
we set $\hat{\theta}_S(y) = y$ (McCullagh and Nelder, 1989). Standardised deviances for
common distributions are:

$$y_i \sim \text{Binomial}(\theta_i, n_i) : D_S(\theta) = 2\sum_i \left\{ y_i \log\left[\frac{y_i/n_i}{\theta_i}\right] + (n_i - y_i)\log\left[\frac{(1-y_i/n_i)}{1-\theta_i}\right] \right\}$$

$$y_i \sim \text{Poisson}(\theta_i) : \qquad D_S(\theta) = 2\sum_i \left\{ y_i \log\left[\frac{y_i}{\theta_i}\right] - (y_i - \theta_i) \right\}$$

$$y_i \sim \text{Normal}(\theta_i, \sigma_i^2) : \quad D_S(\theta) = \sum_i \left[\frac{y_i - \theta_i}{\sigma_i}\right]^2$$

If we denote by D_{Si} the contribution of the ith observation to the standardised deviance, so that $\sum_i D_{Si} = D_S$, then the deviance residual dr_i can
be defined as

$$dr_i = sign_i \sqrt{D_{Si}}$$

where $sign_i$ is the sign of $y_i - E(y_i|\theta)$.

When $y_i = 0$, care is needed when calculating these residuals: for a binomial
distribution with $y_i = 0$ we obtain $dr_i = -\sqrt{-2n_i\log(1-\theta_i)}$ and for the
Poisson $dr_i = -\sqrt{2\theta_i}$.

Suppose we have relatively little information in the prior and so the conditions for the asymptotic normality of θ hold (see §3.6).

$$\theta \sim N(\hat{\theta}, (n\hat{I}(\hat{\theta}))^{-1})$$

where $\hat{\theta}$ is the maximum likelihood estimate under the currently assumed model, and $\hat{I}(\hat{\theta}) = -\frac{d^2}{d\theta^2} \log p(y|\theta)\big|_{\hat{\theta}}$ is the observed Fisher information. Then we may write the saturated deviance as

$$D_S(\theta) = -2\log p(y|\theta) + 2\log p(y|\hat{\theta}_S(y)) = D(\theta) - D(\hat{\theta}) + D(\hat{\theta}) - D(\hat{\theta}_S).$$

Then from the asymptotic normality we obtain

$$\begin{aligned}
D(\theta) - D(\hat{\theta}) &= -2\log p(y|\theta) + 2\log p(y|\hat{\theta}(y)) \\
&\approx p\log 2\pi - \log\left|n\hat{I}(\hat{\theta})\right| + (\theta - \hat{\theta})^T \hat{I}(\hat{\theta})(\theta - \hat{\theta}) - p\log 2\pi + \log\left|n\hat{I}(\hat{\theta})\right| \\
&= (\theta - \hat{\theta})^T \hat{I}(\hat{\theta})(\theta - \hat{\theta})
\end{aligned}$$

which has an approximate χ_p^2 distribution, where $p = dim(\theta)$. In addition, $D(\hat{\theta}) - D(\hat{\theta}_S)$ is a function of the data that from classical GLM theory has, approximately, a χ_{n-p}^2 sampling distribution (McCullagh and Nelder, 1989). Taken together, the posterior expectation of the saturated deviance has the approximate relationship

$$E[D_S(\theta)] \approx p + (n - p) = n.$$

As a rough assessment of the goodness of fit for models for which the saturated deviance is appropriate, essentially Poisson and binomial, we may therefore compare the mean saturated deviance with the sample size.

Example 8.3.4. *Bristol (continued): tenth data*
We examine the Bristol data as in Example 8.3.1, but using the tenth data in Table 8.1 to illustrate analysis of small numbers. If the observed proportion in the ith hospital is denoted $p_i = y_i/n_i$, the corresponding contribution to the standardised deviance is

$$D_{Si} = 2n_i\left[p_i \log(p_i/\theta_i) + (1 - p_i)\log\left(\frac{1 - p_i}{1 - \theta_i}\right)\right]$$

```
for (i in 1:12) {
  y[i]          ~ dbin(theta, n[i])
  prop[i]    <- y[i]/n[i]
  # (extra 0.00001 avoids numerical errors if prop[i] = 0 or 1)
  Ds[i]      <- 2*n[i]*(prop[i]*log((prop[i]+0.00001)/theta)
                + (1-prop[i])*log((1-prop[i]+0.00001)/(1-theta)))
  # sign of deviance residual
```

```
sign[i]     <- 2*step(prop[i] - theta) - 1
dev.res[i] <- sign[i]*sqrt(Ds[i])
}
dev.sat     <- sum(Ds[])
```

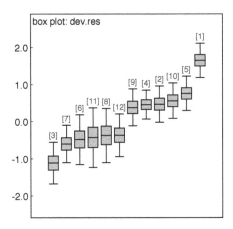

FIGURE 8.5
Deviance residuals for Bristol child heart surgery mortality ("tenth data").

The deviance residuals (Figure 8.5) follow a reasonable pattern between -2 and 2. The saturated deviance has posterior mean 7.5 (95% interval 6.5 to 11.6), and comparison with the sample size of 12 suggests that this model fits adequately.

8.4 Predictive checks and Bayesian *p*-values

8.4.1 Interpreting discrepancy statistics — how big is big?

Residuals can be thought of as examples of statistics which measure the *discrepancy* between the observed data and an assumed model. Although often straightforward to calculate, a problem arises in their calibration — when is an observed discrepancy "large?" In this section we consider a general means of calibrating discrepancies via *p*-values obtained using simulated replicate data, where the discrepancy statistics may be functions of data alone or involve data and parameters.

Ideally models should be checked by comparing predictions made by the model to actual new data. But often we use the same data for model building and checking, in which case special caution is required. We note that although this process is generally referred to as "model checking," we are also checking the reasonableness of the prior assumptions.

8.4.2   Out-of-sample prediction

Suppose we have two sets of data y_f and y_c assumed to follow the same model, where y_f is used to fit the model and y_c for model criticism: for example, we may leave out a random 10% sample of data for validation purposes. The idea is to compare y_c with predictions y_c^{pred} based on y_f and some assumed sampling model and prior distribution, and if the assumptions are adequate the two should look similar. Comparisons are derived from the predictive distribution

$$p(y_c^{\text{pred}}|y_f) = \int p(y_c^{\text{pred}}|y_f, \theta)p(\theta|y_f)d\theta$$

$$= \int p(y_c^{\text{pred}}|\theta)p(\theta|y_f)d\theta$$

which is easy to obtain by simulating θ from its posterior distribution and then simulating $y_c^{\text{pred}}|\theta$.

Ideally we would want to use as much data as possible for model fitting, but also as much data as possible for model criticism. At one extreme, we could take out just one observation $y_c = y_i$, $y_f = y_{\setminus i}$, and then the predictive distribution is $p(y_i^{\text{pred}}|y_{\setminus i})$. Full cross-validation extends this idea and repeats for all i, while 10-fold cross-validation might, for example, remove 10% of the data, see how well it is predicted, and repeat this process 10 times.

At the other extreme we could try and predict *all* of the data, so that $y_c = y$ and the predictive distribution is $p(y)$, conditioned only on the prior. This is the *prior predictive* approach to model criticism (Box, 1980), but it is clearly strongly dependent on the prior information, producing vacuous predictions when this is weak — an impediment to the use of Bayes factors for comparing models (see §8.7).

In practice we often do not split data, and so diagnostics are likely to be conservative.

8.4.3   Checking functions based on data alone

A function $T(y_c)$ is termed a checking or test statistic (Gelman et al., 2004) if it would have an extreme value if the data y_c conflict with the assumed model. A common choice is just $T(y_{ci}) = y_{ci}$ to check for individual outliers.

A check is made whether the observed $T(y_c)$ is compatible with the simulated distribution of $T(y_c^{\text{pred}})$, either graphically and/or by calculating a

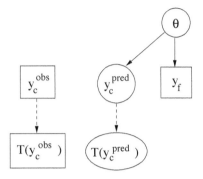

FIGURE 8.6
Graphical representation of predictive model checking, in which observed data y_c is compared to predicted data y_c^{pred} by a checking function T.

Bayesian p-value, which is the predictive probability of getting such an extreme result.

$$p_{\text{Bayes}} = \Pr\left(T(y_c^{pred}) \leq T(y_c)|y_f\right)$$
$$= \int \int I(T(y_c^{pred}) \leq T(y_c))p(y_c^{pred}|\theta)p(\theta|y_f)d\theta dy_c^{pred}.$$

This is easily calculated by simulating data, calculating the discrepancy function, and setting an indicator function for whether the observed statistic is greater than the current simulated value (Figure 8.6).

Example 8.4.1. *Bristol (continued): cross-validation*
Suppose we take Bristol out and predict how many deaths we would expect were there a common mortality risk in all the hospitals.

```
for (i in 2:12) {    # remove Bristol
    ...
}
# predicted number of deaths in centre 1 (Bristol)
y1.pred ~ dbin(theta, n[1])
P.bris <- step(y1.pred-y[1]-0.00001) + 0.5*equals(y1.pred, y[1])
```

Here the *mid p-value* is used for discrete data: $\Pr(y_1^{pred} > y_1) + \frac{1}{2}\Pr(y_1^{pred} = y_1)$.

node	mean	sd	MC error	2.5%	median	97.5%	start	sample
P.bris	0.0	0.0	1.0E-12	0.0	0.0	0.0	1001	10000
y1.pred	16.05	3.81	0.04069	9.0	16.0	24.0	1001	10000

We would predict that Bristol would have between 9 and 24 deaths assuming a common risk model: the probability of getting at least the observed total of 41 under this model is essentially 0.

The following example concerns secondary features of a model, and does not involve a separate dataset for checking.

Example 8.4.2. *Is a sequence of flips of a biased coin real or fake?*
A standard illustration of the non-regularity of random outcomes is to test whether a sequence of flips of a fair coin is a real or fake sequence. Test statistics include the longest run; and the number of switches between heads and tails, and the fake sequence generally will show too short a longest run and too many switches.

Generally an unbiased coin is assumed, but the analysis can be extended to a coin with unknown probability θ of coming up heads. Consider the following data on 30 coin-flips, where 1 indicates a head and 0 a tail.

```
y=c(0,0,0,1,0,1,0,0,0,1,0,0,1,0,1,0,0,0,0,1,1,0,0,0,1,0,0,1,0,1)
```

There are 17 switches between heads and tails, and the maximum run length is 4.

The following code shows how to calculate the number of switches and the maximum run length both for observed and replicate data, where the maximum run length is given by max.run.obs[N]. Mid-p-values are then calculated for both switches and maximum run length, which fully take into account the uncertainty about θ.

```
for(i in 1:N) {
  y[i]              ~ dbern(theta)
  y.rep[i]          ~ dbern(theta)
}
theta ~ dunif(0,1)
switch.obs[1]    <- 0
switch.rep[1]    <- 0
for (i in 2:N) {
  switch.obs[i]  <- 1 - equals(y[i-1], y[i])
  switch.rep[i]  <- 1 - equals(y.rep[i-1], y.rep[i])
}
s.obs            <- sum(switch.obs[])
s.rep            <- sum(switch.rep[])
P.switch         <- step(s.obs-s.rep-0.5)
                    + 0.5*equals(s.obs, s.rep)
run.obs[1]       <- 1
run.rep[1]       <- 1
max.run.obs[1]   <- 1
max.run.rep[1]   <- 1
```

```
for (i in 2:N) {
    run.obs[i]        <- 1 + run.obs[i-1]
                           * equals(y[i-1], y[i])
    run.rep[i]        <- 1 + run.rep[i-1]
                           * equals(y.rep[i-1], y.rep[i])
    max.run.obs[i] <- max(max.run.obs[i-1], run.obs[i])
    max.run.rep[i] <- max(max.run.rep[i-1], run.rep[i])
}
P.run                 <- step(max.run.obs[N]-max.run.rep[N]-0.5)
                           + 0.5*equals(max.run.obs[N], max.run.rep[N])
```

node	mean	sd	MC error	2.5%	median	97.5%	start	sample
theta	0.3442	0.08273	2.597E-4	0.192	0.3408	0.5151	1	100000
s.rep	12.71	3.259	0.01033	6.0	13.0	19.0	1	100000
max.run.rep[30]	6.937	3.047	0.008692	3.0	6.0	15.0	1	100000
P.run	0.1153	0.256	8.294E-4	0.0	0.0	1.0	1	100000
P.switch	0.9094	0.2642	8.941E-4	0.0	1.0	1.0	1	100000

The underlying probability θ of a head is estimated to be around 0.34, and in a true random sequence we would expect between 6 and 19 switches (median 13) and for the maximum run length between 3 and 15 (median 6). The mid-p-values are 0.91 for switches and 0.12 for runs, providing some evidence that the sequence is not from a real coin, considering p-values close to 0 and 1 as extreme.

The power to detect discrepancies from the model is substantially improved if we can choose a checking function whose predictive distribution is independent of unknown parameters in the model. If such a checking function can be found, then there is no need to have a separate dataset to use for checking.

Example 8.4.3. *Newcomb (continued): checking for a low minimum value*
The Newcomb data (Example 8.2.1) show evidence of non-normality. If we particularly wanted to check for low minimum values, we could calculate the checking statistic $T^{\text{obs}} = (y_{[1]} - y_{[N/2]})/(y_{[N/4]} - y_{[N/2]})$, and its posterior predictive replicates, where $y_{[r]}$ is the rth lowest value of y_i (rounding r to the nearest integer if necessary). T^{obs} will be high compared to its replicates if the minimum is lower than expected under the assumed normal model. Since the checking statistic is "standardised," its predictive distribution will be independent of the true mean μ and variance $\sigma^2 = 1/\tau$, and depends only on the normality assumption. To show this, note that the statistic is unchanged if each of the four terms $y_{[r]}$ in the expression is replaced by $(y_{[r]} - \mu)/\sigma$, each of which has a distribution which is independent of μ and σ. Compared to the observed value of 23.7, the predicted distribution for T^{rep} (Figure 8.7) has mean 3.7 and SD 1.1, and the p-value P.T is estimated to be very close to 0, so the observed value is clearly incompatible.

```
for(i in 1:N){
   Y[i]        ~ dnorm(mu,tau)
   Y.rep[i] ~ dnorm(mu,tau)
}
N.50        <- round(N/2)
N.25        <- round(N/4)
Y.rep.min <- ranked(Y.rep[], 1)
Y.rep.50  <- ranked(Y.rep[], N.50)
Y.rep.25  <- ranked(Y.rep[], N.25)
T.rep       <- (Y.rep.min - Y.rep.50)/(Y.rep.25 - Y.rep.50)
P.T         <- step(T.rep - T.obs)
```

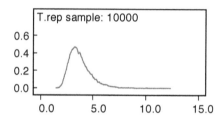

FIGURE 8.7
Posterior density of discrepancy statistic for low minimum value.

Suppose instead we used the sample variance as a checking statistic, which would be higher if the minimum is lower than expected.

```
V.obs       <- sd(Y[])*sd(Y[])
V.rep       <- sd(Y.rep[])*sd(Y.rep[])
P.V         <- step(V.rep - V.obs)
```

Compared to the observed sample variance of 115.5, the predicted sample variance has mean 119 and SD 30, and the p-value P.V is estimated to be around 0.49, so no indication of discrepancy is given. This is not unexpected — the distribution of this statistic depends strongly on σ^2, and the estimate of σ^2 is highly influenced by the outliers. Thus the uncertainty about estimating the parameters reduces the power of the check to detect outliers when the same dataset is used for estimating parameters and checking.

8.4.4   Checking functions based on data and parameters

If we want to check deviations from an assumed parametric form, it may be appropriate to use a discrepancy function $T(y_c, \theta)$ (Gelman et al., 2004) that depends on both the data and parameters, for example, standardised Pearson residuals. We can then simulate $T(y_c^{pred}, \theta)$ and calculate

$$p_{\text{Bayes}} = \Pr\left(T(y_c^{pred}, \theta) \le T(y_c, \theta)|y_f\right)$$
$$= \int\int I(T(y_c^{pred}, \theta) \le T(y_c, \theta)) \times p(y_c^{pred}|\theta)p(\theta|y_f)d\theta dy_c^{pred},$$

which is obtained by seeing how often the discrepancy function based on predicted data is less than the observed discrepancy.

Although ideally model checking should be based on new data, in practice the same data are generally used for both developing and checking the model. This needs to be carried out with caution, as the procedure can be conservative, and we shall see in the next example that the very features we are trying to detect can be influential in fitting the model and so mask their presence.

Example 8.4.4. *Newcomb (continued): checking for skewness*
Instead of specifically focusing on the minimum observation, as described above, we might consider a more generic test of distributional shape such as sample skewness. We could construct a checking function based on data alone and whose distribution under normality does not depend on the unknown mean and variance, such as $T(y) = \sum_i[(y_i - \bar{y})/\text{SD}(y)]^3$, which has observed value -290.

We can compare this with a similar measure of skewness but based on parameters $T(y, \mu, \sigma) = \sum_i[(y_i - \mu)/\sigma]^3$.

```
for (i in 1:N) {
   Y[i]               ~ dnorm(mu, tau)
   Y.rep[i]           ~ dnorm(mu, tau)
   T.data.obs[i] <- pow((Y[i] - mean(Y[]))/sd(Y[]),3)
   T.data.rep[i] <- pow((Y.rep[i] - mean(Y.rep[]))/sd(Y.rep[]),3)
   T.para.obs[i] <- pow((Y[i] - mu)/sigma,3)
   T.para.rep[i] <- pow((Y.rep[i] - mu)/sigma,3)
}
mu                 ~ dunif(-100, 100)
tau                ~ dgamma(0.001, 0.001)
sigma            <- 1/sqrt(tau)
T.data.obs.tot   <- sum(T.data.obs[])
T.data.rep.tot   <- sum(T.data.rep[])
T.para.obs.tot   <- sum(T.para.obs[])
T.para.rep.tot   <- sum(T.para.rep[])
P.data           <- step(T.data.obs.tot - T.data.rep.tot)
P.para           <- step(T.para.obs.tot - T.para.rep.tot)
```

node	mean	sd	MC error	2.5%	median	97.5%	start	sample
T.data.obs.tot	-289.8	0.0	1.0E-12	-289.8	-289.8	-289.8	1	10000
T.data.rep.tot	0.02773	18.65	0.1727	-37.1	-0.01383	37.07	1	10000
T.para.obs.tot	-293.3	81.25	0.8554	-475.1	-285.5	-157.4	1	10000
T.para.rep.tot	-0.444	31.57	0.2758	-62.93	-0.4164	62.88	1	10000

For the data-based measure, the replicate has a mean of around 0 and SD 19, showing a massive discrepancy with the observed value of −290. For the parametric checking function, the observed checking function has a posterior mean of −293 (SD 81), while the predictive check has a mean of around 0 (SD 32), showing a much less extreme (although still strong) discrepancy.

The above examples emphasise that increased power to detect deviations from assumptions can be obtained by careful choice of checking functions that focus on the suspected discrepancy and whose distributions do not depend on unknown parameters, or at least not strongly.

Example 8.4.5. *Dugongs (continued): prediction as model checking*
By adding in an extra line to the code for Example 8.3.3 to generate a replicate dataset at each iteration, we can see whether the observed data are compatible with the predictions.

```
y.pred[i]   ~ dnorm(mu[i], inv.sigma2)
P.pred[i] <- step(y[i] - y.pred[i])
```

Figure 8.8 shows 95% prediction bands for the replicate data, and so an informal comparison can be made by superimposing the observed data.

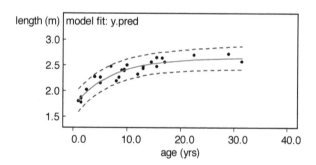

FIGURE 8.8
Prediction bands for replicate data — dugongs example.

The procedure outlined in this example appears attractive but could be conservative if there were considerable uncertainty about the parameters, in which case the predictive distributions would be wide and observations would tend to appear compatible with the model.

Such conservatism arises when the predictive distribution of the checking function depends on parameters which are not very precisely estimated, and so, as we found in Example 8.4.4, we should choose a checking function whose predictive distribution depends primarily on the assumptions being checked. This is illustrated in the next example.

8.4.5 Goodness of fit for grouped data

Classical goodness of fit statistics are based on grouping data into K bins labelled by k and then comparing observed O_k to expected E_k counts under the assumed model: comparison might be via a Pearson chi-squared $X^2 = \sum_k (O_k - E_k)^2 / E_k$ or likelihood ratio statistic $G = 2 \sum_k O_k \log(O_k/E_k)$, both of which have asymptotically χ^2_{K-p} distributions, where p is the number of estimated parameters. We might adopt either of these statistics as checking functions that are based on data and parameters.

If we assume that each of the O_k is a Poisson observation with mean E_k under the null hypothesis, then $G = 2 \sum_k O_k \log(O_k/E_k)$ is a saturated deviance and hence, using the results of §8.3.4, would be expected to have a posterior mean of around K if the null model is true, although this ignores the constraint that $\sum_k O_k = \sum_k E_k$.

Example 8.4.6. *Claims*
An insurance company takes a sample of $n = 35$ customers and adds up the number of claims each has made in the past year to give $y_1, \ldots, y_n$. These form the following distribution, where $m_k = \sum_i I(y_i = k)$ is the number of y_i that equal k, with $\sum m_k = n$.

k	0	1	2	3	4	5
m_k	10	5	10	7	2	1

The total number of claims in the sample is $\sum_k k m_k = 59$, and the average number of claims per customer is $\sum_k k m_k / 35 = 1.69$. Suppose we fit a Poisson model with mean λ. Then we can generate replicate data y^{rep} at each iteration, create replicate aggregate data m^{rep}, and hence provide a null distribution for any checking function.

We consider the classical likelihood ratio goodness-of-fit statistic $G(m, \lambda) = 2 \sum_k m_k \log(m_k/E_k)$, where $E_k = n e^{-\lambda} \lambda^k / k!$ is the expected count in cell k under the fitted model, which will be a random quantity since it depends on the unknown λ. G^{rep} is also calculated for replicate data: these data may extend beyond the range of counts observed in the original dataset.

A Jeffreys prior for the Poisson is adopted, $p_J(\lambda) \propto \lambda^{-\frac{1}{2}}$, approximated by a Gamma(0.5,0.001) distribution.

```
# remember that k = number of claims + 1
for (i in 1:n) {
    y[i]          ~ dpois(lambda)
    y.rep[i]      ~ dpois(lambda)
```

```
    for (k in 1:K) {
      eq[i,k]      <- equals(y[i], k-1)    # needed to construct
                                           # aggregate data
      eq.rep[i,k] <- equals(y.rep[i], k-1)
    }
  }
  for (k in 1:K) {
    m[k]           <- sum(eq[,k])          # aggregate counts
    m.rep[k]       <- sum(eq.rep[,k])
    # log of expected counts
    logE[k]        <- log(n) - lambda + (k-1)*log(lambda)
                     - logfact(k-1)
    # likelihood ratio statistic
    LR[k]          <- 2*m[k]*(log(m[k]+0.00001) - logE[k])
    LR.rep[k]      <- 2*m.rep[k]*(log(m.rep[k]+0.00001)
                     - logE[k])
  }
  G                <- sum(LR[])
  G.rep            <- sum(LR.rep[])
  P                <- step(G - G.rep)
  lambda           ~ dgamma(0.5, 0.001)   # Jeffreys prior
```

node	mean	sd	MC error	2.5%	median	97.5%	start	sample
G	7.96	1.465	0.0137	6.956	7.382	12.1	10001	10000
G.rep	6.473	3.531	0.03121	1.668	5.747	15.09	10001	10000
P	0.701	0.4578	0.004491	0.0	1.0	1.0	10001	10000
lambda	1.7	0.22 1	0.001996	1.291	1.693	2.161	10001	10000

We note that the likelihood ratio checking function $G(m, \lambda)$ has a posterior mean of around 8 — since this can be considered a saturated deviance we would expect a posterior mean of around $K = 6$ and so does not suggest a discrepancy from the null assumption. A more precise analysis is obtained by comparing the distribution of the observed and replicate likelihood ratio statistics — these show substantial overlap, and the p-value of 0.70 does not indicate any problems with the Poisson assumption. The minimum value of $G(m, \lambda)$ over 10,000 iterations is 6.955 — this is also the classical likelihood ratio statistic which may be compared to a χ^2_5 distribution under the null hypothesis, giving an approximate p-value of 0.22 and so similarly not indicating a significant deviation from the Poisson assumption.

However, a simple look at the data might lead us to suspect an excess of zero claims. A more powerful checking function would focus on this single form of discrepancy and would not depend (or only depend weakly) on λ. This requires a little ingenuity.

If data were really from a Poisson distribution, we would expect $T = m_0 m_2 / m_1^2$ to be around $\frac{\Pr(Y=0|\lambda) \Pr(Y=2|\lambda)}{\Pr(Y=1|\lambda)^2} = \frac{e^{-\lambda} \frac{1}{2}\lambda^2 e^{-\lambda}}{\lambda^2 e^{-2\lambda}} = 0.5$. Since this does not depend on λ, we would hope that the distribution of T will depend mainly on the Poisson assumption and be largely independent of the true rate of claims. $T = 4$ with these data.

```
# more powerful invariant test for excess zero counts
T.rep <- m.rep[1]*m.rep[3]/(m.rep[2]*m.rep[2])
P.inv <- step(T - T.rep)
```

T^{rep} has a median of 0.46 and 95% credible interval 0.083 to 2.7. The resulting p-value is 0.99, providing strong evidence of excess zero claims, a discrepancy entirely missed by the more generic goodness-of-fit tests. A "zero-inflated Poisson" model to fit these data is discussed in §11.6.

8.5 Model assessment by embedding in larger models

Suppose we are checking whether an observation y_i can be assumed to come from a distribution $p(y_i|\theta_{i0})$, where this null model is a special case of a larger family $p(y_i|\theta_i)$. Three approaches are possible:

- Fit the more flexible model in such a way as to provide diagnostics for individual parameter values $\theta_i = \theta_{i0}$.

- Fit the full model and examine the posterior support of the null model: this can be difficult when the null model is on the boundary of the parameter space.

- Fit the full and null model and compare them using a single criterion, such as DIC (see §8.6.4, e.g., salmonella example).

Example 8.5.1. *Newcomb (continued): checking normality*
We have seen in §8.2 how a Student's t-distribution can be expressed as a scale mixture of normals, in which if we assume $Y_i \sim N(\mu, \sigma^2)$, and $\lambda_i \sim \chi_4^2$, then if $\tau_i = 4/(\lambda_i\sigma^2)$, then $(Y_i - \mu)\sqrt{\tau_i} \sim t_4$. If we define $s_i = 4/\lambda_i$, large s_is should correspond to outlying observations. These are plotted in Figure 8.9. In WinBUGS and OpenBUGS the data index i corresponding to each box plot can be displayed by right-clicking on the plot and selecting `Properties->Special->show labels`. This will identify the two outliers correctly as the measurements of -2 and -44.

```
Y[i]           ~ dnorm(mu, invsigma2[i])
invsigma2[i] <- tau/s[i]
```

```
s[i]           <- 4/lambda[i]
lambda[i]       ~ dchisqr(4)
```

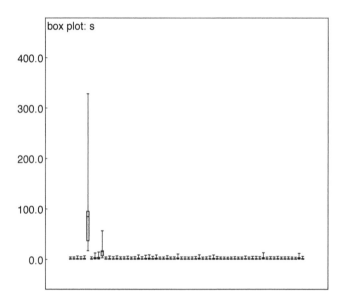

FIGURE 8.9
Outlier diagnostics using the t-distribution for the Newcomb light speed data.

This analysis assumes a t distribution with *known* degrees of freedom ν in order to provide a diagnostic for outlying observations. However, we could consider ν to be unknown and estimate its posterior distribution, noting that the normal assumption is equivalent to the t with infinitely large ν. We could adapt the model of § 4.1.2 but follow Jeffreys' suggestion $p(\nu) \propto 1/\nu$ as a prior for a positive integer (§5.2.8) — this is equivalent to a uniform prior on $\log \nu$. Using a discrete approximation by placing a uniform prior on $\nu = 2, 4, 8, 16, 32, \ldots, 1024$ leads to 86% posterior probability for $\nu = 2$.

8.6 Model comparison using deviances

Given a set of candidate models, we would often like to say something about which is "better," or even "best." This general idea can be made more formal in various ways, and we shall see that the appropriate tool depends on our intended action. In particular, we may be interested in selecting the "correct" model, informal examination of sensitivity to multiple "well-supported" models, making short-term predictions, or explicit weighting of the models.

Classical model choice uses hypothesis testing for comparing nested models, typically the likelihood ratio test. For non-nested models, alternatives include the Akaike Information Criterion (AIC).

$$\text{AIC} = -2\log p(y|\hat{\theta}) + 2p = D(\hat{\theta}) + 2p, \qquad (8.4)$$

where $\hat{\theta}$ is the maximum likelihood (minimum deviance) estimate, p is the number of parameters in the model (dimension of θ), and lower AIC is favoured. The $2p$ term serves to penalise more complex models which would otherwise be favoured by having a lower minimised deviance $-2\log p(y|\hat{\theta})$. Just as in model checking, predictive ability forms a natural criterion for comparing models, and AIC is designed to optimise predictions on a replicate dataset of the same size and design as that currently in hand. We note that, asymptotically, AIC is equivalent to leave-one-out cross-validation (§ 8.4.2) (Stone, 1977). Of course, if an external dataset is available, then we could compare models on how well they predict such independent validation data.

In a Bayesian context it is natural to base a comparative measure on the posterior distribution of the deviance, and many authors have suggested using the posterior mean deviance $\overline{D} = E[D]$ as a measure of fit (§ 8.2). This is invariant to the chosen parameterisation of θ and generally converges well. It can, under certain circumstances, be used as a test of absolute fit of a model (§ 8.3.4). But more complex models will fit the data better and so will inevitably have smaller $\overline{D}$ and so, in analogy to the Akaike criterion, we need to have some measure of "model complexity" to trade off against $\overline{D}$.

However, in Bayesian models with informative prior distributions, which include all hierarchical models, the prior effectively acts to restrict the freedom of the model parameters, so the appropriate "number of parameters" is less clear. We now consider a proposal for assessing the "effective number of parameters."

8.6.1 p_D: The effective number of parameters

Spiegelhalter et al. (2002) use an informal information-theoretic argument to suggest a measure p_D defined by

$$p_D = E_{\theta|y}[-2\log p(y|\theta)] + 2\log p(y|\tilde{\theta}(y))$$
$$= \overline{D} - D(\tilde{\theta}), \qquad (8.5)$$

where $\tilde{\theta}$ is a "good" plug-in estimate of θ. If we take $\tilde{\theta} = E[\theta|y] = \overline{\theta}$, then

$$p_D = \text{"posterior mean deviance - deviance of posterior means."}$$

Suppose we assume that the posterior distribution $p(\theta|y)$ is approximately multivariate normal with mean $\overline{\theta}$; then a Taylor expansion of $D(\theta)$ around $D(\overline{\theta})$, followed by posterior expectation, gives

$$E_{\theta|y}[D(\theta)] \approx D(\overline{\theta}) - E\left[\text{tr}\left((\theta - \overline{\theta})^T L_{\overline{\theta}}''(\theta - \overline{\theta})\right)\right]$$
$$= D(\overline{\theta}) + \text{tr}\left(-L_{\overline{\theta}}'' \, V\right)$$

where $V = E\left[(\theta - \overline{\theta})(\theta - \overline{\theta})^T\right]$ is the posterior covariance matrix of θ, and

$$-L_{\overline{\theta}}'' = -\left. \frac{d^2}{d\theta^2} \log p(y|\theta) \right|_{\overline{\theta}}$$

is the observed Fisher's information evaluated at the posterior mean of θ — this might also be denoted $\hat{I}(\overline{\theta})$ using the notation of §3.6.

Hence

$$p_D \approx \text{tr}\left(-L_{\overline{\theta}}'' V\right). \qquad (8.6)$$

The approximate form given in (8.6) shows p_D might be thought of as the dimensionality p of θ, times the ratio of the information in the likelihood about the parameters as a fraction of the total information in the posterior (likelihood + prior). To elucidate this with an example, suppose we observe $y \sim N(\theta, \sigma^2)$ with σ^2 known and one parameter θ with prior distribution $\theta \sim N(\mu, \omega^2)$. Then the standard normal-normal results (§3.3.2) show that $V = 1/(\sigma^{-2} + \omega^{-2})$ and $-L_{\overline{\theta}}'' = \sigma^{-2}$, the Fisher information $I(\theta)$. Hence $p_D = \sigma^{-2}/(\sigma^{-2} + \omega^{-2})$, the ratio of the information in the likelihood to the information in the posterior. We can also write $p_D = \omega^2/(\sigma^2 + \omega^2)$, known as the intra-class correlation coefficient.

Suppose that the information in the likelihood dominates that in the prior, either through the sample size being large or a "vague" prior being used for θ, and conditions for asymptotic normality apply. Then $\overline{\theta} \approx \hat{\theta}$, the maximum likelihood estimate. Also, $V^{-1} \approx -L_{\overline{\theta}}''$ from the asymptotic results (§3.6), hence $p_D \approx p$, the actual number of parameters.

This could also be derived from the results of §8.3.4 for situations of weak prior information in which we argue that

$$D(\theta) - D(\hat{\theta}) \approx (\theta - \hat{\theta})^T \hat{I}(\hat{\theta})(\theta - \hat{\theta})$$

has a χ_p^2 distribution. Thus, taking expectations of each side reveals that, in these circumstances, $p_D \approx p$.

Example 8.6.1. *Dugongs (continued): effective number of parameters*
Example 8.3.3 has weak prior information and hence we should expect that p_D should be close to the actual number of parameters. We find that $p_D = 3.7$, compared to the four parameters.

8.6.2 Issues with p_D

p_D is an apparently attractive measure of the implicit number of parameters, but has two major problems:

- p_D is not invariant to reparameterisation, in the sense that if the model is rewritten into an equivalent form but in terms of a function $g(\theta)$, then a different p_D may arise since $p(y|\overline{\theta})$ will generally not be equal to $p(y|\overline{g(\theta)})$. This can lead to misleading values in some circumstances and even negative p_Ds.

- An inability to calculate p_D when θ contains a categorical parameter, since the posterior mean is not then meaningful. This renders the measure inapplicable to mixture models (§11.6)—though see §8.6.6 for alternatives.

Example 8.6.2. *Transformed binomial: negative p_D due to severe posterior skewness*
For a binomial parameter θ, consider three different ways of putting a uniform prior on θ, generated by putting appropriate priors on $\psi_1 = \theta, \psi_2 = \theta^5$, and $\psi_3 = \theta^{20}$.

Let $\psi = \theta^n$. If $\theta \sim \mathrm{Unif}(0,1)$, then $\Pr(\theta < t) = t$, and so $\Pr(\psi < t) = t^{1/n}$. Differentiating reveals that $p(\psi) \propto t^{1/n-1}$, which we can identify as a $\mathrm{Beta}(1/n, 1)$ distribution.

Suppose we now observe $n = 2$ observations of which $r = 1$ is a "success." The appropriate code is therefore:

```
r <- 1; n <- 2; a[1] <- 1; a[2] <- 5; a[3] <- 20
for (i in 1:3) {
   a.inv[i] <- 1/a[i]
   theta[i] <- pow(psi[i], a.inv[i])
   psi[i]      ~ dbeta(a.inv[i], 1)
}
r1 <- r; r2 <- r; r3 <- r    # replicate data
r1              ~ dbin(theta[1], n)
```

```
r2                ~ dbin(theta[2], n)
r3                ~ dbin(theta[3], n)
```

In the WinBUGS output, obtained from setting a DIC monitor (after conver-gence) via the Inference -> DIC... dialog box, the mean deviance $\overline{D(\theta)}$ is labelled Dbar, while the plug-in deviance $D(\bar{\theta})$ is labelled Dhat.

	Dbar	Dhat	pD
r1	1.945	1.386	0.559
r2	1.945	1.550	0.395
r3	1.933	2.289	-0.356

The posterior distribution for θ does not depend on the parameter on which the prior distribution has been placed, and the mean deviances Dbar are theoretically invariant to the parameterisation (although even after 500,000 iterations the mean deviance for the third parameterisation has not reached its asymptotic value). WinBUGS, however, uses the posterior of ψ, not θ, to calculate p_D. These are very different, with that corresponding to the third parameterisation being negative.

As suggested in §8.2, WinBUGS (and current OpenBUGS) parameterises the deviance in terms of the stochastic parents of y in the model: i.e., if θ are the parameters that directly appear in the sampling distribution $p(y|\theta)$, but there are stochastic nodes ψ such that $\theta = f(\psi)$ in the model, then Dhat $= D(f(\overline{\psi}))$. It is therefore important to try to ensure that the stochastic parents of the data have approximately normal posterior distributions. The example above is an instance of p_D being negative due to the posterior of ψ being very non-normal, in which case $f(\overline{\psi})$ does not provide a good estimate of θ — in fact it leads to a plug-in deviance that is greater than the posterior mean deviance.

One possibility, illustrated in the salmonella example below, is to change the parameterisation into one for which the posterior is more likely to be symmetric and unimodal, which will generally mean placing prior distributions on transformations that cover the real line. Unfortunately, this may not be the most intuitive scale on which to place a prior distribution.

An alternative is to ignore the p_D calculated by WinBUGS and calculate your own p_D directly in terms of the deterministic parameters $\theta = \theta(\psi)$ so that Dhat $= D(\bar{\theta})$. In the previous example, this would lead to the plug-in deviances all being based on $\bar{\theta}$ and hence no difference between the parame-terisations. As another example, in generalised or nonlinear regression models one might use the posterior means of the linear predictors $\mu = \beta'x$, whose posterior distributions are likely to be approximately normal, rather than higher-level parameters of which β are deterministic functions. For example, in Example 6.3.1 we would calculate the posterior mean of every μ_i, instead of L_0 and L_∞, which may have skewed posteriors. The deviance at the posterior means of these parameters must then be calculated externally to BUGS and then subtracted from the posterior mean deviance (Equation 8.5) to calculate p_D. It is envisaged that this will become the standard way of calculating p_D in future versions of OpenBUGS.

The previous example shows the problems that arise when using the posterior mean as a summary of a highly skewed distribution. The mean is also a very poor summary of a bimodal posterior distribution, and the following example shows this can also lead to negative values of p_D.

Example 8.6.3. *Conflicting ts: negative p_D due to a prior-data conflict*
Consider a single observation y from a t_4 distribution with unit scale and unknown mean μ, which itself has been given a standard t_4 distribution with unit scale and mean 0.

```
y <- 10
y   ~ dt(mu, 1, 4)
mu  ~ dt(0, 1, 4)
```

This extreme value leads to the bimodal posterior distribution shown in Figure 8.10.

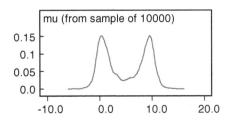

FIGURE 8.10
Posterior distribution resulting from a prior-data conflict.

The mean posterior deviance is $\overline{D(\theta)} = 10.8$: the posterior mean of μ is 4.9, which has a low posterior ordinate, and plugging this into the deviance gives a plug-in deviance $D(\bar{\theta}) = 12.0$. Hence $p_D = -1.2$.

WinBUGS and OpenBUGS do not calculate p_D (or DIC) if θ contains a discrete parameter. If this parameter is categorical, the posterior mean, required by p_D, is not meaningful. If the parameter is discrete but quantitative, however, it could be made continuous, as in the following example.

Example 8.6.4. *Salmonella (continued): parameterising to stabilise p_D*
The negative binomial model for the salmonella data (Example 6.5.2) has a discrete parent parameter r, which means we cannot calculate p_D. We first try mak-

ing the prior distribution on r continuous by using `r ~ dunif(1,max)` instead of the discrete uniform prior adopted previously.

This produces $p_D \approx 1.6$, which seems inappropriately low given there are four parameters in the model: three regression coefficients and r. The problem is as described above: the posterior distribution for r is highly skewed. We can, however, try placing a prior on a parameter that would be expected to have a more symmetric and unimodal posterior distribution, such as $\log r$. Taking Jeffreys' suggestion for a positive integer (§5.2.8), we could make $p(r) \propto 1/r$, or equivalently $p(\log r) \propto$ constant.

```
logr.cont      ~ dunif(0, 10)
log(r.cont) <- logr.cont
r              <- round(r.cont)
```

`logr.cont` has a well-behaved posterior distribution, and p_D now becomes 4.3, close to the true number of four parameters.

8.6.3   Alternative measures of the effective number of parameters

Consider again the situation in which information in the likelihood dominates that in the prior. We have already seen (§8.6.1) that in these circumstances

$$D(\theta) \approx D(\bar{\theta}) + \chi_p^2$$

so that $E(D(\theta)) \approx D(\bar{\theta}) + p$ (leading to $p_D \approx p$ as shown above) and $Var(D(\theta)) \approx 2p$. Thus, with negligible prior information, half the variance of the deviance is an estimate of the number of free parameters in the model — this estimate generally turns out to be remarkably robust and accurate in these situations of weak prior information. This in turn suggests using $p_V = Var(D)/2$ as an estimate of the effective number of parameters in a model in more general situations: this was noted by Gelman et al. (2004). p_V has the great advantage over p_D of being invariant to reparameterisations.

A second alternative to p_D was proposed by Plummer (discussion of Spiegelhalter et al. (2002)): the expected Kullback–Leibler information divergence $E_{\theta_0,\theta_1}(I(\theta_0, \theta_1))$ between the predictive distributions at two different parameter values, where

$$I(\theta_0, \theta_1) = E_{y^{pred}|\theta_0} \left[\log \left(\frac{p(y^{pred}|\theta_0)}{p(y^{pred}|\theta_1)} \right) \right]$$

This is the definition of p_D used in the JAGS implementation of BUGS, computed using the sample average of

$$\log \left(\frac{p(y_0^{pred}|\theta_0)}{p(y_0^{pred}|\theta_1)} \right)$$

over values of θ_0 and θ_1 generated from two parallel MCMC chains, where y_0^{pred} are posterior predictions from the chain for θ_0. This is non-negative, does not depend on the parameterisation, and is identical to the previous p_D in the normal linear model with known variance. Plummer (2008) calls this measure p_D^* and gives an approximate cross-validatory justification.

Some examples of these alternative measures compared to p_D are shown in Table 8.2. They work well in situations where p_D works well, but although they are always positive and invariant to parameterisation, they can also be misleading in extreme situations, for example, with a strong prior-data conflict or severely skewed posteriors.

TABLE 8.2
Comparison of p_D and alternative measures of effective model complexity.

Example	No. Parameters	p_D	p_V	p_D^*
Dugongs (6.3.1, model 3)	4	3.6	4.7	4.7
Transformed Binomial (8.6.2, version 3)	1	-0.4	0.3	0.1
Conflicting ts (8.6.3)	1	-1.2	23.2	3.6
Salmonella (Poisson, 6.5.2)	3	3.0	3.0	3.0
Salmonella (negative binomial, 8.6.4)	4	4.3	4.8	4.6

8.6.4 DIC for model comparison

The measure of fit $\overline{D}$ can be combined with the measure of complexity p_D to produce the *Deviance Information Criterion* (DIC).

$$DIC = \overline{D} + p_D$$
$$= D(\overline{\theta}) + 2p_D$$

This can be seen as a generalisation of Akaike's criterion: for models with weak prior information, $\overline{\theta} \approx \hat{\theta}$, and hence $p_D \approx p$ and DIC $\approx$ AIC. We can therefore expect Akaike-like properties of favouring models that can make good short-term predictions of a repeat set of similar data, rather than attempting to identify a "true" model — see §8.9 for discussion on alternative model comparison criteria.

Example 8.6.5. *DIC for model comparison*
Table 8.3 shows DIC comparisons for previous examples, which can be computed in WinBUGS and OpenBUGS by setting a DIC monitor via the Inference -> DIC... dialog box. Note that this must be set *after* the chains have reached convergence. For the salmonella data, the DIC improves by 8 when the negative binomial model is used instead of the Poisson, despite the increase in complexity.

For Newcomb's light speed data (Example 8.2.1), the t_4 distribution is strongly preferred to the normal. However, if the t_4 distribution is implemented as a scale mixture of normals, as suggested in §8.2, the deviance, and hence the DIC, is on a different scale and therefore cannot be compared to the other two analyses. Here DIC is assessing the ability to predict y_i *conditionally* on the current value of the mixing parameter λ_i — whereas the explicit t_4 model is *integrated* over the λ_i. This is changing the *focus* of the prediction problem — see §10.8 for further discussion.

p_D and DIC are particularly useful for comparing hierarchical models, as discussed in §10.8. For these models it is unclear how much random effects will contribute to the complexity of a model, since the implicit prior constraint of "shrinkage" to the mean (§10.1) simultaneously reduces the model's effective complexity. In Example 10.3.3 (Hepatitis B) each random effect contributes less than half an effective parameter to p_D. DIC also indicates that the hierarchical model is superior to a non-hierarchical model with no child-specific effects.

TABLE 8.3
Examples of DIC for model comparison.

Example	$\overline{D}$	p_D	DIC
Salmonella (Poisson, 6.5.2)	139.2	3.0	142.3
Salmonella (negative binomial, 8.6.4)	125.9	4.3	134.5
Newcomb (normal, 8.2.1)	501.7	2.0	503.8
Newcomb (t_4 distribution)	436.4	2.0	438.4
Newcomb (t_4 distribution as scale mixture)	408.8	15.1	423.8
Hepatitis B (non-hierarchical, 10.3.3)	1132.3	4.0	1136.3
Hepatitis B (hierarchical)	813.8	98.3	912.1

As mentioned above, the minimum DIC is intended to identify the model that would be expected to make the best short-term predictions, in the same spirit as Akaike's criterion. Plummer (2008) gave a more formal justification, showing that DIC approximated a cross-validatory loss but only when the effective number of parameters is much smaller than the number of independent observations.

It is important to note that only *differences between models* in DIC are important, and not *absolute* values. It is difficult to say what constitutes an important difference in DIC. With two simple hypotheses H_0 and H_1, $\exp[(DIC_0 - DIC_1)/2]$ would be a likelihood ratio, and so a DIC difference of 10 would be a likelihood ratio of 148, while a DIC difference of 5 would be a likelihood ratio of 12. By this rough analogy, differences of more than

10 might definitely rule out the model with the higher DIC, and differences between 5 and 10 are substantial. But if the difference in DIC is, say, less than 5, and the models make very different inferences, then it could be misleading just to report the model with the lowest DIC. In this case there is uncertainty about the choice of model. It may then be helpful to use methods of inference which account for this uncertainty, as discussed in §8.8.

We note that DIC can legitimately be negative! A probability density $p(y|\theta)$ can be greater than 1 if on a small standard deviation, and hence a deviance can be negative, and a DIC negative.

8.6.5 How and why does WinBUGS partition DIC and p_D?

WinBUGS (and OpenBUGS) separately reports the contribution to Dbar, pD and DIC for each differently named (scalar, vector, or array) node, together with a Total. This enables the individual contributions from different portions of data to be assessed.

In some circumstances some of these contributions may need to be ignored and removed from the Total. For example, in the following model:

```
for (i in 1:N) {
   Y[i] ~ dnorm(mu, tau)
}
tau    <- 1/pow(sigma, 2)
sigma  ~ dnorm(0, 1)I(0, )
mu     ~ dunif(-100, 100)
```

where Y is observed data, then the DIC tool will give DIC partitioned into Y, sigma, and the Total, where sigma has been constrained to be greater than 0 using the I() construct (Appendix A.2.2). Clearly, in this case, there should be no contribution from sigma, but because of the lower bound specified using the I() notation in the prior, WinBUGS treats sigma as if it were an observed but censored stochastic node when deciding what to report in the DIC table.

In another situation, we might genuinely have censored data, e.g.,

```
Y[i] ~ dnorm(mu, tau)I(Y.cens[i], )
```

where Y is unknown but Y.cens is the observed lower bound on Y (see Example 9.6.1).

WinBUGS has no way of knowing that in the first case, sigma should be ignored in the DIC, whereas in the second case Y should be included. This is as much a problem of how the BUGS language represents censoring, truncation, and bounds using the same notation as it is to do with how DIC is displayed, but it illustrates the ambiguity and how it is the user's responsibility to pick out the relevant bits.

Note that JAGS currently just sums over all data in the model when presenting DIC and p_D^*.

8.6.6   Alternatives to DIC

For models with discrete parameters, such as finite mixture models, the plug-in deviance required by p_D cannot be calculated since the posterior mean of a discrete parameter is either undefined or not guaranteed to be part of the discrete parameter space. Celeux et al. (2006), and their discussants, investigated ways of defining p_D and DIC for mixture and related models, though found them all problematic. For example, a mixture model could be reformulated by integrating over the discrete component membership parameter, as in Example 11.6.1 — however, the resulting DIC may be sensitive to the constraint chosen to identify the components (§11.6).

Plummer (2008) proposed the *penalised expected deviance* as an alternative model comparison measure. Both this and DIC estimate the ability to predict a replicate dataset, but judge this ability by different loss functions. Whereas DIC estimates the deviance for a replicate dataset *evaluated at the posterior expectations* of parameters θ, Plummer's criterion estimates the *expected* deviance for a replicate dataset. Both criteria incorporate a penalty to adjust for the underestimate in the loss ("optimism") due to using the data twice to both fit and evaluate the model. Since it does not require a "plug-in" estimate such as the posterior mean, this criterion can be used with discrete parameters. It is calculated as $\overline{D} + p_{opt}$, where the optimism p_{opt} is estimated from two parallel MCMC chains using importance sampling, as described by Plummer (2008) and provided in JAGS. The importance sampling method is unstable when there are highly influential observations, otherwise $p_{opt} \approx 2p_D^*$. A similar approximation was derived in the context of variable selection by van der Linde (2005).

The *pseudo-marginal likelihood* was proposed by Geisser and Eddy (1979) as a cross-validatory measure of predictive ability,

$$\prod_i p(y_i | y_{\setminus i}) = \prod_i \int p(y_i | \theta) p(\theta | y_{\setminus i}) d\theta,$$

where $y_{\setminus i}$ is all observations excluding y_i. Gelfand and Dey (1994) described an importance sampling method for estimating it based on a single MCMC run, which avoids the need to refit the model with each observation excluded in turn. The full-data posterior density $p(\theta | y)$ is used as a proposal distribution to approximate the leave-one-out posterior $p(\theta | y_{\setminus i})$. Given an MCMC sample $\theta^{(1)}, \dots, \theta^{(T)}$ from the posterior of θ, the importance weights are then $w_{it} = p(\theta^{(t)} | y_{\setminus i}) / p(\theta^{(t)} | y) \propto 1/p(y_i | \theta^{(t)})$, and the estimate of $p(y_i | y_{\setminus i})$ is the *harmonic mean* of $p(y_i | \theta^{(t)})$ over the posterior sample:

$$p(y_i | y_{\setminus i}) \approx \sum_t w_{it} p(y_i | y_{\setminus i}, \theta^{(t)}) / \sum_t w_{it}$$
$$= T / \sum_t (1/p(y_i | \theta^{(t)}))$$

Thus, the quantity $1/p(y_i|\theta^{(t)})$ is monitored during MCMC sampling, and the estimate of $p(y_i|y_{\backslash i})$ is the reciprocal of its posterior mean. The individual $p(y_i|y_{\backslash i})$ are called *conditional predictive ordinates* (CPOs) and may also be used as outlier diagnostics. Again, this method may be unstable, particularly if some of the CPOs are large (common in hierarchical models) and may require a large MCMC sample for a precise estimate. However, unlike DIC, it does not depend on plug-in estimates or on the model parameterisation.

8.7 Bayes factors

Traditional Bayesian comparison of models M_0 and M_1 is based on hypothesis tests using the *Bayes factor*. The posterior odds of model M_0 compared to M_1 are given by

$$\frac{p(M_0|y)}{p(M_1|y)} = \frac{p(M_0)}{p(M_1)} \frac{p(y|M_0)}{p(y|M_1)}$$

where

$$\frac{p(y|M_0)}{p(y|M_1)} = \frac{\int p(y|\theta_0)p(\theta_0)d\theta_0}{\int p(y|\theta_1)p(\theta_1)d\theta_1} = B_{01}$$

is known as the Bayes factor for M_0 compared to M_1. In other words,

posterior odds of M_0 = Bayes factor $\times$ prior odds of M_0.

The Bayes factor B_{01} quantifies the weight of evidence in favour of the null hypothesis H_0: "M_0 is true." If both models (hypotheses) are equally likely a priori, then their relative prior odds is 1 and B_{01} is the posterior odds in favour of model M_0 (Jeffreys (1939), p. 275, Gelman et al. (2004), p. 185).

The Bayes factors are in some sense similar to a likelihood ratio, except that the likelihood is *integrated* instead of maximised over the parameter space. As with AIC, there is no need for models to be nested, although unlike AIC, the objective is the identification of the 'true' model (Bernardo and Smith, 1994). Jeffreys (1939) provided a table relating the size of the Bayes factor to the "strength of evidence."

$p(y|M_r)$ is the *marginal likelihood* or *prior predictive probability* of the data, and it is important to note that this will depend crucially on the form of the prior distribution. A simple example will show that Bayes factors require informative prior distributions under each model. Consider a scalar θ so that the relevant term for the Bayes factor is $p(y) = \int p(y|\theta)p(\theta)d\theta$. Suppose θ is given a uniform prior, so that $p(\theta) = 1/(2c)$; $\theta \in [-c, c]$. Then $p(y) = \frac{1}{2c}\int_{-c}^{c} p(y|\theta)d\theta \propto \frac{1}{c}$ for large c. Therefore $p(y)$ can be made arbitrarily small by increasing c.

TABLE 8.4
Calibration of Bayes factors provided by Jeffreys.

Bayes factor range	Strength of evidence in favour of H_0 and against H_1
> 100	Decisive
32 to 100	Very strong
10 to 32	Strong
3.2 to 10	Substantial
1 to 3.2	"Not worth more than a bare mention"
	Strength of evidence against H_0 and in favour of H_1
1 to 1/3.2	"Not worth more than a bare mention"
1/3.2 to 1/10	Substantial
1/10 to 1/32	Strong
1/32 to 1/100	Very strong
< 1/100	Decisive

Suppose we are comparing models with weak prior information. Schwarz's Bayesian Information Criterion (BIC) is:

$$\text{BIC} = -2\log p(y|\hat{\theta}) + p\log n,$$

where $\hat{\theta}$ is the maximum likelihood estimate. The difference $\text{BIC}_0 - \text{BIC}_1$ gives an approximation to $-2\log B_{01}$. Kass and Wasserman (1995) show that this approximation has error $O_p(n^{-1/2})$ under a prior distribution which carries information equivalent to a single observation — the *unit-information* prior.

Alternatively, the posterior probability of model r, among a set of models indexed by k, is approximated by

$$p(M_r|y) = \exp(-0.5\text{BIC}_r)/\sum_k \exp(-0.5\text{BIC}_k) \tag{8.7}$$

Example 8.7.1. *Paul the psychic octopus*
In the 2010 football World Cup competition, Paul "the psychic octopus" made 8 predictions of the winners of football matches and got all $y = 8$ right. Our analysis will ignore the possibilities of draws, assume there was no bias or manipulation in the experiment, and ignore selection effects arising from Paul only becoming famous due to the first correct predictions, in the face of competition from numerous other wildlife. We assume a binomial model with probability θ of a correct prediction.

Rather naively, we could set up two simple hypotheses: H_0 representing that the predictions are just chance, so that $\theta = 0.5$; H_1 representing Paul having 100% predictive ability, so that $\theta = 1$. Since these are simple hypotheses with no unknown parameters, the Bayes factor is just the likelihood ratio $p(y|H_0)/p(y|H_1) = 1/2^8 = 1/256$, which from Table 8.4 represents "decisive" evidence against H_0 by Jeffreys criteria. However, the posterior odds against Paul being psychic also depend on the prior odds $p(H_0)/p(H_1)$ of Paul not having any psychic abilities (or knowledge of football), which it is reasonable to assume are so huge that this likelihood ratio makes little impact!

It may be more sensible to compare H_0 with an alternative hypothesis H_1 that Paul has *some* psychic ability, represented by a prior distribution on $\theta|H_1$. Naively this would be uniform on 0.5 to 1, but we introduce some scepticism by restricting it to be less than 0.55. So we are both sceptical of any effect existing at all, and even if it did exist, sceptical of a large effect. The code then essentially follows that of the biased coin example in §5.4.

```
q[1] <- 0.5; q[2] <- 0.5          # prior assumptions
r <- 8; n <- 8                    # data
r              ~ dbin(theta[pick], n)   # likelihood
pick           ~ dcat(q[])
theta[1] <- 0.5                   # if random (assumption 1)
theta[2]  ~ dunif(0.5, 0.55)      # if psychic
psychic  <- pick - 1              # 1 if psychic, 0 otherwise
```

```
node    mean    sd      MC error 2.5% median 97.5% start sample
psychic 0.6012 0.4896 0.001601 0.0   1.0    1.0   1     100000
```

The posterior probability of psychic abilities is now 0.6, corresponding to posterior odds $p(H_0|y)/p(H_1|y) = 0.4/0.6 = 0.66$. If the prior odds are taken as 1, this means that the Bayes factor is 0.66 in favour of psychic abilities, but again the prior odds against psychic abilities should realistically be much larger.

8.7.1 Lindley–Bartlett paradox in model selection

We have already seen that the Bayes factor depends crucially on the prior distribution within competing models. We now use a simple example to show how this can lead to an apparent conflict between using tail areas to criticise assumptions, and using Bayes factors — a conflict that has become known as the *Lindley–Bartlett paradox*.

Suppose we assume $Y_i \sim N(\theta, 1)$; we want to test $H_0 : \theta = 0$ against $H_1 : \theta \neq 0$. Then the sufficient statistic is $\overline{Y}$ with distribution $\overline{Y} \sim N(\theta, 1/n)$. For H_0, $p(\overline{y}|H_0) = \sqrt{\frac{n}{2\pi}} \exp[-n\overline{y}^2/2]$. For H_1, assume $p(\theta) = 1/(2c)$; $\theta \in [-c, c]$, $\theta \neq 0$, then

$$p(\overline{y}|H_1) = \frac{1}{2c} \int_{-c}^{c} \sqrt{\frac{n}{2\pi}} \exp[-n(\overline{y} - \theta)^2/2] \, d\theta \approx \frac{1}{2c}.$$

Hence the Bayes factor is

$$B_{01} = \frac{p(\overline{y}|H_0)}{p(\overline{y}|H_1)} = \sqrt{\frac{n}{2\pi}} \exp[-n\overline{y}^2/2] \times 2c.$$

From a classical hypothesis-testing perspective, we would declare a "significant" result if $\bar{y} > 1.96/\sqrt{n}$. At this critical value, the Bayes factor is $\sqrt{\frac{n}{2\pi}} \exp[-1.96^2/2] \times 2c$. Hence

- For fixed n, we can give H_0 very strong support by increasing c

- For fixed c, we can give H_0 very strong support by increasing n

So data that would just *reject* H_0 using a classical test will tend to *favour* H_0 for (a) diffuse priors under H_1 and (b) large sample sizes.

8.7.2   Computing marginal likelihoods

Computing Bayes factors for a generic model is challenging outside simple conjugate situations, as reviewed by Han and Carlin (2001) and Ntzoufras (2009). There is no easy method which works for all models specified in BUGS. BIC gives an approximation, as described above, though this essentially implies a default "unit information" prior for the parameters and does not allow user-specified priors. Other methods are based either on

- directly computing the marginal likelihood for each model, or

- considering the model choice as a discrete parameter and jointly sampling from the model and parameter space (§8.8.2).

For computing the marginal likelihood $p(y)$ for a particular model M, *harmonic mean* and related estimators are also sometimes used:

$$p(y) \approx \left(\frac{1}{T} \sum_{t=1}^{T} \left\{ \frac{g(\theta^{(t)})}{p(y|\theta^{(t)})p(\theta^{(t)})} \right\} \right)^{-1}$$

where $g()$ is an importance sampling density chosen to approximate the posterior. Although this is temptingly easy to program in BUGS by monitoring the term inside the braces and taking the reciprocal of its posterior mean, it is impractical in all but the simplest of models, since $p(y|\theta^{(t)})$ will frequently be very small; thus unfeasibly long runs would be required to stably estimate the posterior mean — indeed it may never converge (Neal, 2008). *Bridge sampling* or *path sampling* estimators (Gelman and Meng, 1998) are more effective, though usually require problem-specific tuning. Similarly, methods by Chib (1995) and Chib and Jeliazkov (2001) were shown to be effective by Han and Carlin (2001), but to implement these in BUGS would need substantial problem-specific programming, including access to the underlying source code (Ntzoufras, 2009).

Jointly sampling from the model and parameter space is a generally more reliable method of obtaining posterior probabilities of models in a BUGS context, particularly for comparing models with different sets of predictor variables, and techniques to do this in BUGS are reviewed in §8.8.2.

8.8   Model uncertainty

Neglecting uncertainty about the choice of model has been called a "quiet scandal" in statistical practice (Breiman, 1992) – see, for example, Draper (1995) and Burnham and Anderson (2002) for discussions. Drawing conclusions on the basis of a single selected model can conceal the possibility that other plausible models would give different results. Sensitivity analysis is recommended as a minimum, and this section discusses methods to formally incorporate model uncertainty in conclusions.

8.8.1   Bayesian model averaging

Posterior model probabilities $p(M_r|y)$ can be used to do "model averaging" to obtain predictions which account for model uncertainty. If we need to predict $\tilde{Y}$, and the predictive distribution assuming M_r is $p(\tilde{y}|y, M_r)$, then the "model-averaged" prediction is

$$p(\tilde{y}|y) = \sum_i p(\tilde{y}|y, M_r) p(M_r|y)$$

where

$$p(M_r|y) = p(M_r)p(y|M_r) / \sum_k \{p(M_k)p(y|M_k)\}$$

However, as discussed in §8.7.2, the marginal likelihood $p(y|M_r)$ involved in this definition is not, in general, straightforward to calculate in BUGS. We now describe techniques to accomplish model-averaged predictions without needing to calculate marginal likelihoods.

8.8.2   MCMC sampling over a space of models

We could consider the model choice as an additional parameter: specify prior probabilities for the model choice m and model-specific parameters θ_m, and sample from their joint posterior distribution $p(m, \theta_m|y)$, thus computing the posterior model probabilities. Any predictions are automatically averaged over the competing models.

Reversible jump MCMC   However, if we are choosing between models with different numbers of parameters, then the *dimension* of the space changes as the currently chosen model changes. The *reversible jump* MCMC algorithm was devised by Green (1995) to allow sampling over a space of varying dimension. The Jump add-on to WinBUGS (Lunn et al., 2009c)[†] performs reversible

---

[†]This is under development for OpenBUGS.

jump for variable selection in linear and binary regression and selecting among polynomial splines with different numbers of knots. See the manual included with Jump for further details and worked examples. It could be extended in the future to select within other classes of models, such as mixture models, for which specialised programming is currently required to implement reversible jump MCMC.

Product space search   In reversible jump MCMC, a value for θ_m is only sampled if the sampler is currently visiting model m. Carlin and Chib (1995) described an alternative MCMC method for sampling from $p(m, \theta_m|y)$, where values of θ_m are sampled for all m, whatever the currently chosen model. This requires a *pseudoprior* to be specified for each θ_m conditionally on the model *not* being m. While this is less efficient than reversible jump, it enables standard MCMC algorithms, available in BUGS, to be used. It can suffer from poor mixing unless the pseudopriors and priors on the models are chosen carefully. In practice, each model can be fitted independently and the resulting posteriors used to choose pseudopriors for a joint model. See the `Pines` example for BUGS (available from the BUGS web site or distributed with OpenBUGS) or Carlin and Louis (2008) for further details.

Variable selection priors   There are several methods specifically for variable selection in regression models, including stochastic search variable selection (George and McCulloch, 1993) and Gibbs variable selection (Dellaportas et al., 2002). The general idea is that there is a vector of covariate effects β and a vector I of the same length containing 0/1 indicators for each covariate being included in the model. β is then given a "spike and slab" prior (Mitchell and Beauchamp, 1988). This is a mixture of a probability mass $p(\beta_j|I_j = 0)$ concentrated around zero, representing exclusion from the model, and a relatively flat prior $p(\beta_j|I_j = 1)$ given that the variable is included:

$$p(\beta_j) = p(I_j = 1)p(\beta_j|I_j = 1) + p(I_j = 0)p(\beta_j|I_j = 0).$$

In BUGS, an example is

```
beta[j] <- b[pick[j]]      # effect of jth covariate
b[1]       ~ dnorm(0, tau)   # "spike": tau is large, b[1] <- 0
b[2]       ~ dnorm(0, eps)   # "slab": precision eps is small
pick[j] <- I[j] + 1
I[j]       ~ dbern(p[j])
```

where `p[j]` is the prior probability that the jth covariate is included, assuming these probabilities are independent. Thus the posterior probabilities of including each covariate arise naturally as the posterior mean of each `I[j]`. The methods differ in how exactly they define the priors. For more details on these methods and their implementation in BUGS, see O'Hara and Sillanpää

(2009) and Ntzoufras (2009) — while programming is generally straightforward, their efficiency and accuracy can depend on the choice of prior and parameterisation.

8.8.3 Model averaging when all models are wrong

Bayesian model averaging involves choosing and computing prior and posterior probabilities on models, interpreted as beliefs in their truth. Bernardo and Smith (1994) showed decision-theoretically that this provides optimal prediction or estimation under an "$\mathcal{M}$-closed" situation — in which the true process which generated the data is among the list of candidate models.

However, often one does not believe any of the models are true — an "$\mathcal{M}$-open" situation. Typically the truth is thought to be more complex than any model being considered. Model averaging is more difficult in this case, though some suggestions have been made. For example, substituting AIC or DIC for BIC in Equation (8.7) gives "Akaike weights" (Akaike, 1979) or DIC weights for averaging models, which measure their predictive ability, rather than their probability of being true. Using DIC in this way is attractively simple, though this method has not been formally assessed or justified. The resulting probabilities are difficult to interpret, though Burnham and Anderson (2002) suggest they represent posterior model probabilities under an implicit prior which favours more complex models at larger sample sizes.

Bootstrapping DIC A more interpretable way of averaging models without invoking a "true model" is to *bootstrap* the model selection process. Assuming independent data points, we resample from the data, choose the best-fitting model according to some criterion, repeat the process a large number of times, and average the resulting predictions. Buckland et al. (1997) used this method with AIC in a frequentist context, and Jackson et al. (2010a) used it with DIC for averaging Bayesian models. The resulting model probabilities $p(M_r|y)$ are the proportion of samples to which model r fits best according to the criterion. These are not Bayesian posterior probabilities, but rather *frequentist* probabilities, under sampling uncertainty, that the model will give the best predictions among those being compared.

Resampling and refitting would often be impractical for Bayesian models fitted by MCMC, which are typically intensive to compute. Therefore Jackson et al. (2010a), following Vehtari and Lampinen (2002), adapted a "Bayesian bootstrap" method which only requires one model fit and no resampling. Instead of sampling with replacement from the data vector y, the Bayesian bootstrap samples *sets of probabilities* q_i that the underlying random variable Y takes the value of each sample point $y_1, \ldots, y_n$. In one bootstrap iteration, samples $q_i^{(rep)}$ of q_i are drawn from a "flat" Dirichlet$(1, \ldots, 1)$ distribution. This is the posterior distribution of *the sampling distribution of Y*, which is assumed to be a discrete distribution over the observed values. This posterior is obtained by combining the sample $y_1, \ldots, y_n$ with an improper prior (Rubin, 1981).

The bootstrap replicate of a sample statistic (e.g., the mean) that can be expressed as $\sum_i f(y_i)$ is the weighted sum $n \sum_i q_i^{(rep)} f(y_i)$. Since the DIC can be decomposed into a sum over observations i, $DIC(y|M_r) = \sum_{i=1}^{n} DIC_i$, where $DIC_i = 2\overline{D(y_i|\theta)} - D(y_i|\hat{\theta})$, the bootstrap replicate of the DIC is

$$DIC(y|M_r)^{(rep)} = n \sum_{i=1}^{n} q_i^{(rep)} DIC_i$$

The sample of replicate DICs for each competing model can be used to give a bootstrap "confidence interval" surrounding the DIC for each model and probabilities that each model is best among the candidates.

Implementing this in BUGS requires the contribution to the posterior deviance from each data point to be monitored explicitly, similar to the method of deviance residuals (§8.3.4). For example, in a normal model:

```
for (i in 1:n) {
  y[i]      ~ dnorm(mu[i], tau)
  dev[i] <- log(2*pi) - log(tau) + pow(y[i] - mu[i], 2)*tau
  ...
}
```

The deviance of the observation y[i] evaluated at the posterior means of mu[i] and tau is subtracted from the posterior mean of dev[i] to produce DIC_i. The replicates can then be computed outside BUGS, using random samples of Dirichlet(1,...,1) variables (created, e.g., by BUGS). Note that the resulting model-averaged posterior has no Bayesian interpretation, since two sampling models for the data are used simultaneously — it is best viewed as a computational approximation to resampling.

8.8.4 Model expansion

Instead of averaging over a *discrete* set of models, a more flexible framework for model uncertainty is to work within a single model that encompasses all reasonable possibilities. This is recommended, for example, by Gelman et al. (2004). Model uncertainty is then considered as a choice over a *continuous* space of models. Support for different model choices is assessed by examining posterior distributions of parameters in the larger model, as in §8.5, and the model is checked to ensure that it gives predictions which agree with observations, as in §8.4.

The class of Bayesian nonparametric models illustrated in §11.8, for example, can reasonably be thought to "include the truth" in most practical situations. However, these do not naturally represent many model choice situations — a common example is whether to include or exclude a covariate in a regression. The encompassing flexible model would then be the one which includes all covariates being considered. In §8.8.2, we described flexible models of this type, where the prior distributions for the covariate effects had "spikes" at zero representing the possibility that the covariate is not included.

Smooth distributions are often a more realistic expression of prior belief than the mixture priors of this kind implied by discrete model averaging. Giving privilege to an effect of zero would not make sense if all potential predictors are thought to have non-zero, though perhaps inconsequentially small, effects. On the other hand, routinely using very vague priors for all potential effects would often lead to identifiability problems or poor predictive ability. Weakly informative priors might then be used, which typically "shrink" the effect towards zero. Gelman et al. (2008), for example, recommend a default Cauchy prior for logistic regression. For a review and comparison of such "shrinkage" priors for linear regression, see O'Hara and Sillanpää (2009), and for binary and survival regression see Rockova et al. (2012).

8.9 Discussion on model comparison

Broadly, there are two rather different approaches to Bayesian model comparison — one based on Bayes factors (or BIC) and the other on DIC, AIC, or similar measures. We can contrast the approaches under the following headings:

- *Assumptions.* The Bayes factor approach attempts to identify the "correct" model and implicitly assumes that such a thing exists and is in the families of distributions being considered. Posterior probabilities of models rely on it being meaningful to place probabilities on models. DIC/AIC makes no such assumption and only seeks short-term predictive ability.

- *Prior distributions.* Bayes factors require proper prior distributions (although these could be unit-information priors, as in BIC), which are not required for DIC/AIC.

- *Computation.* Bayes factors are notoriously difficult to compute in MCMC, requiring problem-specific programming or approximation; computation of DIC is generally straightforward.

- *Model uncertainty.* Model averaging to account for uncertainty about model choice is natural within a Bayes factor approach, provided one is willing to specify and interpret prior and posterior probabilities on models. Otherwise, DIC or Akaike weights, or bootstrapping, could be used for model averaging, though the theoretical justification is weak.

Working within an expanded model is a more flexible approach to model uncertainty which does not require averaging over a discrete set of models.

- *The "focus" of the analysis.* When dealing with hierarchical models, different model comparison methods can be related to which aspect of the model is of primary interest (§10.8.1).

The first issue is the most important: the situations in which it is reasonable to assume that any particular model is "true" appear very limited. We would therefore argue that the use of Bayes factors is restricted to domains where competing models correspond to clear, identifiable hypotheses that could in principle be proven to be "correct." Examples might include genetics applications where an individual is assumed to either carry some gene or not, or studies of supernatural abilities (as in Example 8.7.1) where the existence of any ability, however small, would be remarkable.

In either approach, we would recommend thorough criticism of model assumptions, as described in the first half of this chapter, and if there is model uncertainty, addressing it either formally or through clear sensitivity analyses.

8.10   Prior-data conflict

Bayesian analysis has traditionally focused on "turning the Bayesian handle," combining a prior distribution with a likelihood to produce a posterior distribution. But what if the likelihood and the prior seem in conflict, in the sense that they support almost non-intersecting areas of the parameter space? A naive approach would just charge on regardless, but this can lead to absurd results: for example, if we assume a normal observation $y \sim N(\theta, 1)$ with standard normal prior $\theta \sim N(0, 1)$, then an observation $y = 10$ will lead to a posterior distribution $\theta \sim N(5, 0.5)$, which is tightly situated around $\theta = 5$, a value supported neither by prior nor data (we note that if we instead assumed Student's t distributions we would obtain a bimodal posterior distribution, as in Example 8.6.3). This has been nicely ridiculed by Stephen Senn's characterisation of a Bayesian as someone who, suspecting a donkey and catching a glimpse of what looks like a horse, strongly concludes he has seen a mule (Senn (1997), p. 46, Spiegelhalter et al. (2004), p. 63).

There are two broad approaches to handling conflict: "identification" and "accommodation." Throughout this discussion we generally assume that the data are given priority and, in the case of conflict, it is the prior distribution that is called into question and discarded if necessary. However, the techniques can be easily adapted to give priority to the prior and discard divergent data, essentially adapting techniques previously used for identifying outliers.

8.10.1   Identification of prior-data conflict

This approach considers the current prior as a null hypothesis and checks whether the data fit the prior model or not. It is essentially a p-value argument as described for model checking (§8.4), in which an observed summary statistic t_0 is compared to a predictive distribution $p_0(t) = \int p(t|\theta)\,p(\theta)d\theta$, but using predictions arising from the prior rather than from the posterior distribution. The aim is to identify conflict and then one can decide whether to question the prior or the data.

As a simple example, assume the prior is $\theta \sim N(\mu, \sigma^2/n_0)$ and the sampling distribution is $Y_m \sim N(\theta, \sigma^2/m)$. Then the predictive distribution is $Y_m \sim N(\mu, \sigma^2/m + \sigma^2/n_0)$ and so the predictive p-value is

$$\Pr(Y_m < y_m) = \Phi\left(\frac{y_m - \mu}{\sigma\sqrt{\frac{1}{n_0} + \frac{1}{m}}}\right).$$

We note that this is also the tail area associated with a standardised test statistic contrasting the likelihood and the prior: i.e., suppose we assumed $Y_m \sim N(\theta_1, \sigma^2/m)$ and interpret the prior distribution as resulting from an observation $\mu \sim N(\theta_2, \sigma^2/n_0)$, then a classical test of the null hypothesis $H_0 : \theta_1 = \theta_2$ would be based on

$$z_m = \frac{y_m - \mu}{\sigma\sqrt{\frac{1}{n_0} + \frac{1}{m}}},$$

a measure of *conflict* between data and prior.

We use one-sided p-values throughout, identifying both high and low values as "interesting."

Example 8.10.1. *Surgery (continued): assessing prior-data conflict*
In Example 1.1.1 we considered a prior distribution for a mortality rate that could be expressed as a Beta(3,27), which has a mean of 10%. In Example 2.7.1 we then assumed that 20 operations were to take place and obtained the predictive probability of the number of successes. Suppose, however, that after the first five operations there had been two deaths, that is, 40% mortality — is this grounds for deciding that the prior distribution was "wrong"?

We can calculate the predictive distribution for Y either in its beta-binomial closed form or by Monte Carlo simulation. Since this predictive distribution is discrete, we assume a mid-p-value, $\Pr(Y > y) + \frac{1}{2}\Pr(Y = y)$.

```
r.obs <- 2
```

```
theta   ~ dbeta(3, 27)
r         ~ dbin(theta, 5)    # sampling distribution
P         <- step(r-r.obs-0.5) + 0.5*equals(r, r.obs) # mid-p-value
```

The mean of P is 0.054, suggesting some evidence of conflict with the prior distribution.

8.10.2   Accommodation of prior-data conflict

Suppose that instead of simply identifying conflict, we wanted to automatically accommodate it: we assume that in the case of conflict we would want to reject or downweight the prior distribution. A natural way of modelling this is to imagine competing priors, perhaps drawn from disagreeing experts. One prior might represent our current opinion and be given substantial prior weight, while an alternative could represent a weak prior covering a wider range of alternatives: this is a natural application of mixture priors (§5.4) in which the idea is essentially to get the data to "choose" between the alternatives.

Example 8.10.2. *Surgery (continued): mixture of priors*
Our prior for the underlying mortality risk in the previous example was Beta(3,27). But suppose a claim was made that the procedure was much more dangerous than this; in fact the mortality rate could be around 50%. Such a prior opinion might be represented by a Beta(3,3) distribution, which is symmetric with mean 0.5 and standard deviation $=\sqrt{0.5 \times 0.5/7} = 0.19$. Suppose, as above, out of the first five operations there are two deaths — what should we now believe about the true mortality rate? What do we expect to happen over the next 10 operations?

A crucial input is the relative belief in the two competing prior distributions, prior 1: $\theta \sim$ Beta(3, 27) or prior 2: $\theta \sim$ Beta(3, 3). We shall take them as initially equally plausible, corresponding to $q_1 = \Pr(\text{prior } 1) = 0.5$. The code shows how a "pick" formulation is used to select the appropriate parameters for the prior distribution.

```
model {
   theta        ~ dbeta(a[pick], b[pick])
   pick         ~ dcat(q[1:2])
   q[1]         <- 0.50
   q[2]         <- 0.50
   q.post[1] <- equals(pick, 1)    # = 1 if prior 1 picked
   q.post[2] <- equals(pick, 2)    # = 1 if prior 2 picked
   r            ~ dbin(theta, n)    # sampling distribution
   r.pred       ~ dbin(theta, m)    # predictive distribution
}
```

node	mean	sd	MC error	2.5%	median	97.5%	start	sample
q.post[1]	0.2416	0.4281	0.004438	0.0	0.0	1.0	1001	50000
q.post[2]	0.7584	0.4281	0.004438	0.0	1.0	1.0	1001	50000
y.pred	3.789	2.328	0.0177	0.0	4.0	8.0	1001	50000
theta	0.3786	0.1843	0.00164	0.07455	0.3872	0.721	1001	50000

Given these early results, there is now a 76% probability that the "sceptical" prior is appropriate and that this is a high-risk operation, and we would now expect 4 (95% interval 0 to 8) deaths out of the next 10 operations. Such a formulation may naturally lead to a bimodal posterior distribution for θ, as shown in Figure 8.11.

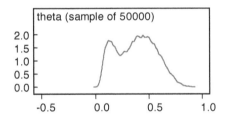

FIGURE 8.11
Posterior distribution for surgery mortality using a mixture of priors.

The formulation above works well when both priors are from the same family of distributions. Alternatively, we could follow the approach of Example 8.7.1 and model

```
y          ~ dbin(theta[pick], n)
theta[1] ~ dbeta(3, 27)
theta[2] ~ dbeta(3, 3)
```

which generalises easily to different parametric forms for the competing prior distributions.

This idea can be extended from a small set of possible prior distributions to a continuous mixture, and in so doing we can provide a "robust" prior that will have some influence if the data and prior agree, and otherwise will be overwhelmed. Essentially this can be implemented by adopting a "heavy-tailed" prior distribution that supports a wide range of possibilities but has little influence in the extremes. For example, if we have a normal sampling distribution $Y \sim N(\mu, 1)$, but the prior distribution is a Student's t distribution, then in the case of conflict the prior is essentially ignored (Dawid, 1973) as the long tails of the prior "accommodate" the conflicting observation.

Using the ideas introduced in §8.2 and §8.5, we can express the t distribution as a normal distribution whose unknown precision is drawn from a chi-squared distribution. Specifically, suppose we thought that a reasonable prior distribution was normal with precision 1, but we wished to express some doubt about this assumption. If we take $\mu \sim N(0, 1/\lambda)$, where $\lambda = X_k^2/k$ and $X_k^2 \sim \chi_k^2$, we are essentially assuming a t_k prior distribution for μ.

Example 8.10.3. *Prior robustness using a t prior distribution*
Suppose we assume $Y \sim N(\mu, 1)$, a prior mean $E[\mu] = 0$, and we want to build in prior robustness by assuming a t_k distribution for μ. We shall illustrate this with $k = 1, 2, 4, 10, 50, 1000$; $k = 1$ corresponds to a very heavy-tailed Cauchy distribution with no mean or variance, while $k = 1000$ is essentially a normal distribution. We construct these t distributions as scale mixtures of normals, as in Example 8.2.1.
Suppose we then observe a single data point $y = 4$, apparently conflicting with the prior mean of 0.

```
y.obs <- 4
df[1] <-1; df[2] <- 2; df[3] <- 4
df[4] <- 10; df[5] <- 50; df[6] <- 1000
#########################################
for (i in 1:6) {
   y[i]              <- y.obs        # replicate data
   y[i]              ~ dnorm(mu[i], 1)
   mu[i]             ~ dnorm(0, lambda[i])
   lambda[i]         <- X[i]/df[i]    # precision is chi-square/df
   X[i]              ~ dchisqr(df[i])
   # compare with prior distributions
   mu.rep[i]         ~ dnorm(0, lambda.rep[i])
   lambda.rep[i] <- X.rep[i]/df[i]
   X.rep[i]          ~ dchisqr(df[i])
}
```

node	mean	sd	MC error	2.5%	median	97.5%	start	sample
lambda[1]	0.2091	0.3384	0.004393	0.00338	0.1054	1.072	1001	10000
lambda[2]	0.3047	0.3576	0.004708	0.0155	0.1954	1.252	1001	10000
lambda[3]	0.4709	0.3992	0.005804	0.05558	0.3619	1.536	1001	10000
lambda[4]	0.6988	0.3487	0.004382	0.2132	0.6339	1.56	1001	10000
lambda[5]	0.9299	0.1895	0.001948	0.5993	0.9165	1.333	1001	10000
lambda[6]	0.9961	0.045	3.477E-4	0.9095	0.9951	1.086	1001	10000
mu[1]	3.449	1.072	0.01359	1.321	3.456	5.482	1001	10000
mu[2]	3.212	1.065	0.01376	1.135	3.198	5.316	1001	10000
mu[3]	2.859	1.035	0.01576	0.9113	2.833	4.97	1001	10000
mu[4]	2.444	0.8902	0.01013	0.7911	2.42	4.277	1001	10000
mu[5]	2.084	0.7478	0.007878	0.6675	2.071	3.575	1001	10000
mu[6]	2.004	0.7031	0.00638	0.6264	1.996	3.401	1001	10000

The estimate of μ based on the Cauchy ($k = 1$) is hardly influenced by the prior and a low value for λ is estimated. The normal ($k = 1000$) has the posterior mean mid-way between the data and the prior — an implausible conclusion whichever is true — and estimates λ to be almost 1.

We could think of a prior t distribution as a sensitivity analysis when we are unsure of a reasonable prior variance for a parameter with a normal prior. Assuming a t prior leads to the data taking preference if there is apparent "conflict" with the prior mean, since if the data and the prior mean are very different, this will tend to support the assumption of a large prior variance and so tends to assume that λ is small.

9

Issues in Modelling

A strength of the Bayesian graphical modelling techniques of BUGS is the way they can represent the typical complexities of real data. This chapter explains various generic issues encountered in data analysis and how they can be addressed in BUGS. For example, data commonly include missing values and measurement errors. A realistic model may need to account for censoring, truncation, grouping, rounding, or constraints on parameters, or use a sampling or prior distribution not already included in BUGS. We also discuss prediction, controlling "feedback" in graphical models, classical bootstrap estimation, and expressing uncertainty surrounding "ranks" or positions in a league table. Each of the techniques we describe may be deployed as part of *any* model in BUGS, with typically only a few extra lines of code.

9.1 Missing data

Missing data are common and there is an extensive literature covering a wide variety of methods for dealing with the problem. Comprehensive textbooks on the topic include Little and Rubin (2002), Molenberghs and Kenward (2007), and Daniels and Hogan (2008). Missing values in BUGS are denoted by NA in the data set, and from a Bayesian perspective, these are treated as additional unknown quantities for which a posterior distribution can be estimated. Hence the Bayesian approach makes no fundamental distinction between missing data and unknown model parameters. We just need to specify an appropriate joint model for the observed and missing data and model parameters, and BUGS will generate posterior samples of the model parameters and missing values in the usual way using MCMC.

The appropriateness of a particular missing data model is dependent on the mechanism that leads to the missing data and the pattern of the missing data. It also makes a difference whether we are dealing with missing responses or missing covariates (or both). Following Rubin (1976), missing data are generally classified into three types: missing completely at random (MCAR), missing at random (MAR), and missing not at random (MNAR). Informally, MCAR occurs when the probability of missingness does not depend on observed or unobserved data, in the less restrictive MAR it depends only on the observed data, and when neither MCAR nor MAR holds, the data are MNAR. Under an MCAR or MAR assumption, it is not usually necessary to specify a model for the missing data *mechanism* in order to make valid inference about parameters of the observed data likelihood, in which case the missing data mechanism is termed *ignorable*. In the case of MNAR missingness, the fact that a given value is missing tells us something about what that value might have been. In this case the missing data mechanism is *informative* and we must specify a model for it. There are two main approaches to this, using either a pattern mixture model or a selection model (Daniels and Hogan, 2008). In either case, the parameters of such a model cannot be uniquely inferred from the data, and so informative priors or parameter constraints are typically required. Inferences can thus be sensitive to the choices made — see Best et al. (1996) and Mason et al. (2012) for detailed discussions in the case of selection models. In the following examples we illustrate how to implement some specific models for missing response or missing covariate data in BUGS that make different assumptions about the missing data mechanism. A comprehensive discussion of a wide range of Bayesian missing data models can be found in Daniels and Hogan (2008).

9.1.1   Missing response data

For *ignorable* missing response data, we can chose to remove it from the data set, but often this is inconvenient. If we simply denote the value as missing (NA) in the dataset, then BUGS will automatically generate values from its posterior predictive distribution — see §9.2 — and inferences on the parameters will be as if we had deleted that response.

Example 9.1.1. *Growth curve (continued): ignorable missing response data mechanism*
We look again at the growth data from a single rat previously considered as an example of regression analysis (Example 6.1.1). We assume that the final datapoint (actually 376 g) is missing. If we assume that the chance of a value being missing does not directly depend on the true underlying weight, then an identical regression model can be adopted and only the data file changed.

```
for (i in 1:5) {
  y[i]        ~ dnorm(mu[i], tau)
  mu[i]       <- alpha + beta*(x[i] - mean(x[]))
}
alpha         ~ dflat()
.....
list(y = c(177,236,285,350,NA), x = c(8,15,22,29,36))
```

node	mean	sd	MC error	2.5%	median	97.5%	start	sample
alpha	290.3	7.029	0.06709	279.4	290.4	300.8	4001	10000

```
beta     8.104 0.8881 0.007908 6.79   8.11    9.355 4001  10000
sigma2  188.1 2915.0 58.29     5.72   29.08   926.0 4001  10000
mu[5]   403.8 16.74  0.158     377.6  404.0   427.7 4001  10000
y[5]    403.6 20.52  0.2122    371.5  404.0   434.2 4001  10000
```

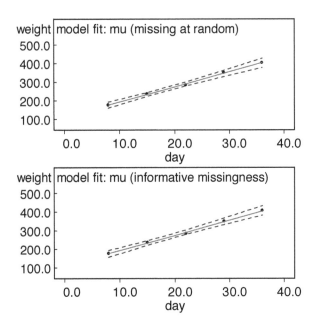

FIGURE 9.1
Model fits for rat 9's data with final observation missing. The plotted point (•) corresponding to the missing value y[5], at 36 days, is the posterior mean. The interval plotted at x[5] is for mu[5], not y[5]. Top: y[5] missing at random. Bottom: informative missingness for y[5].

The model fit is shown at the top of Figure 9.1. The estimated value for the missing data point y[5] lies on the fitted line; the 95% credible interval is wider than that for mu[5] (see table above), since it allows for the uncertainty about mu[5] as well as for the sampling error σ ($= \tau^{-1/2}$), and uncertainty about σ.

Example 9.1.2. *Growth curve (continued): informative missing response data mechanism*
Again we assume the final observation is missing, but now we also assume that the chance of an observation being missing depends on the true weight: specifically

the odds of being missing increase by 2% for every extra gram of weight. The data now has to include an indicator `miss[]` for whether an observation is missing or not, and our assumption about the missing data mechanism is specified using a logistic selection model for the probability of missingness, with b=log(1.02).

```
for (i in 1:5) {
  y[i]              ~ dnorm(mu[i], tau)
  mu[i]             <- alpha + beta*(x[i] - mean(x[]))

  # selection model for missing data mechanism
  miss[i]           ~ dbern(p[i])
  logit(p[i])       <- a + b*(y[i]-250)
}
a                   ~ dlogis(0, 1)
b                   <- log(1.02)
.....
list(y = c(177,236,285,350,NA), x = c(8,15,22,29,36),
     miss = c(0,0,0,0,1))
```

Here we specify a logistic prior for a, equivalent to a uniform prior on the probability of an observation with a true value of 250 g being missing (see §5.2.5).

node	mean	sd	MC error	2.5%	median	97.5%	start	sample
a	-1.973	1.082	0.01081	-4.262	-1.913	0.01912	4001	10000
alpha	291.4	13.13	0.3402	280.5	290.6	303.5	4001	10000
beta	8.228	1.314	0.03709	7.035	8.133	9.677	4001	10000
sigma2	433.0	4537.0	177.5	5.755	30.73	1063.0	4001	10000
mu[5]	406.6	27.89	0.8487	382.7	404.6	435.0	4001	10000
y[5]	408.4	38.23	1.368	378.6	404.9	447.1	4001	10000

The assumption about the missing data mechanism has raised the estimated missing weight from 404 to 408 g, and the posterior mean gradient from 8.10 to 8.23. The fact that the final data point is missing suggests that it has a larger value than it might have had if it were missing at random. Note that the missing weight is no longer estimated on the fitted line but slightly above it — see Figure 9.1 (bottom). In practice, we would want to examine sensitivity to different assumptions about the missing data mechanism, particularly to the value of b. It is also possible to treat b as random and specify a prior distribution for it, although posterior learning about the parameters of the selection model is heavily dependent on model assumptions (see Mason et al. 2012). Note that with more complex examples, reasonably tight priors and careful choice of initial values for the parameters of the selection model may be needed to avoid convergence problems and possible crashes of the MCMC sampling algorithms.

9.1.2 Missing covariate data

In the case of missing covariates, again NA can be specified for each missing value. However, the difference between missing responses and missing covariates is that we would not have specified a prior distribution or model for the covariates if they had been fully observed. Hence we must introduce a model or prior for the missing values, even if the missing data mechanism can be assumed to be ignorable. One option is to specify a prior distribution with common unknown parameters for both observed and unobserved values of the relevant covariate and then priors for the parameters of this distribution. The observed values of the covariate will contribute to the estimation of the unknown parameters, which, in turn, will inform about the missing values. This approach is only valid if the missing and observed covariate values can be assumed to be *exchangeable*, which is a mathematical formalisation of the assumption that a group of quantities is *similar* in some sense. Such assumptions are further discussed in Chapter 10. We illustrate this approach below.

Example 9.1.3. *Dugongs (continued): ignorable missing covariate mechanism*
We look again at version 3 of the non-linear growth curve model for the lengths of 27 dugongs, previously considered as an example of regression analysis (Example 6.3.1). In the current example we assume that the ages of four of the dugongs were not recorded. In specifying a prior distribution for the missing ages we should bear in mind that the growth curve is only meaningful for non-negative values of age. Here we constrain each unknown age to be positive by assuming that all ages arise from a log-normal distribution (dlnorm() — see Appendix C.2) with common unknown parameters. A more appropriate assumption might be a truncated normal distribution, truncated at zero. Note, however, that we would not be able to use the I(,) construct to truncate the distribution, since the parameters of that distribution would be unknown — see §9.6.2 and Appendix A.2.2 for further details, including alternative ways of specifying truncated distributions.

```
for(j in 1:27) {
  y[j]        ~ dnorm(mu[j], tau)
  mu[j]      <- alpha - beta*pow(gamma, x[j])
  # prior on covariate
  x[j]        ~ dlnorm(mu.x, p.x)
}
...
# priors on mean and precision of covariate model
mu.x        ~ dunif(-10, 10)
p.x        <- 1/pow(sd.x, 2)
sd.x        ~ dunif(0, 10)
...
list(x = c(1.0,1.5,1.5,NA,2.5,4.0,5.0,5.0,NA,8.0,8.5,9.0,
          9.5,9.5,10.0,12.0,12.0,13.0,NA,14.5,15.5,15.5,
```

```
        16.5,17.0,NA,29.0,31.5),
y = c(1.80,1.85,1.87,1.77,2.02,2.27,2.15,2.26,2.47,
      2.19,2.26,2.40,2.39,2.41,2.50,2.32,2.32,2.43,
      2.47,2.56,2.65,2.47,2.64,2.56,2.70, 2.72,2.57))
```

node	mean	sd	MC error	2.5%	median	97.5%	start	sample
alpha	2.649	0.06845	0.001528	2.527	2.645	2.797	4001	50000
beta	0.9604	0.07883	9.338E-4	0.8136	0.9584	1.116	4001	50000
gamma	0.8661	0.03251	7.006E-4	0.7926	0.8707	0.9152	4001	50000
mu.x	2.096	0.2087	9.875E-4	1.689	2.094	2.512	4001	50000
sd.x	1.048	0.1722	0.001139	0.7735	1.027	1.444	4001	50000
sigma	0.0955	0.01605	1.407E-4	0.07019	0.09341	0.1325	4001	50000
x[4]	1.395	0.7006	0.004034	0.3363	1.299	3.035	4001	50000
x[9]	16.7	18.08	0.1805	6.831	12.77	54.61	4001	50000
x[19]	17.52	22.87	0.2601	6.821	12.88	59.72	4001	50000
x[25]	38.5	39.62	0.3918	12.3	28.05	127.9	4001	50000

There is considerable uncertainty regarding the "true" values of the missing co-variates, particularly for dugongs 9, 19, and 25. The actual values of the four missing ages are 1.5, 7, 13, and 22.5, and all of these are included within the estimated 95% credible intervals, although for dugong 9 this is only just the case. This reflects the influence of the response value, y, on the posterior distribution of each missing x — dugong 9 was long for its age (y[9] = 2.47) and hence values of age somewhat larger than the actual value are consistent with the fitted model. In fact, dugongs 9 and 19 had identical lengths, and so the posterior distributions for x[9] and x[19] are identical to within sampling error.

If there are other fully observed covariates in the regression model of in-terest, the previous approach will not account for correlation between these and the covariate being imputed. In this case, a better option is to specify a regression model to impute the missing covariates as a function of other co-variates. This model may include covariates not in the main model of interest and is similar in spirit to the two-stage multiple imputation (MI) approach of Rubin (1987). As with standard MI, variables that are predictive of both the missing covariate itself *and* of the missing data mechanism should be included in the imputation model. When multiple covariates have missing values, it is also important to reflect the dependence structure of the covariates in the imputation model.

Example 9.1.4. *Birthweight: regression model for imputing missing covariates*
Trihalomethanes (THM) are a chemical byproduct of the treatment process used to disinfect the public water supply in the UK. Molitor et al. (2009) analyse the association between THM levels in domestic tap water and the risk of giving birth to a low birthweight (< 2.5 kg) baby. They use data from the UK National Births Register on maternal age, baby's gender and birthweight, and use the mother's

residential postcode to link modelled estimates of average THM levels in the water supply zone of residence to each birth. Maternal smoking and ethnicity are known risk factors for low birthweight and are potential confounders of the THM effect due to their spatial patterning, which correlates with spatial variations in THM levels. Smoking and ethnicity are not recorded in the birth register but are available for a subset of the mothers who participated in a national birth cohort study. Molitor et al. (2009) build a full Bayesian model to impute the missing smoking and ethnicity indicators for mothers in the birth register who did not participate in the cohort study and to simultaneously estimate the regression of low birthweight on THM levels adjusted for confounders.

Here we use simulated data that mimics a slightly simplified version of this problem. The model of interest is a logistic regression of the low birthweight indicator lbw on binary indicators of THM level > 60 μg/L (THM), male baby (male), non-white maternal ethnicity (eth), maternal smoking during pregnancy (smk), and deprived local area (dep). smk and eth are recorded for 20% of mothers but are missing for the remaining 80%; all other variables are fully observed. To impute the missing covariate values, we build a bivariate regression model for smk and eth assuming correlated errors. Since these are both binary indicators, we use multivariate probit regression (Chib and Greenberg, 1998) in which smk and eth are equal to thresholded values of a bivariate normal latent variable Z,

$$Z_i = (Z_{i1}, Z_{i2})' \sim \text{MVN}(\mu_i, \Omega), \quad i = 1, \ldots, n;$$
$$smk_i = I(Z_{i1} \geq 0); \quad eth_i = I(Z_{i2} \geq 0),$$

where Ω must be in correlation form for identifiability reasons. The elements of $\mu_i = (\mu_{i1}, \mu_{i2})'$ are modelled as independent linear functions of covariates, which include the other variables in the regression model for lbw plus area-level measures of the proportion of the population who smoke (area.smk) and who are non-white (area.eth). Unlike standard multiple imputation, it is *not* necessary to include the response variable from the regression model of interest (lbw in this case) in the covariate imputation model, since information about lbw is automatically propagated via feedback from the assumed regression model of lbw on smk and eth. Likelihood information about the observed values of smk and eth is included in the imputation model by specifying bounds on Z_i such that $Z_{i1} \in (-\infty, 0)$ if $smk_i = 0$, $Z_{i1} \in [0, \infty)$ if $smk_i = 1$, $Z_{i2} \in (-\infty, 0)$ if $eth_i = 0$, and $Z_{i2} \in [0, \infty)$ if $eth_i = 1$. If smk_i and eth_i are missing, the corresponding bounds on Z_{i1} and Z_{i2} are set to $(-\infty, \infty)$. In BUGS this is done using the I(lower,upper) notation (§9.6) and including vectors giving values of the lower and upper bounds in the data file. Since a value of $\pm\infty$ cannot be specified in the data file, we instead use an arbitrarily large value relative to the scale of the latent Z variable (say ± 10). Initial values for the parameters of the regression model of interest and for the imputation model need to be chosen carefully to ensure that they both provide compatible information about the missing covariate values; strongly conflicting initial values can cause the MCMC samplers in BUGS to crash.

```
for (i in 1:n) {
```

```
  lbw[i]              ~ dbern(p[i])
  logit(p[i])         <- beta[1] + beta[2]*THM[i] +
                         beta[3]*male[i] + beta[4]*dep[i] +
                         beta[5]*smk[i] + beta[6]*eth[i]
}
for (k in 1:6) {
  beta[k]             ~ dnorm(0, 0.0001)
}
for (k in 2:6) {
  OR[k]               <- exp(beta[k])
}
# multivariate probit covariate imputation model
for (i in 1:n) {
  Z[i,1:2]            ~ dmnorm(mu[i,1:2],
                           Omega[1:2,1:2])I(lo[i,1:2],up[i,1:2])
  mu[i,1]             <- delta[1,1] + delta[2,1]*THM[i] +
                         delta[3,1]*male[i] + delta[4,1]*dep[i] +
                         delta[5,1]*area.smk[i] +
                         delta[6,1]*area.eth[i]
  mu[i,2]             <- delta[1,2] + delta[2,2]*THM[i] +
                         delta[3,2]*male[i] + delta[4,2]*dep[i] +
                         delta[5,2]*area.smk[i] +
                         delta[6,2]*area.eth[i]
}
for (i in 1:Nmis) { # Data file is ordered so subjects
                    # 1,...,Nmis have missing values
  smk[i]              <- step(Z[i,1]) # thresholded value of Z[i,1]
  eth[i]              <- step(Z[i,2]) # thresholded value of Z[i,2]
}
Sigma[1,1]          <- 1
Sigma[2,2]          <- 1
Sigma[1,2]          <- corr
Sigma[2,1]          <- corr
corr                  ~ dunif(-1, 1)
Omega[1:2, 1:2]     <- inverse(Sigma[,])
for (k in 1:6) {
  delta[k,1]          ~ dnorm(0, 0.0001)
  delta[k,2]          ~ dnorm(0, 0.0001)
}
```

node	mean	sd	MC error	2.5%	median	97.5%	start	sample
OR[2]	1.178	0.121	0.002129	0.956	1.172	1.425	1001	20000
OR[3]	0.8095	0.08177	0.001518	0.6607	0.8052	0.9821	1001	20000
OR[4]	1.007	0.101	0.001919	0.8243	1.001	1.221	1001	20000
OR[5]	2.915	0.5202	0.01434	2.033	2.867	4.057	1001	20000
OR[6]	4.116	0.7181	0.01833	2.897	4.053	5.716	1001	20000

```
corr  -0.3039 0.05583 0.002219 -0.4096 -0.3046 -0.1904 1001  20000
```

There is a small excess risk of low birthweight for mothers with high THM levels in their tap water supply, although the posterior 95% credible interval just includes the null odds ratio. Maternal smoking and non-white ethnicity confer substantially increased risks of low birthweight, although the wide credible intervals for these effects reflect uncertainty due to the high proportion of missing values. Analysis of the complete cases only produced a far more uncertain estimate of the THM effect (mean of OR[2] = 1.13, 95% CI 0.76 to 1.64), whilst analysis of the full data excluding the confounders smk and eth from the regression model produced an upwardly biased estimate of the THM effect (mean of OR[2] = 1.44, 95% CI 1.21 to 1.70).

9.2 Prediction

There are a number of reasons why we may want to predict an unknown quantity Y^{pred}. We may want to "fill in" missing or censored data (§9.1) or predict replicate datasets in order to check the adequacy of our model (§8.4). Finally, we may simply want to make predictions about the future.

If we were working within a classical paradigm, it would not be straightforward to make full predictions after fitting a statistical model. Although point predictions of a future quantity Y^{pred} may be easy, obtaining the appropriate full predictive distribution for Y^{pred} is challenging, as one needs to account for three components: uncertainty about the expected future value $E[Y^{\mathrm{pred}}]$, the inevitable sampling variability of Y^{pred} around its expectation, and the uncertainty about the size of that error, as well as the correlations between these components. Fortunately, it is so trivial to obtain such predictive distributions using MCMC that it can be dealt with very briefly.

Suppose we have a model $p(y^{\mathrm{pred}}|\theta)$ and a fully specified prior distribution $p(\theta)$. We have already seen in §2.7 how Monte Carlo methods can produce the predictive distribution of a future quantity as $p(y^{\mathrm{pred}}) = \int p(y^{\mathrm{pred}}|\theta)p(\theta)d\theta$ by simply including y^{pred} in the model and treating it as an unknown quantity. The same principle applies if instead we are using MCMC methods with a posterior distribution $p(\theta|y)$ (Chapter 4). In the example below we point out that, rather than explicitly include the quantities to be predicted in the model description, it may be easier to just expand the dataset to include missing data indicated as NA.

Example 9.2.1. *Dugongs (continued): prediction*
Consider again the growth model for dugongs from Example 6.3.1. Suppose we
want to project the length of dugongs beyond the currently observed age range,
say at ages 35 and 40 years. We could explicitly include, as quantities in the model,
the *expected* lengths of all dugongs at those ages, as well as the *observable* lengths
of specific future dugongs. Assuming

$$y_j \sim \text{Normal}(\mu_j, \sigma^2 = \tau^{-1}), \quad \mu_j = \alpha - \beta\gamma^{x_j}$$

for the observed data, we add the code

```
mu35 <- alpha - beta*pow(gamma, 35)
mu40 <- alpha - beta*pow(gamma, 40)
y35    ~ dnorm(mu35, tau)
y40    ~ dnorm(mu40, tau)
```

Alternatively, it will generally be easier to leave the model description unmodified
and instead expand the data file with two dugongs of the appropriate ages but
with missing lengths:

```
list(N = 29,
     x = c(1.0, 1.5, 1.5, ..., 29.0, 31.5, 35, 40),
     y = c(1.80, 1.85, 1.87, ..., 2.72, 2.57, NA, NA))
```

Posterior predictive summaries for the quantities of interest are given by

node	mean	sd	MC error	2.5%	median	97.5%	start	sample
mu[28]	2.638	0.05949	0.002778	2.528	2.635	2.762	1001	10000
mu[29]	2.642	0.06291	0.002959	2.528	2.638	2.775	1001	10000
y[28]	2.637	0.1153	0.002654	2.408	2.638	2.865	1001	10000
y[29]	2.642	0.1179	0.003107	2.413	2.64	2.881	1001	10000

The intervals around mu[28] and mu[29] reflect uncertainty concerning the fitted
parameters α, β, and γ, as is the case for the other elements of mu. Intervals
around the missing ys additionally reflect the sampling error σ and uncertainty
about the value of σ. The model fit and predictive intervals for mu[28] and
mu[29] are shown together in Figure 9.2. Widening of the intervals towards the
right-hand side of the plot has nothing to do with the fact that the right-most
intervals are predictive; this is simply due to greater uncertainty in the fitted curve
for older animals.

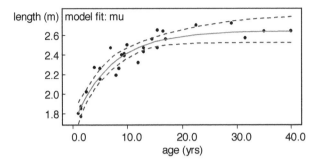

FIGURE 9.2
Model fit for observed dugongs data, with 95% posterior predictive intervals for the expected dugong length at ages 35 and 40 years. The points plotted (•) at 35 and 40 years are the posterior median values of `y[28]` and `y[29]`, which are specified as missing. The points coincide exactly with the predictive medians for `mu[28]` and `mu[29]`.

9.3 Measurement error

Errors in measurement can occur for both responses and covariates. The former case is straightforward, since standard statistical models can be thought of as encompassing errors in measurement and we can select the appropriate response distribution. When covariates are measured with error, there are two possible models: *classical* and *Berkson*.

The more common, classical model is represented in the graph shown in Figure 9.3(a), in which the observed covariate x is assumed conditionally independent of the response y given the "true" underlying covariate value z. The covariate may be categorical or continuous, but in either case needs to be provided with a prior distribution, with parameters ψ, say, which may or may not be known. In addition, an error model is assumed with parameters ϕ, which in order to be identifiable will need to be either assumed known or estimable from a subset of data in which both x and z are observed.

Example 9.3.1. *Cervix: case-control study with errors in covariates*
Carroll et al. (1993) consider the problem of estimating the odds ratio of a disease d in a case-control study where the binary exposure variable is measured with error. Their example concerns exposure to herpes simplex virus (HSV) in women with invasive cervical cancer $(d = 1)$ and in controls $(d = 0)$. Exposure to HSV is measured by a relatively inaccurate western blot procedure x for 1929 of the 2044 women, whilst for 115 women, it is also measured by a refined or "gold standard" method z. The data are given in Table 9.1. They show a substantial

(a)   (b)

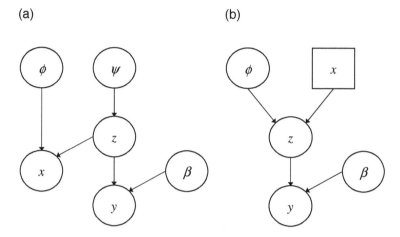

FIGURE 9.3

DAGs depicting classical and Berkson measurement error models. The response variable y is regressed on "true" covariates z, with regression coefficients β. (a) Classical model: the observed covariates x are assumed dependent on the true values z via an error model with parameters ϕ; a prior distribution with parameters ψ is specified for z. (b) Berkson error model: the true covariates z are assumed dependent on the observed values x via an error model with parameters ϕ.

amount of misclassification, as indicated by low sensitivity and specificity of x in the "complete" data. The degree of misclassification is also significantly higher for the controls than for the cases ($p = 0.049$ by Fisher's exact test).

A (prospective) logistic model is fitted to the case-control data as follows

$$d_i \sim \text{Bernoulli}(p_i), \quad \text{logit}(p_i) = \beta_0 + \beta z_i, \quad i = 1, \ldots, 2044,$$

where β is the log odds ratio of disease d. Since the relationship between d and z is only directly observable in the 115 women with "complete" data, and because there is evidence of differential measurement error, the following parameters are required in order to estimate the misclassification model:

$$\phi_{11} = \Pr(x = 1 | z = 0, d = 0)$$
$$\phi_{12} = \Pr(x = 1 | z = 0, d = 1)$$
$$\phi_{21} = \Pr(x = 1 | z = 1, d = 0)$$
$$\phi_{22} = \Pr(x = 1 | z = 1, d = 1)$$
$$\psi = \Pr(z = 1)$$

BUGS code for the model is as follows:

```
for (i in 1:n) {
```

TABLE 9.1

Case-control data for cervix example.

Complete data				Incomplete data			
d	z	x	Count	d	z	x	Count
1	0	0	13	1	—	0	318
1	0	1	3	1	—	1	375
1	1	0	5	0	—	0	701
1	1	1	18	0	—	1	535
0	0	0	33				
0	0	1	11				
0	1	0	16				
0	1	1	16				

```
    d[i]              ~ dbern(p[i])
    logit(p[i]) <- beta0 + beta*z[i]
    z[i]              ~ dbern(psi)
    x[i]              ~ dbern(phi[z1[i], d1[i]])
    z1[i]            <- z[i] + 1
    d1[i]            <- d[i] + 1
  }
  for (j in 1:2) {
    for (k in 1:2) {
      phi[j, k]   ~ dunif(0, 1)
    }
  }
  psi               ~ dunif(0, 1)
  beta0             ~ dnorm(0, 0.0001)
  beta              ~ dnorm(0, 0.0001)
```

where the z1 and d1 variables are created because phi[] must be indexed with
1s and 2s, as opposed to 0s and 1s, and functions are not allowed as indices.
The data can be specified "long-hand" with three entries (d, z, x) for each of
the 2044 individuals. Alternatively the individual-level data can be "constructed"
from Table 9.1 via the following additional code:

```
  for (j in 1:8) {
    for (i in offset[j]:offset[j+1]-1) {
      d[i] <- d.com[j]; x[i] <- x.com[j]; z[i] <- z.com[j]
    }
  }
  for (j in 9:12) {
    for (i in offset[j]:offset[j+1]-1) {
      d[i] <- d.inc[j-8]; x[i] <- x.inc[j-8]
    }
  }
```

where d.com, x.com, and z.com denote the complete data, d.inc and x.inc denote the incomplete data, and offset contains the cumulative counts of individuals within each category: offset=c(1, 14, 17, 22, 40, 73, 84, 100, 116, 817, 1352, 1670, 2045). Posterior summaries are given in the table below.

node	mean	sd	MC error	2.5%	median	97.5%	start	sample
beta	0.6213	0.3617	0.01924	-0.09153	0.6188	1.345	1001	20000
beta0	-0.9059	0.199	0.01045	-1.321	-0.8996	-0.5283	1001	20000
phi[1,1]	0.3177	0.05309	0.00199	0.2109	0.3186	0.4199	1001	20000
phi[1,2]	0.2212	0.08055	0.003301	0.07556	0.2188	0.3884	1001	20000
phi[2,1]	0.5691	0.06352	0.002116	0.4428	0.5683	0.6941	1001	20000
phi[2,2]	0.7638	0.06187	0.002506	0.6409	0.7646	0.8806	1001	20000
psi	0.4923	0.04304	0.001771	0.4057	0.4929	0.5771	1001	20000

From this output we can estimate that the chance of falsely identifying HSV using a western blot is 32% in controls and 22% in cases, while the chance of missing a true HSV is 44% in controls and 24% in cases. Accounting for this misclassification results in a substantially de-attenuated estimate of the exposure log-odds ratio, although the increased uncertainty means that this is no longer statistically significant (posterior mean and 95% CI for beta = 0.62 (−0.09, 1.35) compared to 0.45 (0.27, 0.63) if covariate misclassification is ignored).

Example 9.3.2. *Dugongs (continued): measurement error on age*
Recalling again Example 6.3.1, we now assume that the observed age x_j is an imperfect measure of the true age z_j, with measurement standard deviation 1. We consider the model

$$y_j \sim \text{Normal}(\mu_j, \sigma^2), \quad \mu_j = \alpha - \beta\gamma^{z_j}, \quad x_j \sim \text{Normal}(z_j, 1),$$

with $\alpha, \beta \sim \text{Uniform}(0, 100)$, $\gamma \sim \text{Uniform}(0, 1)$, and $\log\sigma \sim \text{Uniform}(-10, 10)$. In addition, a prior distribution for each z_j is required. In the absence of prior knowledge we assume $z_j \sim \text{Uniform}(0, 100)$ for $j = 1, \ldots, n$. BUGS code for the model is given by

```
for(j in 1:n) {
  y[j]        ~ dnorm(mu[j], tau)
  mu[j]       <- alpha - beta*pow(gamma, z[j])
  x[j]        ~ dnorm(z[j], 1)
  z[j]        ~ dunif(0, 100)
}
alpha         ~ dunif(0, 100)
beta          ~ dunif(0, 100)
gamma         ~ dunif(0, 1)
tau           <- 1/sigma2
```

```
log(sigma2) <- 2*log.sigma
log.sigma    ~ dunif(-10, 10)
```

The model fit is shown in Figure 9.4. Note that the "true" ages z_j, $j = 1, \ldots, n$, are estimated such that the model fit is improved. This is reflected by a posterior median value for σ^2 of 0.0078, which is reduced from 0.0094 (when no measurement error was assumed — see Example 6.3.1). Figure 9.5 shows the posterior distribution of the difference between each observed and "true" age, calculated by adding the code:

```
for (j in 1:n) {resx[j] <- x[j] - z[j]}
```

Where the fit is not improved by estimating z[j] away from the observed value x[j], the distribution of resx[j] is approximately standard normal. We can see that, particularly in the first half of the dataset, there is considerable adjustment of the ages being entered into the regression equation.

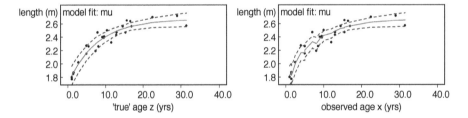

FIGURE 9.4

Model fits from analysis of dugongs data assuming a classical measurement error model for the observed ages. Left-hand side: model fit plotted against posterior median "true" ages, z_j, $j = 1, \ldots, n$. Right-hand side: model fit plotted against observed ages x_j, $j = 1, \ldots, n$.

Berkson errors arise in situations where the observed covariates are expected to be less variable than the "true" values, perhaps because the observed values are aggregated. This can occur when the covariates measure environmental exposure, say, such as levels of air pollution. The observed values may be summary measures for geographical areas, each, perhaps, taken at a single site or summed/averaged over the area. The *actual* exposures of individuals within those areas would then be expected to be more variable than the recorded values. This leads to a measurement error model in which the true covariate value depends on the observed value, rather than the other way round as in classical measurement error (see the right-hand side of Figure 9.3). The following example illustrates the use of a Berkson model for air pollution data.

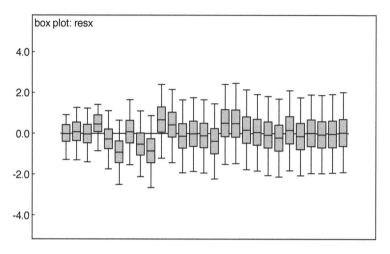

FIGURE 9.5

Box plot summarising posterior distributions of the difference between each observed and "true" age for the dugongs example with covariate measurement error.

Example 9.3.3. *Air pollution: Berkson measurement error*

Whittemore and Keller (1988) examine data regarding the potential effects of exposure to nitrogen dioxide (NO_2) on respiratory illness. One hundred and three children are categorised as having respiratory illness or not and of being exposed to one of three different levels of NO_2 in their bedrooms:

Respiratory illness (y)	Bedroom NO_2 level in ppb (x)			
	< 20	20–40	40+	Total
Yes	21	20	15	56
No	27	14	6	47
Total (n)	48	34	21	103

The three exposure categories (indexed by j) are thought of as a surrogate for "true exposure," and the nature of the measurement error relationship is known precisely from a separate calibration study:

$$z_j = \alpha + \beta x_j + \epsilon_j, \quad j = 1, 2, 3,$$

where x_j equals 10, 30, and 50 for $j = 1$, 2, and 3, respectively, and z_j is interpreted as the "true average value" of NO_2 in group j. In addition, from the calibration study we assume $\alpha = 4.48$, $\beta = 0.76$, and $\epsilon_j \sim \text{Normal}(0, 81.14)$, $j = 1, 2, 3$. We wish to fit the following logistic regression:

$$y_j \sim \text{Binomial}(p_j, n_j), \quad \text{logit}(p_j) = \theta_1 + \theta_2 z_j, \quad j = 1, 2, 3,$$

where n_j is the total number of children in group j.

```
for (j in 1:3) {
    y[j]              ~ dbin(p[j], n[j])
    logit(p[j]) <- theta[1] + theta[2]*z[j]
    z[j]              ~ dnorm(mu[j], 0.01232)
    mu[j]            <- alpha + beta*x[j]
}
theta[1]          ~ dnorm(0, 0.0001)
theta[2]          ~ dnorm(0, 0.0001)
```

node	mean	sd	MC error	2.5%	median	97.5%	start	sample
theta[1]	-0.8096	0.8559	0.03736	-2.889	-0.6557	0.3502	12001	20000
theta[2]	0.04207	0.03144	0.001313	-0.002226	0.03667	0.1212	12001	20000
z[1]	12.8	8.299	0.2011	-3.881	12.98	28.81	12001	20000
z[2]	27.43	7.474	0.08438	12.95	27.37	42.39	12001	20000
z[3]	41.43	8.56	0.1437	25.35	41.23	58.56	12001	20000

Note that the effect of NO_2 exposure on the chances of developing a respiratory illness is almost, but not quite, significant in this analysis: the 95% credible interval for θ_2, which represents the log odds ratio for a unit increase in "true exposure" only just includes zero.

9.4 Cutting feedback

In the dugongs example 9.3.2 above the true ages are estimated such that they improve the fit of the line. In many cases, this is exactly what we would want: the information in the measured ages regarding the values of the true ages is supplemented by feedback from the response data (dugong lengths) due to the assumed relationship between length and age. However, there are situations in which we might wish to infer the values of missing variables based solely on the observed values of those variables. In other words, we may wish to ignore, or "cut" the feedback from the response data. We can achieve this using the cut() function in WinBUGS,* as illustrated in the following example.

Example 9.4.1. *Cutting feedback*
Consider the simple linear regression presented in Figure 9.6(a). In this case the values of the variable plotted on the x-axis are assumed known. However, suppose we know that they are measured with error and that the standard deviation of those errors is 1.5. We denote the response variable by y_i, $i = 1, \ldots, n$, and

---
*or OpenBUGS. There is no "cut" function currently in JAGS.

the modelled and observed values of the independent variable by z_i and x_i, $i = 1, \ldots, n$, respectively. (Note that $x_i = i$, $i = 1, \ldots, n$.) One option is to assume:

$$y_i \sim \text{Normal}(\mu_i, \sigma^2), \quad \mu_i = a + bz_i, \quad x_i \sim \text{Normal}(z_i, 1.5^2),$$

with appropriate priors on a, b, σ, and each z_i. This would allow estimation of the z_is to be influenced by feedback from the y_is. To cut this feedback, we assume instead

$$z_i = \text{cut}(z_i^*), \quad x_i \sim \text{Normal}(z_i^*, 1.5^2),$$

with appropriate priors on the z_i^*s, e.g., $z_i^* \sim \text{Uniform}(-100, 100)$, $i = 1, \ldots, n$. The cut(.) function here makes a copy of the variable passed as an argument but otherwise severs the link between argument and result, z_i^* and z_i, respectively, in this case. Hence z_i always has the same value as z_i^* but z_i^* is isolated from the y_is and cannot be influenced by them. The following BUGS code fits models with and without feedback as well as the model in which $z_i = x_i$, $i = 1, \ldots, n$. In order to fit multiple models simultaneously we must make multiple copies of the dataset $\{y_i, x_i, i = 1, \ldots, n\}$, as shown below.

```
model {
  for (m in 1:3) {
    for (i in 1:n) {
      y.copy[m, i] <- y[i]
      x.copy[m, i] <- x[i]
      y.copy[m, i]   ~ dnorm(mu[m, i], tau[m])
      mu[m, i]       <- a[m] + b[m]*z[m, i]
    }
    a[m]             ~ dnorm(0, 0.0001)
    b[m]             ~ dnorm(0, 0.0001)
    tau[m]           <- 1/pow(sigma[m], 2)
    sigma[m]         ~ dunif(0, 100)
  }
  for (i in 1:n) {
    z[1, i]          <- x.copy[1, i]
    x.copy[2, i]     ~ dnorm(z[2, i], 0.4444)
    z[2, i]          ~ dunif(-100, 100)
    z[3, i]          <- cut(z.star[i])
    x.copy[3, i]     ~ dnorm(z.star[i], 0.4444)
    z.star[i]        ~ dunif(-100, 100)
  }
}
```

Model fits for the models with and without feedback are shown in Figure 9.6(b) and Figure 9.6(c), respectively. Note that with feedback the z_is are estimated so that the regression line fits as well as possible. One danger of allowing this to happen is that the estimates may become implausible. Indeed, note that in this example several of the z_is are not ordered when feedback is allowed, e.g., posterior median estimates for z_3, z_4, and z_5 are 2.90, 5.59, and 3.40, respectively, although there is considerable overlap between the 95% credible intervals: (0.844, 4.81), (2.85, 7.10), and (1.70, 5.98), respectively. It may be known that they must be ordered, however. An ordering constraint could be applied in such cases, as discussed in §9.7.2, but there are numerous situations in which an obvious constraint to ensure plausibility does not exist. The reader is referred to Lunn et al. (2009a) for further discussion on difficulties with feedback. Without feedback, point estimates for the z_is are approximately equal to the measured values, $x_i = i$, $i = 1, \ldots, n$, but considerable uncertainty is acknowledged and this is propagated into the regression analysis, manifesting as a wider credible interval for the model fit compared to when the observed x_is are assumed error-free.

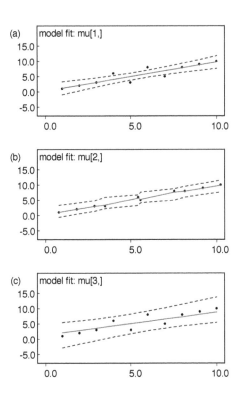

FIGURE 9.6
Model fits for linear regression data, under various assumptions for the independent variable: (a) x_is assumed error-free, i.e., $z_i = x_i$; (b) fully Bayesian model (with feedback); and (c) model without feedback. In (b) and (c), the model fits are plotted against posterior mean estimates of the z_is.

9.5   New distributions

9.5.1   Specifying a new sampling distribution

Suppose we wish to use a sampling distribution that is not included in the list of standard distributions, in which an observation y_i contributes a likelihood term L_i. One possibility is the "zeros trick" based on the following.

We invent a set of observations $z_i = 0$, each of which is assumed to be drawn from a Poisson(ϕ_i) distribution. Each then has a likelihood contribution $\exp(-\phi_i)$, and so if ϕ_i is set to $-\log(L_i)$, we will obtain the correct likelihood contribution. (Note that ϕ_i should always be > 0 as it is a Poisson mean, and so we may need to add a suitable constant to ensure that it is positive.) The BUGS code will look like the following:

```
const     <- 10000    # arbitrary, ensures phi[i] > 0
for (i in 1:n) {
  z[i]    <- 0
  z[i]     ~ dpois(phi[i])
  phi[i] <- -log(L[i]) + const
  L[i]    <- ...
}
```

L_i is set to a function of y_i and θ proportional to the likelihood $p(y_i|\theta)$. This trick allows arbitrary sampling distributions to be used and is particularly suitable when, say, dealing with truncated distributions (§9.6.2).

A new observation from the distribution, denoted y_{n+1}, can be predicted by including it as an additional, but missing, observation in the data file and assigning it an improper uniform prior, e.g., y[n+1] $\sim$ dflat(), defining z_i and ϕ_i in the same way as before for $i = n + 1$. The missing observation is essentially assumed to be an unknown parameter with a uniform prior, but also with a likelihood term corresponding to the sampling distribution.

Note that the DIC (§8.6) for data from distributions specified using the zeros trick, as reported by the WinBUGS or OpenBUGS DIC tool, is calculated with respect to z_i, not y_i. Example 11.6.2 explains how to transform this to the scale of y_i, so it can be compared with the DICs of models for y_i which are specified using built-in sampling distributions.

Example 9.5.1. *A clumsy way of modelling the normal distribution*
We use the "zeros trick" to model a normal distribution with unknown mean μ and unknown standard deviation σ, including predicting a new observation. We

have seven observed values and one missing value as follows: y = c(-1, -0.3, 0.1, 0.2, 0.7, 1.2, 1.7, NA).

```
for (i in 1:8) {
   z[i]    <- 0
   z[i]     ~ dpois(phi[i])
   phi[i] <- log(sigma) + 0.5*pow((y[i] - mu)/sigma, 2)
}
y[8]       ~ dflat()
sigma      ~ dunif(0, 100)
mu         ~ dunif(-100, 100)
```

We must provide an initial value for y[8], via y = c(NA, NA, NA, NA, NA, NA, NA, 0), say, otherwise BUGS will try to generate one from the improper prior and crash.

node	mean	sd	MC error	2.5%	median	97.5%	start	sample
mu	0.365	0.4758	0.006864	-0.5948	0.3693	1.316	4001	10000
sigma	1.18	0.481	0.01139	0.6216	1.067	2.415	4001	10000
y[8]	0.3499	1.355	0.03415	-2.345	0.3564	3.095	4001	10000

Whilst the results match those that would be obtained in a standard analysis using y[i] ~ dnorm(mu, tau); tau <- 1/pow(sigma,2), this is an inefficient procedure, particularly for the prediction, and so a long run is necessary. The MC error for the prediction is 0.03 using the zeros trick and 0.01 for the same number of iterations in the equivalent standard analysis.

An alternative to the "zeros trick" is the "ones trick." Here we invent a set of observations equal to 1 instead, and assume each to be Bernoulli distributed with probability p_i. By making each p_i proportional to L_i (i.e., by specifying a scaling constant large enough to ensure $p_i < 1$ for all i), the required likelihood term is provided:

```
const  <- 10000    # arbitrary, ensures p[i] < 1
for (i in 1:n) {
   z[i] <- 1
   z[i]  ~ dbern(p[i])
   p[i] <- L[i]/const
}
```

We will illustrate use of the "ones trick" in §9.6.2, where we consider how to specify truncated sampling distributions.

9.5.2 Specifying a new prior distribution

Suppose we want to use a prior distribution for θ that does not belong to the standard set. Then we can use the "zeros trick" (see above) at the prior level combined with an improper uniform prior for θ. A single Poisson observation equal to zero, with mean $\phi = -\log(p(\theta))$, contributes a term $\exp(-\phi) = p(\theta)$ to the likelihood for θ; when this is combined with a "flat" prior for θ the correct prior distribution results. This is essentially the same process as predicting from a new distribution covered in the previous section. Summary BUGS code is

```
z     <- 0
z       ~ dpois(phi)
theta ~ dflat()
phi   <- expression for -log(desired prior for theta)
```

For example, if we wished to produce a standard normal prior, we would use

```
phi   <- 0.5*pow(theta, 2)
```

It is important to note that this method produces high auto-correlation, poor convergence, and large MC errors, so it is computationally slow and long runs are necessary. Initial values also need to be specified as the `dflat()` prior cannot be sampled from using `gen.inits`.

New sampling distributions and new prior distributions can also be specified in WinBUGS via the WinBUGS Development Interface (WBDev). This can give big computational savings and clearer BUGS code, at the cost of "lower-level" programming in Component Pascal — see § 12.4.8 for more details. There are similar but less well-documented capabilities in OpenBUGS and JAGS; see Chapter 12.

9.6 Censored, truncated, and grouped observations

9.6.1 Censored observations

A data point is a censored observation when we do not know its exact value, but we do know that it lies above or below a point c, say, or within a specified interval. The most common application is in survival analysis (§11.1), but here we consider general measurement problems. There are two strategies within BUGS:

1. In general, in WinBUGS we can use the I(,) construct (§A.2.2), which specifies a restricted range within which the unknown quantity lies. The unknown quantity is then simply treated as a model parameter. Note that in OpenBUGS the C() function is preferred (see §12.5.1) and JAGS uses a different syntax altogether (see §12.6.2).

2. Each exact observation y contributes $p(y|\theta)$ to the likelihood of θ, whereas an observation censored at c provides a contribution of $\Pr(Y >$

$c|\theta)$ or $\Pr(Y < c|\theta)$. Hence, if the distribution function can be expressed in BUGS syntax, then we can use either the "ones trick" or the "zeros trick" (§9.5) to directly specify the contribution of the censored observations to the likelihood.

Example 9.6.1. *Censored chickens*

Suppose we weigh nine chickens, with a scale that only goes up to 8 units, so that if the scale shows 8 it means that the chicken weighs *at least* 8 units, which we denote 8+. The weights are 6, 6, 6, 7, 7, 7, 8+, 8+, 8+. The population of chickens is assumed to have weights that are normally distributed with mean μ and standard deviation 1 unit. (This is not intended to be a realistic example — all the observed weights are integer-valued and would more realistically be modelled as roundings of a true continuous-valued weight, as in Example 9.6.3). If the 8+ weighings were exactly 8, and μ was assigned a locally uniform prior, then the posterior distribution for μ would be Normal$(7, 1/9)$. WinBUGS code accounting for the censoring via method 1 above is

```
model {
    for (i in 1:6) {y[i] ~ dnorm(mu, 1)}        # uncensored data
    for (i in 7:9) {y[i] ~ dnorm(mu, 1)I(8,)}   # censored data
    mu                     ~ dunif(0, 100)
}
data:
    list(y = c(6,6,6,7,7,7,NA,NA,NA))
```

node	mean	sd	MC error	2.5%	median	97.5%	start	sample
mu	7.193	0.3478	0.003604	6.515	7.19	7.875	1001	10000
y[7]	8.571	0.4809	0.005356	8.018	8.446	9.805	1001	10000

We note that the posterior mean of μ is greater than 7 since the censored observations have been estimated to be between 8.0 and 9.8 — see Figure 9.7 for the posterior density of censored observation y[7] (y[8] and y[9] also have the same posterior). The posterior standard deviation of μ is 0.35, slightly greater than if the data had been exact rather than censored — we can think of the effective sample size having been reduced from 9 to $1/0.3478^2 = 8.3$.

Now consider the second method outlined above, making use of the "zeros trick." Each censored observation provides a term $\Pr(Y > 8|\mu)$ to the likelihood of μ, which is equal to $\Pr(Y - \mu > 8 - \mu) = 1 - \Phi(8 - \mu) = \Phi(\mu - 8)$, where $\Phi(.)$ is the cumulative distribution function of the standard normal distribution, available in BUGS via the syntax phi(.):

```
    for (i in 1:6) {y[i] ~ dnorm(mu, 1)}
    for (i in 1:3) {
        zeros[i]              <- 0
        zeros[i]            ~ dpois(p[i])
```

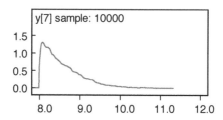

FIGURE 9.7
Posterior distribution for censored observation y[7] in the "censored chickens" example.

```
   p[i]                   <- -log(phi(mu-8))
 }
 mu                       ~ dunif(0, 100)
```

```
node mean  sd    MC error 2.5%  median 97.5% start sample
mu   7.192 0.348 0.003687 6.519 7.185  7.882 1001  10000
```

We obtain largely the same results as with method 1, with similar Monte Carlo standard errors. Method 2 is computationally more efficient, as the censored observations are integrated out before analysis. However, method 1 is more generally applicable, as it does not require the distribution function to be known.

9.6.2   Truncated sampling distributions

A sampling distribution is truncated if for some reason we never observe cases above or below a specified point, although in the permissible range of observations the data follow a standard distribution. The sampling distribution must therefore be normalised to condition on lying in the permissible range, say $Y < c$, so that the likelihood contribution of an observation y is $p(y|\theta)/\Pr(Y < c|\theta)$. This will not generally be of standard form, and so either a new distribution has to be defined (§12.4.8) or the "ones"/"zeros" trick used (§9.5).

It is very important to realise that the I(,) construct is not appropriate for truncated distributions with unknown parameters, since the generated likelihood term will ignore the truncation and be incorrect (see Appendix A.2.2). However, the I(,) construct can be used when specifying truncated prior distributions with no unknown parameters — see Examples 5.3.2 and 6.3.1.

Example 9.6.2. *Truncated chickens*
In Example 9.6.1, suppose that any chicken weighing 8 or more units is sent back to get more exercise, so the distribution of chicken weights is right-truncated at

8. Therefore we only hear about the six chickens that weighed 6 or 7 units, and each of these provides a likelihood contribution of $\exp[-(y_i - \mu)^2/2]/\Phi(c - \mu)$. So using the "ones trick":

```
for (i in 1:6) {
  z[i] <- 1
  z[i]    ~ dbern(p[i])
  p[i] <- exp(-0.5*(y[i] - mu)*(y[i] - mu))/phi(8 - mu)
}
mu        ~ dunif(0, 100)
```

node	mean	sd	MC error	2.5%	median	97.5%	start	sample
mu	6.737	0.4965	0.00498	5.819	6.72	7.75	1001	10000

Although we don't know how many chickens were weighed and returned, knowledge of this truncation has raised the estimated population mean from the sample mean of 6.5 to 6.74. The posterior standard deviation of 0.5 means that the effective sample size is $1/0.5^2 = 4$, so the truncation has considerably reduced the precision. Note that in this situation chickens weighing 8 or more units are not included in the data collection process, whereas in the censoring example (Example 9.6.1), while they may not be fully observed, they may still be present.

In JAGS, the T(,) construct may simply be used to truncate the sampling distribution in the following way. JAGS computes the appropriate normalising constant internally.

```
y[i] ~ dnorm(mu, 1)T(,8)
```

A similar general facility is planned in OpenBUGS but is only partially implemented currently.

As may be apparent from the examples above, there are subtle differences between truncation and censoring. Truncation is appropriate when values outside a given range are actually impossible. Censoring, on the other hand, is appropriate when values beyond that range are possible *in principle*, but have not been observed due to the nature of the measurement device/method — that is, they may be observable using a different method.

9.6.3 Grouped, rounded, or interval-censored data

If observations are either grouped into categories or rounded, to the nearest integer, say, the information provided by an observation is that it lies in a particular interval, say (*lower, upper*). This can be treated as interval censoring and handled by assuming we have a true but unobserved quantity z which contributes to the likelihood for θ but we only know it lies between *lower* and *upper*. This is specified using I(lower,upper).

Example 9.6.3. *Grouped chickens*
Suppose all nine chickens in Example 9.6.1 have been weighed and reported as 6, 6, 6, 7, 7, 7, 8, 8, 8, but we know that when the scales report 7 units, say, the true weight z could be anything between 6.5 and 7.5.

```
for (i in 1:9) {
   lower[i] <- y[i] - 0.5
   upper[i] <- y[i] + 0.5
   z[i]        ~ dnorm(mu, 1)I(lower[i], upper[i])
}
mu              ~ dunif(0, 100)
```

```
node mean    sd      MC error 2.5%  median 97.5% start sample
mu    7.001 0.3496 0.00364   6.322 7.003  7.692 1001  10000
z[1]  6.08  0.2778 0.002909  5.543 6.113  6.483 1001  10000
z[4]  6.993 0.2829 0.002573  6.527 6.991  7.474 1001  10000
z[7]  7.922 0.2759 0.002482  7.517 7.885  8.458 1001  10000
```

The posterior mean is 7, as might be expected from symmetric data, but the effective sample size is reduced from 9 to $1/0.35^2 = 8.2$ by the grouping. The true weight for a chicken reported as 6 is estimated to be 6.08, slightly "shrunk" towards the mean by the assumption that the weights in the population are normally distributed — see Figure 9.8.

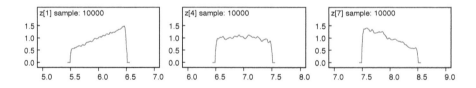

FIGURE 9.8
Posterior density estimates for "true" chicken weights in the "grouped chickens" example.

9.7   Constrained parameters

9.7.1   Univariate fully specified prior distributions

A single parameter θ may be subject to a range constraint such as $\theta > 0$. Provided the distribution of θ does not contain any unknown parameters, then this can be accommodated by using the I(,) construct (§9.6). For example, a standard normal variable constrained to be positive is expressed as

```
theta ~ dnorm(0,1)I(0,)
```

See Examples 5.3.2 and 6.3.1. An alternative approach, which better generalises to more complex constraints, is essentially an extension of the "ones trick." We introduce an auxiliary observation z taking the value 1. This is assumed to arise from a Bernoulli distribution whose parameter takes the value 1 if the constraint is obeyed, and 0 otherwise. When sampling θ, only values that obey the constraint, and therefore provide non-zero likelihood, will be accepted, as illustrated in the example below.

Example 9.7.1. *Half-normal*
The code below shows the half-normal distribution being generated in two different ways.

```
theta[1]     ~ dnorm(0,1)I(0,)
theta[2]     ~ dnorm(0,1)
z            <- 1
z            ~ dbern(constraint)
constraint <- step(theta[2])
```

The results show the substantially increased Monte Carlo error associated with the auxiliary data method:

node	mean	sd	MC error	2.5%	median	97.5%	start	sample
theta[1]	0.8009	0.6106	0.006119	0.03056	0.6696	2.253	4001	10000
theta[2]	0.7897	0.5973	0.01643	0.02286	0.6743	2.206	4001	10000

9.7.2   Multivariate fully specified prior distributions

Order constraints on a series of parameters can be expressed using the I(,) construct, provided the prior distribution does not contain unknown parameters. For example, to order a[1]<a[2]<a[3]:

```
a[1] ~ dnorm(0, 0.001)I(, a[2])
a[2] ~ dnorm(0, 0.001)I(a[1], a[3])
a[3] ~ dnorm(0, 0.001)I(a[2], )
```

Or as in Example 7.3.1, a[2] and a[3] could be defined by adding positively distributed increments to a[1] and a[2], respectively. In JAGS and Open-BUGS, the elements of an unconstrained vector can also be sorted using the sort() function, for example, b[1:3] <- sort(a[]).

The auxiliary data method can be used to impose more complex constraints, as the following example shows.

Example 9.7.2. *Doughnut: bivariate normal with a hole in it*
Suppose we assume $\theta_i \sim \text{Normal}(0, 1)$, $i = 1, 2$, but with the curious constraint that $\theta_1^2 + \theta_2^2 > 1$. This can be generated using the following code.

```
theta[1]      ~ dnorm(0, 1)
theta[2]      ~ dnorm(0, 1)
z             <- 1
z             ~ dbern(constraint)
constraint <- step(theta[1]*theta[1] + theta[2]*theta[2] - 1)
```

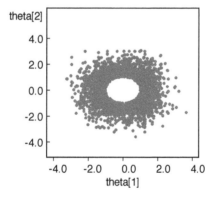

FIGURE 9.9
Scatterplot showing 5000 samples from a bivariate normal distribution subject to the constraint $\theta_1^2 + \theta_2^2 > 1$.

A scatterplot of 5000 simulations (Figure 9.9) shows a bivariate normal distribution with a hole in the centre.

Additive constraints on parameters can be easily imposed by reparameterisation. For example, if we require a set of parameters β_i, $i = 1, \ldots, n$, to sum to 0, we can define each as $\beta_i = b_i - \bar{b}$, where the b_is are independent with known prior distributions.

Example 9.7.3. *Bristol (continued): sum-to-zero constraint*
We consider Example 8.3.1 as a logistic model:

$$y_i \sim \text{Binomial}(\theta_i, n_i), \quad \text{logit}\,\theta_i = \alpha + \beta_i, \quad i = 1, \ldots, 12.$$

We allow each centre i to have its own effect parameter β_i, but with the commonly imposed constraint that those parameters add to 0, in order to ensure identifiability.

```
for (i in 1:12) {
  y[i]             ~ dbin(theta[i], n[i])
  logit(theta[i]) <- alpha + beta[i]
  beta[i]         <- b[i] - mean(b[])
  b[i]             ~ dunif(-10,10)
}
alpha              ~ dunif(-10,10)
```

Figure 9.10 shows a box plot of the beta[i]s from the above model. These are identifiable, due to the sum-to-zero constraint, but the individual b[i]s are not. Non-identifiable parameters can actually be introduced to improve convergence in hierarchical models — see §10.5.

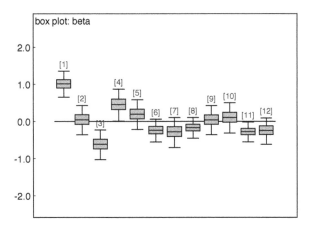

FIGURE 9.10
Posterior summaries for centre effects beta[i] in the Bristol example.

9.7.3   Prior distributions with unknown parameters

All of the parameter constraints considered thus far have taken a standard
prior distribution and truncated it. Those standard priors have also had fixed
parameters. Unfortunately, when the prior parameters are unknown, the I()
method for truncation does not work correctly in BUGS — it is simply a trick
for restricting the values sampled for a given node and only works in simple settings. In particular, the normalising constant for the truncation is not
accounted for when computing the likelihood contribution of the truncated
parameter(s) to the full conditional distribution(s) of the prior parameter(s).
In such cases, we have several alternative options, but as this issue is particularly relevant to hierarchical models, discussion of these is deferred until
§10.2.2.

9.8   Bootstrapping

As suggested in Chapter 2, BUGS can, in principle, be used to perform any
statistical procedure based on random sampling, which need not necessarily be
a Bayesian analysis. For example, we can implement classical nonparametric
bootstrap estimation as follows.

Example 9.8.1. *Bootstrapping in BUGS: the Newcomb data*
The Newcomb data have previously been considered as an illustration of model
elaboration for non-normal data (see Examples 8.2.1, 8.4.3, and 8.4.4). Carlin
and Louis (2008) point out that, given the outliers, a more robust analysis might
consider inference on the median rather than the mean of the distribution. If we
wish to avoid a parametric assumption about the shape of the distribution, then
we can adopt the basic bootstrap procedure of taking a series of repeat samples
with replacement and calculating the sample mean and median for each of these
repeats. This can be easily carried out in BUGS using the code below.

```
for (i in 1:N) {
  p[i]      <- 1/N          # set up uniform prior on 1 to N
}
for (j in 1:N){
  pick[j]   ~ dcat(p[])   # pick random number between 1 and N
  Yboot[j] <- Y[pick[j]]  # set jth bootstrap observation
}
mean       <- mean(Yboot[])
# now find median of bootstrap sample: this is halfway
# between observation N/2 and N/2+1...
n1         <- N/2
```

```
n2              <- n1 + 1
median          <- (ranked(Yboot[], n1) + ranked(Yboot[], n2))/2
```

We note how the discrete uniform prior distribution is set up and the use of the dcat distribution to select a random observation. We monitor one of the bootstrap elements Yboot[1]; its density in Figure 9.11 provides an approximate sampling distribution for the data. The median has a discrete distribution with median 27 and 95% interval 26.0 to 28.5 (the true value for the speed of light would give 33, well outside the 95% bootstrap interval).

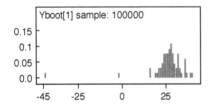 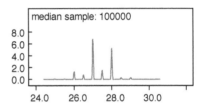

FIGURE 9.11
Empirical distributions for Yboot[1] and median based on 100,000 simulations.

9.9 Ranking

There is increasing attention to the profiling of schools, hospitals, and so on, often resulting in institutions being ranked into a league table similar to sports teams or competitors. Generally the rank is treated as a descriptive statistic, but we can also think of the observed rank as an imperfect measure of the "true rank" and perform statistical inference. This can be useful when we want to assess the probability, for example, that a treatment that currently looks best is truly the best treatment being examined.

The observed rank is a highly unreliable summary statistic since it can be very sensitive to small changes in the data. Bayesian methods can provide posterior interval estimates for ranks for which WinBUGS and OpenBUGS contain "built-in" options.[†] rank(x[], i) returns the rank of the ith element

---

[†]In JAGS, rank(x[]) transforms a vector x into a vector of ranks, so that the equivalent of rank(x[],i) is y <- rank(x[]); y[i]. The equivalent of ranked(x[], i) is y <- sort(x[]); y[i].

of x. equals(rank(x[],i),1) = 1 if the ith element of x has the lowest value,
and 0 otherwise; the mean is the probability that the ith element has the lowest
value. ranked(x[], i) returns the value of the ith-ranked element of x. The
Rank option of the Inference menu monitors the rank of each element of a
specified vector.

Example 9.9.1. *Bristol (continued): ranking*
Consider again the child heart surgery mortality rates introduced in Example 8.3.1.
Ignoring Bristol, we consider whether the variability in mortality rates between
hospitals allows any confident ranking. We assume the mortality in each hospital
is $p_i, i = 1, ..., N$, with $N = 11$, which are assumed to have independent Jeffreys
Beta(0.5,0.5) priors. We would like to assess the true rank of each hospital and
the probability that each has the highest or lowest mortality.

```
for (i in 1:N) {
  numbers1toN[i]  <- i
  p[i]            ~ dbeta(0.5, 0.5)
  r[i]            ~ dbin(p[i], n[i])
  hosp.rank[i]    <- rank(p[], i)          # rank of hospital i
  prob.lowest[i]  <- equals(hosp.rank[i], 1) # =1 if hosp i is lowest
  prob.highest[i] <- equals(hosp.rank[i], N) # =1 if hosp i is highest
}
hosp.lowest       <- inprod(numbers1toN[], prob.lowest[])
                                         # index of lowest hosp
hosp.highest      <- inprod(numbers1toN[], prob.highest[])
                                         # index of highest hosp
```

The rank function produces the rank of each hospital at each iteration so that,
for example, hosp.rank[i] = 1 if hospital i currently has the lowest mortality
rate p[i]. prob.lowest[i] will then be 1 at that iteration, and so the mean
of prob.lowest[i] will provide the probability that hospital i is the "safest"
hospital.

We note the "which is min/max" trick used to pick out the index of, say, the
lowest hospital: prob.lowest[] is a vector of zeros except for a 1 in the position
of the hospital with the lowest p at that iteration; by taking an inner-product
$\sum_i i*$prob.lowest[i], hosp.lowest takes on the value equal to the index of
the hospital currently ranked lowest.

The upper panel of Figure 9.12 shows that there are substantial posterior probabilities (0.71 and 0.68, respectively) of hospital 2 being the "safest" (with the
lowest mortality) and hospital 3 being the "least safe": these are hospitals 3 and 4
in the original data table in Example 8.3.1. The lower panel illustrates that there
is considerable uncertainty regarding each hospital's rank, however.

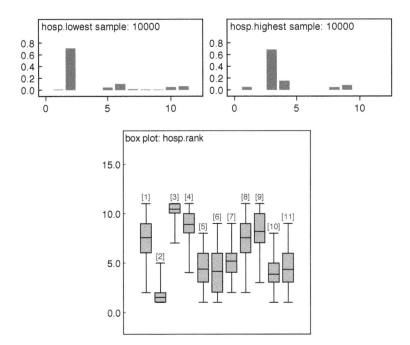

FIGURE 9.12

Top: Posterior histograms for the hospital with the lowest (left) and highest (right) mortality. Bottom: Box plot comparing hospital-specific posterior ranks.

10

Hierarchical models

We are often interested in making inferences on many parameters corresponding to different "units," for example, different patients, geographical areas, schools, hospitals, etc. The main goal is often to quantify the degree of similarity across units so that we can make predictions about "new" units. Suppose, for example, that a specific operation is carried out in a number of hospitals, labelled A, B, C, D, etc. Further suppose that the observed mortality rates for that operation in hospitals A, B, and C are 0.1, 0.19, and 0.14, respectively. What would you predict for hospital D? What information did you use to come up with that prediction? Most people would predict a value that is similar, in some sense, to the other values. We tend to recognise that it is unlikely that all hospitals have the *same* underlying mortality rate, due to employing different surgeons and having different catchment areas, for example, but we also tend to assume that knowing something about the other hospitals tells us *at least something* about the one of interest. Our natural inclination, therefore, is towards an assumption somewhere between the unit-specific parameters being identical and being entirely independent.

10.1 Exchangeability

An assumption of "similarity," in the sense that the units' "labels" convey no information, is related to the assumption of "exchangeability" made about observations (§3.6.2), which was shown to be equivalent to assuming the observations were independent and identically distributed from a distribution with unknown parameters, where those parameters are given a prior distribution. Similarly, under broad conditions (de Finetti (1931); see also Bernardo and Smith (1994) for an overview) exchangeability of the unit-specific parameters, θ_i, $i = 1, ..., N$, can be shown to be mathematically equivalent to assuming that they arise from a common "population" distribution whose parameters are unknown and assigned appropriate prior distributions. Therefore the θ_is are similar but not identical.

For example, suppose $\theta_i \sim N(\mu, \omega^2)$, $i = 1, ..., N$, with $\mu \sim N(.,.)$ and $\omega \sim \text{Uniform}(.,.)$, say. The data y_i are a group of observations for each unit i, and these are assumed to be generated conditionally on the corresponding

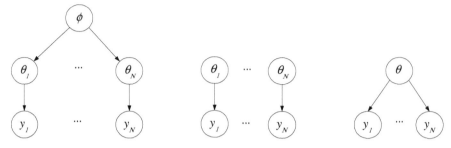

FIGURE 10.1

A basic hierarchical model (left), independent-parameters model (middle), and identical-parameters model (right).

unit-specific parameter θ_i. This is illustrated as a directed acyclic graph in Figure 10.1 (left). In this model, we learn about θ_i not only through *direct* information from y_i, but also through *indirect* information which comes from the remaining $y_j : j \neq i$, via the population distribution, which is parameterised by ϕ.

To contrast this with the alternative assumptions alluded to above, an "independent-parameters" assumption would involve assigning prior distributions with fixed values directly to each θ_i (Figure 10.1, middle) — here the information about θ_i comes only from y_i. An "identical-parameters" assumption would set $\theta_i = \theta$, $i = 1, ..., N$ (Figure 10.1, right). If we denote, collectively, the unknown parameters of the population distribution by ϕ, an exchangeability assumption is an assumption of *conditional independence*, given ϕ.

The posterior distribution for such a model is still proportional to the likelihood multiplied by the prior, but the prior distribution, of all unknown parameters, is decomposed into an exchangeability assumption for the unit-specific parameters and a prior for the population parameters:

$$p(\theta_1, ..., \theta_N, \phi) = p(\phi) \prod_{i=1}^{N} p(\theta_i|\phi)$$

Hence the prior is "hierarchical," that is, it is specified in "layers." Such models are thus generally referred to as *hierarchical models*, but they may also be called *multilevel* or *mixed effects* models. The term "mixed effects" stems from the combination of *fixed effects* and *random effects* in a "classical" statistical model: the population mean parameters, e.g., μ, are referred to as fixed effects, whereas the exchangeable parameters θ_i, $i = 1, ..., N$, or the corresponding "residuals" $\theta_i - \mu$, $i = 1, ..., N$, are known as random effects. Of course, in a Bayesian model, the population mean parameters are also, strictly speaking, "random," because they are assigned a prior probability distribution. Note that a finite *mixture model* or *latent class model* (§11.6) is a special case of a hierarchical model in which $p(\theta_i|\phi)$ is a categorical distribution, so that the units are classified into a finite number of groups.

Example 10.1.1. *Bristol (continued): hierarchical model*
We fit a hierarchical model to the data from Example 8.3.1 on mortality rates from child heart surgery in English hospitals. The outlying data from Bristol are excluded, leaving 11 hospitals.

Denote the underlying "true" mortality rates by θ_i, $i = 1, ..., 11$. The sampling model is then given by $p(y|\theta) = \prod_{i=1}^{11} p(y_i|\theta_i) = \prod_{i=1}^{11} \text{Binomial}(\theta_i, n_i)$, where y and θ denote $\{y_1, ..., y_{11}\}$ and $\{\theta_1, ..., \theta_{11}\}$, respectively. If we wish to assume that the θ_is are "similar" then one option is $\theta_i \sim \text{Beta}(a, b)$, $i = 1, ..., 11$, with a and b unknown and assigned appropriate priors. Previously when specifying beta priors, we have simply chosen fixed values for a and b to express our uncertainty regarding the relevant proportion parameter. However, now we are faced with expressing uncertainty about a and b themselves, which are somewhat unintuitive. Instead we could specify priors for the mean and standard deviation, as defined in §5.3.1, and define a and b as deterministic functions of those. Another option is to use a link function in combination with a normal population distribution, for example, $\text{logit } \theta_i \sim \text{Normal}(\mu, \omega^2)$, $i = 1, ..., N$, with appropriate priors for μ and ω, say. Here we use this latter option with vague uniform priors for μ and ω:

```
for (i in 1:11) {
   y[i]                   ~ dbin(theta[i], n[i])
   logit(theta[i]) <- logit.theta[i]
   logit.theta[i]    ~ dnorm(mu, inv.omega.squared)
}
inv.omega.squared <- 1/pow(omega, 2)
omega                    ~ dunif(0, 100)
mu                       ~ dunif(-100, 100)
```

The results are summarised in Figure 10.2, where posterior distributions for each θ_i are compared with those obtained under an assumption of independence. There are several phenomena to notice here. Firstly note that the credible intervals for the exchangeable θ_is are narrower. If assuming independence, one may as well analyse each hospital's data independently; there is nothing to gain from simultaneous analysis. Under an assumption of exchangeable θ_is, however, each posterior *borrows strength* (or precision) from the others, via their joint influence on the estimation of the underlying population parameters. In other words, the information about each hospital's mortality rate is supplemented with information about the population distribution of mortality rates. One side effect of this borrowing of strength is that the uncertainty about mortality rates is spread more evenly across the hospitals, as can be observed from the more uniform width of credible intervals. Another side effect of borrowing of strength is known as "shrinkage to the mean." Units contribute to the estimation of the population parameters in proportion to the amount of data they provide. In turn, unit-specific estimates for units with relatively few data are supplemented more, in relative terms, by the information available on the population parameters. Because of sampling variation, there is a tendency for unit-specific estimates, under independence, to be more extreme when data are few. Hence the more extreme values tend to get pulled in towards the population mean (because the data-rich units, which typically give less extreme estimates, contribute more to locating that population mean).

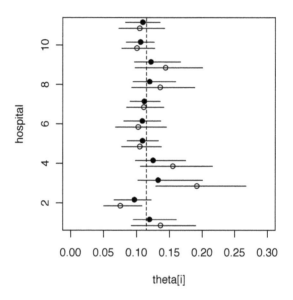

FIGURE 10.2
Posterior summaries for hospital-specific mortality rates (θ_i) from analysis of surgery data. Ninety-five percent credible intervals and posterior medians are shown both for hierarchical analysis, assuming exchangeable/similar θ_is ($\bullet$), and for independent θ_is ($\circ$). The vertical dashed line represents the posterior median population mean mortality rate.

The effects of fitting a hierarchical model, observed in Figure 10.2 above — borrowing of strength, global smoothing of uncertainty, and shrinkage to the mean — are general phenomena, that is, they can be observed to some degree in all hierarchical analyses. Shrinkage to the mean can initially appear somewhat undesirable, but it is simply the similarity assumption overriding the effects of sampling variation. With sufficient data for a given unit, there is negligible shrinkage.

10.2   Priors

10.2.1   Unit-specific parameters

In Example 10.1.1 above we allude to various options being available for the type of population distribution specified for the unit-specific parameters θ_i, $i = 1, ..., N$. In many settings we will want to explore whether the values of unit-specific parameters are related to various covariates. Had we chosen a beta population distribution in Example 10.1.1, this would have been cumbersome, since the underlying mean and variance must be constrained appropriately. It is often much easier to transform the unit-specific parameters to a scale on which they can take any real value and then specify a normal population distribution, say, for the transformed parameters. We may then specify any appropriate model for the relationship between the mean and covariates of interest, and the variance is also free to take any positive value.

Example 10.2.1. *Bristol (continued): hospital-level covariate*
Recalling Example 10.1.1, let x_i, $i = 1, ..., N$, denote the observed values of some covariate that might predict hospitals' mortality rates, such as the typical condition of patients who tend to be admitted. We transform to a logistic scale, as before, and specify

$$\text{logit}\,\theta_i \sim \text{Normal}(\mu_i, \omega^2), \quad \mu_i = \alpha + \beta x_i, \quad i = 1, ..., N$$

```
for (i in 1:11) {
   y[i]                 ~ dbin(theta[i], n[i])
   logit(theta[i]) <- logit.theta[i]
   logit.theta[i]    ~ dnorm(mu[i], inv.omega.squared)
   mu[i]              <- alpha + beta*x[i]
}
```

with appropriate priors for α, β, and ω.

Note that whenever we apply a covariate model to unit-specific parameters, we are making exchangeability assumptions regarding the residual differences between the parameters and the covariate model, rather than the unit-specific parameters themselves.

10.2.2   Parameter constraints

In many cases we will wish to impose constraints, such as positivity, on unit-specific parameters. However, in hierarchical models the unit-specific parameters have prior distributions whose parameters are also unknown. This means that the tricks outlined in §9.7 for applying constraints do not work. In particular, we cannot use the I(,) construct or the auxiliary data method (see Example 9.7.2, for example) for constraining unit-specific parameters. Fortunately, there remain several basic options, as outlined below.

- Choose an appropriately constrained standard distribution: For example, gamma or log-normal for positivity, or beta for proportion parameters. Of course, this has the downside that the resulting model may be difficult to interpret, or that covariate modelling is cumbersome, as mentioned in the previous section. One particularly awkward situation is when a set of parameters must sum to a specific value, e.g., the probabilities associated with a multinomial distribution must sum to 1. In principle, choosing a Dirichlet prior for the probability vector ensures that the elements always sum to 1. However, no current implementation of BUGS allows learning about the parameters of a Dirichlet distribution, which would be required if the model were hierarchical (with unit-specific probability vectors). Fortunately, there is a workaround for this outlined in Example 10.3.4.

- Reparameterise the model: We could choose a new set of parameters, for which a normal population distribution (with unknown population parameters) would seem appropriate, and use these as the basis for our hierarchical model. We may then "construct" the constrained parameters from the unconstrained set. As a very simple example, which we have already seen in Example 10.2.1, we can construct a proportion parameter from a normally distributed parameter by taking the inverse-logit (or *expit*) transformation. A more complex example is the unit-specific probability vectors of a multinomial distribution, alluded to above. In this case, we could specify

$$\phi_{ir} = \frac{\exp(\theta_{ir})}{\sum_{s=1}^{p} \exp(\theta_{is})}, \quad r = 1, ..., p,$$

where i indexes units, r indexes the p probabilities for each unit, and the θ_i vectors are assumed to arise from a p-dimensional multivariate normal distribution. Use of this approach is also illustrated in Example 10.3.4.

- Specify a new distribution that applies the relevant constraint: This can be achieved in WinBUGS using the WinBUGS Development Interface (WBDev — see § 12.4.8). Thus a truncated distribution, say, can be specified for the unit-specific parameters in such a way that the normalising constant is fully accounted for in computing the full conditional distributions of the population parameters. This can also be achieved via the "ones" or "zeros" trick (§9.5). Although these make use of auxiliary data, they work correctly as long as the new distribution is fully specified, including any normalising constants, say.

10.2.3 Priors for variance components

Although, ostensibly, we have numerous options in terms of specifying a prior for the variance of unit-specific parameters, we must be careful, if attempting to be vague relative to the data, that our choice does not influence the resulting posterior too much, or even render it improper. Improper posteriors can be avoided by ensuring that the prior is proper, but an apparently vague prior for a random effects variance component, even if proper, can have undesirable consequences. There is rarely an issue with the residual variance, because the likelihood contribution from the observed data will invariably support values for the residual variance well away from zero, but variances of unobserved parameters require considerable care. Obvious choices include proper approximations to the Jeffreys prior: either the conjugate gamma prior on the precision scale, $\omega^{-2} \sim \text{Gamma}(\epsilon, \epsilon)$ with ϵ small, or the uniform prior on the log standard deviation scale, $\log \omega \sim \text{Uniform}(-A, A)$ with A large. However, the resulting posteriors can be very sensitive to choices of ϵ and A. As ϵ tends away from 0 and A tends away from ∞, the prior becomes inappropriately biased away from $\omega = 0$ (Gelman, 2006) when we usually want to allow for the possibility of no between-group variability. Note that $\sigma^{-2} \sim \text{Gamma}(\epsilon, \epsilon)$ is less problematic as a prior for a *sampling* variance σ^2, where zero variance is usually implausible.

Instead, Gelman (2006) recommends using a uniform prior on the scale of the standard deviation, over a large (or semi-infinite) range, or a half-normal distribution with large variance. When a more informative prior is required, Gelman (2006) recommends working with the half-t family of distributions, including the half-Cauchy as a special case. The half-Cauchy is a non-standard distribution in BUGS, but it can be specified by noting that if z is a normal random variable with mean zero and standard deviation B, and γ is a χ_1^2 random variable, then $z/\sqrt{\gamma}$ is Cauchy distributed, with location zero and scale B, e.g.

```
omega          <- abs(z)/sqrt(gamma)
z               ~ dnorm(0, inv.B.squared)
inv.B.squared <- 1/pow(B, 2)
gamma           ~ dgamma(0.5, 0.5)
```

gives a half-Cauchy distribution for ω. The scale B is the median of the half-Cauchy distribution, which means that prior beliefs are easily expressed. Half-Cauchys also assign considerable weight to the value zero and have very heavy tails, which means that both zero and large random-effects variances (ω^2) can be accommodated if the likelihood is indicative of such. We will revisit the half-Cauchy in §10.4.

Covariance matrices In multivariate settings, i.e. when there are multiple unit-specific parameters for each unit — see Example 10.3.2 — we may extend the above ideas to prior specification for the variance of each set of random effects. We may, additionally, wish to estimate the covariances between each pair of random effects. The variance-covariance matrix must always be positive-semidefinite, however, and this is not a straightforward constraint to apply. A Wishart(R, k) prior for the precision matrix, where R and k denote inverse-scale matrix and degrees of freedom, respectively, ensures positive-semidefiniteness, but can lead to sensitivity, as in the case of its univariate, gamma counterpart. The least informative, proper Wishart prior, obtained by setting k equal to the number of parameters, is actually not that uninformative, especially as the number of parameters increases, although it will usually suffice. In light of this, it is advisable, whether attempting to be informative or not, to set R equal to k multiplied by some prior guess at the value of the between-unit covariance matrix, since the prior mean of the precision matrix is given by kR^{-1}. If attempting to be informative, Wishart priors are somewhat restrictive, since once the mean is set, the level of uncertainty is controlled by k alone. Hence choosing k to match prior beliefs for one parameter may lead to implausible values for others. O'Malley and Zaslavsky (2005) propose a more flexible, *scaled* inverse-Wishart prior for covariance matrices. The reader is referred to Gelman and Hill (2007), pp. 284–287, 376–380, for further discussion including how to implement such priors in BUGS. Other alternative priors tend to be parameterised in terms of decompositions such as the spectral or Cholesky decomposition. Barnard et al. (2000) propose decomposing the covariance matrix into a vector of standard deviations and a correlation matrix, with (typically) independent priors. This is particularly attractive because specifying priors for standard deviations and correlations is relatively intuitive. The correlation matrix is required to be positive definite, but the authors present methods for ensuring this. BUGS is yet to be equipped with such specialised sampling tools, however.

For low-dimensional problems, multivariate-normal random-effects distributions might be best specified in terms of conditional distributions, for example,

$$\begin{pmatrix} \theta_{i1} \\ \theta_{i2} \end{pmatrix} \sim \text{MVN}_2 \left(\begin{pmatrix} \mu_1 \\ \mu_2 \end{pmatrix}, \begin{pmatrix} \omega_1^2 & \rho\omega_1\omega_2 \\ \rho\omega_1\omega_2 & \omega_2^2 \end{pmatrix} \right)$$

is equivalent to

$$\theta_{i1} \sim \text{Normal}(\mu_1, \omega_1^2),$$

$$\theta_{i2}|\theta_{i1} \sim \text{Normal}\left(\mu_2 + \frac{\omega_2}{\omega_1}\rho(\theta_{i1} - \mu_1), (1 - \rho^2)\omega_2^2 \right). \tag{10.1}$$

Any prior that constrains $\omega_1 > 0$, $\omega_2 > 0$, and $0 \leq \rho < 1$ is then permissible. Some ways of modelling correlations among multivariate spatially correlated quantities are discussed in §11.3.

10.3 Hierarchical regression models

The exchangeability ideas discussed above are easily extended to regression analysis. This is illustrated in the following examples, both of which specify a *Generalised Linear Mixed Model* (GLMM) – a GLM in which the linear predictor contains random effects (Breslow and Clayton, 1993). Use of non-linear forms for the regression function is also handled straightforwardly, as illustrated in Example 10.4.1.

Example 10.3.1. *Salmonella (continued): hierarchical model*
We first return to the "salmonella" example of §6.5. Initially we modelled the data as arising from a Poisson distribution as follows:

$$y_{ij} \sim \text{Poisson}(\mu_i), \quad \log \mu_i = \alpha + \beta \log(x_i + 10) + \gamma x_i, \quad i = 1, ..., 6, j = 1, ..., 3,$$

but the model was incapable of predicting the level of variability apparent in the observed data. One option was to fit a negative binomial model instead (see Example 6.5.2). Another option is to allow the log-mean of the Poisson distribution to be "adjusted" by some amount, λ_{ij}, say, for each observation:

$$y_{ij} \sim \text{Poisson}(\mu_{ij}), \quad \log \mu_{ij} = \alpha + \beta \log(x_i + 10) + \gamma x_i + \lambda_{ij}.$$

By allowing the mean to vary on an observation-by-observation basis, more variability can be accounted for. We might think it reasonable now to assume the λ_{ij}s are similar/exchangeable. Hence $\lambda_{ij} \sim \text{Normal}(\mu_\lambda, \omega_\lambda^2)$, $i = 1, ..., 6$, $j = 1, ..., 3$, except that estimation of μ_λ would be confounded with that of α, and so we set $\mu_\lambda = 0$ and specify a prior for ω_λ only, e.g., $\omega_\lambda \sim \text{Uniform}(0, 100)$. The BUGS code for this model and predictive distributions for the number of colonies at each dose is

```
for (i in 1:6) {
  for (j in 1:3) {
    y[i,j]               ~ dpois(mu[i,j])
    log(mu[i,j])        <- log.fit[i] + lambda[i,j]
    lambda[i,j]          ~ dnorm(0, inv.omega.lambda.squared)
  }
  log.fit[i]            <- alpha + beta*log(x[i] + 10)
                            + gamma*x[i]
  log(fit[i])          <- log.fit[i]
  y.pred[i]             ~ dpois(mu.pred[i])
  log(mu.pred[i])      <- log.fit[i] + lambda.pred[i]
  lambda.pred[i]        ~ dnorm(0, inv.omega.lambda.squared)
}
alpha                   ~ dnorm(0, 0.0001)
```

```
beta                          ~ dnorm(0, 0.0001)
gamma                         ~ dnorm(0, 0.0001)
omega.lambda                  ~ dunif(0, 100)
inv.omega.lambda.squared <- 1/pow(omega.lambda, 2)
```

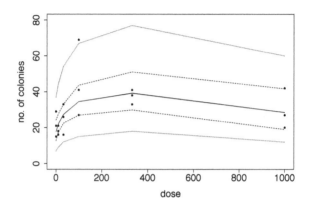

FIGURE 10.3

Posterior median model fit (—), 95% credible intervals (– –), and 95% prediction intervals (.....) from hierarchical regression analysis of salmonella data.

The model fit and predictive intervals are shown in Figure 10.3. Note that, in order to generate unit-specific predictions in a hierarchical model, it is necessary to also generate a predicted random effect (`lambda.pred[i]` in this case) — this is known as "mixed prediction" or "mixed replication" (see §10.7 and Figure 10.8). Posterior summaries for the main parameters are given in the table below. Note that the model fit and parameter estimates are very similar to those previously obtained from the negative binomial model – see Example 6.5.2. A quantitative measure of the model fit is given by the posterior mean deviance in the table below (110.6). Corresponding figures for the non-hierarchical Poisson model and the negative binomial model are 139.3 and 130.8, respectively. Clearly the hierarchical model fits better, but is the improved fit worth the expense of 19 additional parameters (ω_λ and one λ_{ij} for each observation)? Note, however, that the random effects λ_{ij} do not count, collectively, as 18 "whole" parameters, due to the correlation among them induced by the exchangeability assumption. To determine the "effective" number of parameters we can calculate DIC, which gives $p_D = 13.4$, meaning that the random effects count as ~10 parameters. In the non-hierarchical Poisson and negative binomial models, 3 and 4 parameters are used, respectively. Adding posterior mean deviance and the effective number of parameters gives the DIC for each model: 124, 142, and 135 for the hierarchical Poisson, non-hierarchical Poisson, and negative binomial models, respectively, suggesting that the hierarchical Poisson model is superior. We note that the hierarchical Poisson and the negative binomial both can be considered as dealing with extra-Poisson variability and will generally give very similar inferences. Both give similar inferences for the regression parameters. However, as detailed in §10.8.1, we might expect somewhat different deviances, as they have a different "focus" (see §10.8.1), just as a t distribution has a different deviance depending on whether it is expressed directly or as a scale mixture of normals (§8.2). As we are only interested in the regression coefficients, not the random effects, in this case the negative binomial model may be favoured in spite of the lower DIC in the hierarchical Poisson model.

node	mean	sd	MC error	2.5%	median	97.5%	start	sample
alpha	2.147	0.3758	0.01461	1.384	2.152	2.881	1001	100000
beta	0.3173	0.1026	0.004063	0.1168	0.3166	0.5267	1001	100000
deviance	110.6	6.065	0.02823	100.7	109.9	124.3	1001	100000
gamma	-9.96E-4	4.612E-4	1.601E-5	-0.001936	-9.912E-4	-8.523E-5	1001	100000
omega.lambda	0.2813	0.0836	9.769E-4	0.1438	0.2719	0.4726	1001	100000

Example 10.3.2. *Hepatitis B immunisation*
Hepatitis B (HB) is endemic in Africa. This example concerns a program of childhood vaccination introduced in Gambia. The effectiveness of the program depends on the duration of immunity afforded by vaccination. Our dataset comes from a study in which 106 children were immunised against HB and followed up. The level of immunity is measured by a quantity known as *anti-HB titre*. This was measured at the time of vaccination (baseline) and on either two or three follow-up occasions for each child. We wish to construct a model useful for predicting an individual child's protection against HB after vaccination. A similar study in Senegal (Coursaget et al., 1991) found that anti-HB titre is proportional to the reciprocal of time since vaccination, and so we work on the basis that the relationship between log-anti-HB-titre and log-time-since-vaccination should be approximately linear. The raw data are depicted in Figure 10.4; the observations for each child are joined by straight lines. We use the baseline log-titre as a covariate since this will be available for a given new individual, and it clearly gives an indication of subsequent titre values. Let y_{0i} and y_{ij} denote the log-baseline and jth follow-up log-titre measurement, respectively, for individual i. In addition, let t_{ij} denote the log-time to which y_{ij} corresponds. Specification of a non-hierarchical model might begin

$$y_{ij} \sim \text{Normal}(\psi_{ij}, \sigma^2), \quad \psi_{ij} = \alpha + \beta(t_{ij} - \bar{t}) + \gamma(y_{0i} - \overline{y_0}),$$

where $\bar{t}$ and $\overline{y_0}$ denote the empirical means of the t_{ij}s and y_{0i}s, respectively. We might, however, think it unrealistic to assume the same intercept and slope for all individuals. We can assume instead that different individuals have distinct intercepts and slopes, but that all slopes and all intercepts are similar/related. For example,

$$\psi_{ij} = \alpha_i + \beta_i(t_{ij} - \bar{t}) + \gamma(y_{0i} - \overline{y_0}),$$
$$\alpha_i \sim \text{Normal}(\mu_\alpha, \omega_\alpha^2), \quad \beta_i \sim \text{Normal}(\mu_\beta, \omega_\beta^2),$$

with μ_α, ω_α, μ_β, and ω_β assigned appropriate priors. Another option is to make a multivariate exchangeability assumption, which allows for correlation between intercepts and slopes — an individual with a large intercept might be more likely to have a small slope, say. For example, we might assume that each pair (α_i, β_i) arises from a bivariate normal population distribution with unknown population mean vector and unknown population covariance matrix. We might specify this directly, assuming multivariate normal and inverse-Wishart priors for the mean and covariance, respectively, or indirectly via conditional distributions as in (10.1). (Note that it is not feasible to estimate a distinct γ_i for each individual, as there is only one baseline measurement.) We will return to this example shortly, after discussing how the data might be formatted.

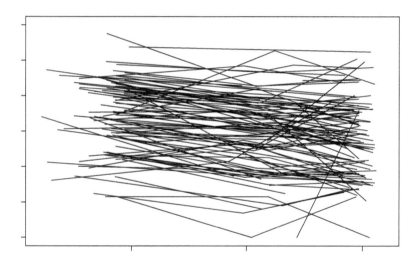

FIGURE 10.4
Raw hepatitis B data: log anti-HB titre vs log time since vaccination. Each child's data are joined together by straight lines.

10.3.1   Data formatting

Hierarchical models are invariably used for fitting data from multiple "units." The fact that those units generally do not provide the same amount of data presents somewhat of a data formatting problem for BUGS. The natural thing to do would be to specify hierarchical data in a matrix, with each row comprising a specific unit's data. Such a matrix would be what we call a "ragged array," due to each row having a different length. We can only specify regular arrays in BUGS, as described in §12.4.2, and so here we present the various options available.

1. "Pad out" the ragged array with missing values, so that it becomes regular. We first choose a maximum row length corresponding to the unit with the most data and fill in the missing values with NAs. The model is then specified as if the data were balanced, as illustrated in the next example, and the missing values are "imputed" from their predictive distributions (see §9.1). This has no impact on the inferences themselves, but can create a lot of unnecessary computation and can be cumbersome to set up, especially if the data are very imbalanced. Hence we present two further alternatives.

• Firstly, format the entire dataset as a single vector, with data from each unit forming contiguous elements and units arranged one after another, and then either:

2. construct a vector of indices at which each unit's data starts in the data vector. This is called the "offsets" approach and is illustrated in the example below. Or,

3. construct a vector of the same length as the data vector which specifies the unit to which each observation corresponds. We may then use the "nested indexing" feature of BUGS, which, again, is illustrated below.

Example 10.3.3. *Hepatitis B (continued)*
The follow-up HB data form a ragged array, as shown in the table below. In this example we illustrate three different approaches to specifying a hierarchical regression model.

Child	log-titre y_{ij}			log-time t_{ij}		
1	4.997	8.028		6.541	6.963	
2	6.830	4.905	6.295	5.841	6.529	6.982
3	3.951	4.356		7.026	10.00	
4		...			...	

Method 1, "padding out":

```
for (i in 1:N) {
  for (j in 1:3) {
    y[i,j]     ~ dnorm(psi[i,j], inv.sigma.squared)
    psi[i,j] <- alpha[i] + beta[i]*(t[i,j] - tbar)
```

```
                    + gamma*(y0[i] - y0bar)
    }
    alpha[i]     ~ dnorm(mu.alpha, inv.omega.alpha.squared)
    beta[i]      ~ dnorm(mu.beta, inv.omega.beta.squared)
  }

  list(N = 106,
       y = structure(
  .Data = c(4.997,8.028,NA,6.830,4.905,6.295,3.951,4.356,NA,...),
  .Dim =   c(106, 3)),
       t = structure(
  .Data = c(6.541,6.963,0,5.841,6.529,6.982,7.026,10.00,0,...),
  .Dim =   c(106, 3)), ...)
```

Note that we can specify any appropriate log-time for the missing data, in this case 0, since the missing values themselves are of no interest.

Method 2, "offsets":

```
for (i in 1:N) {
  for (j in offset[i]:(offset[i+1]-1)) {
    y[j]     ~ dnorm(psi[j], inv.sigma.squared)
    psi[j] <- alpha[i] + beta[i]*(t[j] - tbar)
             + gamma*(y0[i] - y0bar)
  }
  alpha[i]  ~ dnorm(mu.alpha, inv.omega.alpha.squared)
  beta[i]   ~ dnorm(mu.beta, inv.omega.beta.squared)
}

list(N = 106,
     y = c(4.997,8.028,6.830,4.905,6.295,3.951,4.356,...),
     t = c(6.541,6.963,5.841,6.529,6.982,7.026,10.00,...),
     offset = c(1,3,6,8,...), ...)
```

Note that the length of offset[] must be N+1, with offset[N+1] = Nobs + 1, where Nobs is the total number of observations.

Method 3, "nested indexing":

```
for (j in 1:Nobs) {
  y[j]       ~ dnorm(psi[j], inv.sigma.squared)
  psi[j]   <- alpha[child[j]] + beta[child[j]]*(t[j] - tbar)
             + gamma*(y0[child[j]] - y0bar)
}
```

```
for (i in 1:N) {
   alpha[i]  ~ dnorm(mu.alpha, inv.omega.alpha.squared)
   beta[i]   ~ dnorm(mu.beta, inv.omega.beta.squared)
}

list(N = 106, Nobs = 288,
       y = c(4.997,8.028,6.830,4.905,6.295,3.951,4.356,...),
       t = c(6.541,6.963,5.841,6.529,6.982,7.026,10.00,...),
       child = c(1,1,2,2,2,3,3,...), ...)
```

For this example, because there is a maximum of only three observations per child, we used the "padding out" method and specified vague Normal$(0, 100^2)$ priors for μ_α, μ_β, and γ, Uniform$(0, 100)$ priors for ω_α and ω_β, and a Uniform$(-10, 10)$ prior for $\log \sigma$. To specify a bivariate exchangeability assumption instead, we could replace the definitions of alpha[i] and beta[i] in the examples above with

```
alpha[i]      <- theta[i,1]
beta[i]       <- theta[i,2]
theta[i,1:2]  ~ dmnorm(mu[], inv.Omega[,])
```

and specify priors for mu[] and inv.Omega[,] via, for example,

```
mu[1:2]               ~ dmnorm(m[], T[,])
Omega.inv[1:2,1:2]  ~ dwish(R[,], k)
```

with appropriate values for m, T, R, and k, e.g., $m = (0, 0)'$, $T = 100^{-2}I_2$, $R = ks^2I_2$, and $k = 2$, where s is some guess at the between-child standard deviation of regression coefficients, and I_2 is the 2×2 identity matrix. Alternatively, we could specify bivariate normal random effects indirectly via conditional distributions:

```
for (i in 1:N) {
   alpha[i] ~ dnorm(mu.alpha, inv.omega.alpha.squared)
   beta[i]  ~ dnorm(mean.beta, prec.beta)
}
mean.beta <- mu.beta + omega.beta*rho*
                       (alpha[i] - mu.alpha)/omega.alpha
prec.beta <- inv.omega.beta.squared/(1 - pow(rho, 2))
```

with, for example,

```
mu.alpha ~ dnorm(0, 0.0001); mu.beta ~ dnorm(0, 0.0001)
omega.alpha ~ dunif(0, 100); omega.beta ~ dunif(0, 100)
rho ~ dunif(0, 1)
```

The resulting posterior distributions, assuming univariate exchangeability, are summarised and compared with those from the equivalent non-hierarchical model in the table below.

Hierarchical model		Non-hierarchical model	
Parameters	Posterior median (95% interval)	Parameters	Posterior median (95% interval)
μ_α	6.14 (5.84, 6.43)	α	6.13 (5.94, 6.33)
μ_β	−1.08 (−1.34, −0.796)	β	−1.05 (−1.48, −0.608)
γ	0.671 (0.504, 0.837)	γ	0.674 (0.561, 0.787)
σ^2	0.987 (0.791, 1.23)	σ^2	2.98 (2.54, 3.54)
ω_α^2	2.02 (1.46, 2.82)		
ω_β^2	0.0726 (3.64E-4, 0.608)		
DIC	912	DIC	1136
p_D	98.4	p_D	4.03

First, note the strong agreement for the intercept and slope parameters between the hierarchical and non-hierarchical models. This might be expected but is largely due to the linear nature of both models — fitting a straight line through the pooled data gives the same result as averaging parameters after fitting a straight line to each individual. If the model were nonlinear, this would be unlikely to occur. Second, note how the residual variability has been reduced from around 3 to around 1: this suggests that allowing for variation in child-specific αs and βs explains approximately two thirds of the initially unexplained variance. Note also that the vast majority of variability between individuals is between intercepts rather than slopes. Indeed, we might have expected this, given the apparent similarity of individual slopes in Figure 10.4. The DIC figures show that the hierarchical model is vastly superior, even after taking account of an additional ∼94 (effective) parameters. The actual number of parameters used in the hierarchical model is 218, including 212 random effects. This suggests that each random effect is worth less than half a parameter.

Example 10.3.4. *Students' goals: hierarchical categorical/multinomial models*
Four hundred and seventy eight students in grades 4–6 (aged 7–13) from 9 schools in Michigan, USA were asked whether good grades, popularity, or sporting ability was most important to them (Chase and Dummer, 1992). Summarising the data, 52% of the 227 boys in the study and 52% of the 251 girls chose good grades. The difference between boys and girls appears in their preference for sports versus popularity — while 26% of boys preferred sports (22% popularity), 36% of girls chose popularity (12% sports).

We use Bayesian models to quantify uncertainty about the underlying probabilities and how they differ between girls and boys, and between schools. We model the response Goals for each student i as a categorical outcome with probabilities p_{ik}, where $k = 1,2,3$ indicates sporting ability, popularity, or grades, respectively. Using nested indexing, the variable School[i] is an integer from 1 to nschool=9. Gender[i] is 1 for a boy and 0 for a girl.

Dirichlet/gamma model   First, ignoring gender effects for illustration, we might model between-school variations in these probabilities by assuming the probabilities are a constant vector p_j for each student in school j, and model $p_j \sim \text{Dirichlet}(\alpha)$ (see §7.2). However, no current implementation of BUGS allows inference on the parameters of a Dirichlet distribution ddirch. A convenient workaround is to exploit the relation between the Dirichlet and gamma distributions (Appendix C.4) — if $p_{jk} = q_{jk}/\sum_k q_{jk}$ and $q_{jk} \sim \text{Gamma}(\alpha_k, 1)$, then the vector $p_j \sim \text{Dirichlet}(\alpha)$. Thus the school-specific probabilities in the following hierarchical model are drawn from a Dirichlet distribution with parameters a[].

```
for (i in 1:npupil) {
   Goals[i]   ~ dcat(p[School[i],])
}
for (j in 1:nschool) {
   for (k in 1:3) {
     p[j,k] <- q[j,k]/sum(q[j,])
     q[j,k]   ~ dgamma(a[k], 1)
   }
}
for (k in 1:3) {
   a[k]        ~ dgamma(1, 0.001)
   p.pop[k] <- a[k]/sum(a[]) # population mean of p[,k]
}
```

However, the resulting chains fail to converge in this example due to high posterior correlations between the three a[k].

Multinomial logistic models   Convergence can be achieved using a multinomial logistic model, as in §7.2.3, with $\log(q_{jk}) \sim N(\mu_j, \sigma^2)$ for $k = 1, 2$ and $\log(q_{jk}) = 1$ for the baseline category $k = 3$ (a preference for good grades, the commonest outcome). The parameters μ_1, μ_2, and σ^2 express the mean preferences and their variability between schools in a different way and have lower posterior cross-correlations than the Dirichlet parameters a[]. We use this model to investigate the difference between girls and boys as well as between schools.

Three variants are used, which all include gender effects. We estimate a posterior distribution for or.boy — the *odds ratio* of preferring sport for a boy compared to a girl — under each model, assuming this effect is common between schools. The probabilities of each other choice are transformed to log-odds relative to choosing good grades, $k = 3$. In the first model, for example, b[1] and b[2] are the *log odds* of sports and popularity, respectively, compared to grades, for a girl.

No difference between schools

```
for (i in 1:npupil) {
   Goals[i]        ~ dcat(p[i,])
   for (k in 1:3) {
```

```
   p[i,k]        <- q[i,k]/sum(q[i,])
   log(q[i,k]) <- a[i,k]
 }
 a[i,1]          <- b[1] + b.boy*Gender[i]
 a[i,2]          <- b[2]
 a[i,3]          <- 0
}
b[1]               ~ dnorm(0, 0.0001)
b[2]               ~ dnorm(0, 0.0001)
b.boy              ~ dnorm(0, 0.0001)
or.boy          <- exp(b.boy)
```

Boys are significantly more likely than girls to prefer sporting ability. The posterior median odds ratio is 2.67, the 95% credible interval is (1.67, 4.36), and the DIC for model comparison (§8.6.4) is 961.

Independent school effects As above, but with

```
   a[i,1] <- b[School[i], 1] + b.boy*Gender[i]
   a[i,2] <- b[School[i], 2]
   a[i,3] <- 0
}
for (j in 1:nschool) {
  b[j,1] ~ dnorm(0, 0.0001)
  b[j,2] ~ dnorm(0, 0.0001)
}
```

A DIC of 956 suggests that allowing for differences between schools improves the model. This also modifies the gender odds ratio, since there are also some differences between schools in the proportion of boys and girls in the data — the posterior median is now 2.96 (1.78, 5.01).

Hierarchical school effects As above, but replacing the independent vague normal priors on the log odds of sport and of popularity for girls for each school by a bivariate normal exchangeable prior, parameterised using conditional distributions, as described in §10.2.3.

```
for (j in 1:nschool) {
  b[j,1]    ~ dnorm(mu[1], tau[1])
  mub2[j] <- mu[2] + s[2]/s[1]*cor*(b[j,1] - mu[1])
  b[j,2]    ~ dnorm(mub2[j], taub2)
}
vb2        <- (1 - cor*cor)*v[2]
tau[1]    <- 1/v[1]
taub2     <- 1/vb2
for (k in 1:2) {
```

```
    mu[k]      ~ dnorm(0, 0.0001)
    v[k]       <- s[k]*s[k]
    s[k]       ~ dunif(0, 100)
  }
  cor          ~ dunif(0, 1)
```

A DIC of 951 suggests that allowing *exchangeable* school effects improves the model even further. We are "borrowing strength" from other schools to estimate each school's effect — this model assumes they are not entirely the same, but neither are they entirely unrelated. The increase in the odds of preferring sports for a boy compared to a girl now has a posterior median of 2.86 (1.75, 4.76).

In Figure 10.5, the school and gender-specific *estimated probabilities* of preferring sports, under the independent and hierarchical models, are compared with the raw *proportions observed* to prefer sports. For schools with few children in the data (e.g., Elm with only 5 boys and 16 girls), both of the models produce more stable estimates which borrow information from the rest of the data. In the hierarchical model, the school-specific probabilities are shrunk towards the global average. Uncertainty about the probabilities is expressed as 95% credible intervals.

10.4 Hierarchical models for variances

In the Hepatitis B model of Example 10.3.2 we assumed that all residuals have a common variance. However, it is reasonable to suspect that residual variances for different units might differ. It also seems reasonable to suppose that, although different, these variances are similar in some sense, and so we may wish to assume they are exchangeable. One approach would be to model the log-standard deviations as arising from a normal population distribution, with unknown mean and variance. Another option, which allows standard deviations to be modelled on the natural scale, is to assume they arise from a half-Cauchy population distribution, with unknown scale parameter. This latter option is a natural model in cases where some variances may be outlying. It also facilitates the specification of prior beliefs, since the only parameter represents the population median standard deviation. Both of these approaches are illustrated in the following example, which also demonstrates the ease with which hierarchical regression models are extended from linear, as in Example 10.3.2, to nonlinear.

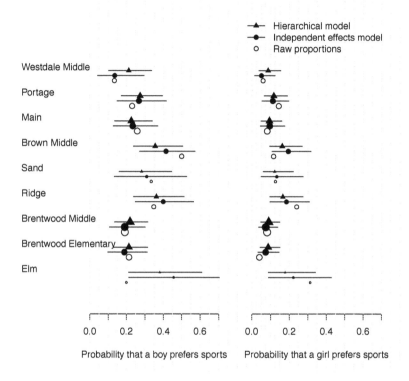

FIGURE 10.5
Probabilities of preferring sports achievement to popularity or grades, for students in nine schools: posterior medians and 95% credible intervals. Plot symbols are scaled in proportion to the number of students in the school/gender.

Example 10.4.1. *Cadralazine: hierarchical model for variances*
Ten cardiac failure patients were each given a 30 mg intravenous bolus dose of cadralazine. The data comprise plasma drug concentrations (mg/L) measured at various times between 2 and 32 hours post dose. Let y_{ij} denote the jth concentration measurement, taken at time t_{ij}, for individual i (there are around six observations for each individual). The time-course of many drugs, following intravenous input, can be modelled as an exponential decay, as follows:

$$y_{ij} \sim \text{Normal}(\psi_{ij}, \sigma_i^2), \quad \psi_{ij} = \frac{D}{V_i} \exp\left(-\frac{CL_i}{V_i} t_{ij}\right),$$

where D denotes the dose (30 mg), and CL_i and V_i denote patient-specific parameters known as the *clearance* and *volume of distribution*, respectively. Note that we are now assuming residuals from different individuals have different variances σ_i^2. Both the clearance and volume must be positive and so the following exchangeability assumption is appropriate for these patient-specific parameters:

$$\theta_i = (\log CL_i, \log V_i)' \sim \text{MVN}_2(\mu, \Omega), \quad i = 1, ..., 10,$$

with appropriate priors for μ and Ω:

$$\mu \sim \text{MVN}_2 \left(\begin{pmatrix} \log 2.9 \\ \log 15 \end{pmatrix}, \begin{pmatrix} 100^2 & 0 \\ 0 & 100^2 \end{pmatrix} \right), \quad \Omega^{-1} \sim \text{Wishart} \left(\begin{pmatrix} 0.08 & 0 \\ 0 & 0.08 \end{pmatrix}, 2 \right),$$

which are vague but represent prior guesses that clearance and volume are 2.9 L/hr and 15 L, respectively, and that their coefficients of variation throughout the population are around 20% ($0.08 = 2 \times 0.2^2$). For the patient-specific residual standard deviations, we could log-transform and assume a normal population distribution:

$$\log \sigma_i \sim \text{Normal}(\mu_\sigma, \omega_\sigma^2), \quad i = 1, ..., 10,$$

with $\mu_\sigma \sim \text{Normal}(0, 100^2)$ and $\omega_\sigma \sim \text{Uniform}(0, 100)$, say. Alternatively, an appropriate model on the natural scale would be

$$\sigma_i \sim \text{half-Cauchy}(B), \quad i = 1, ..., 10,$$

where B is the population median residual standard deviation. We can be vague about B, e.g., $B \sim \text{Uniform}(0, 100)$, but the parameter's direct interpretability means that it is straightforward to be informative. BUGS code for the vague half-Cauchy model is as follows:

```
for (i in 1:10) {
   for (j in offset[i]:(offset[i+1]-1)) {
      y[j]                       ~ dnorm(psi[j], inv.sigma.squared[i])
      psi[j]                     <- D*exp(-CL[i]*time[j]/V[i])/V[i]
   }
   CL[i]                         <- exp(theta[i, 1])
   V[i]                          <- exp(theta[i, 2])
   theta[i, 1:2]                 ~ dmnorm(mu.theta[], inv.Omega[,])
   sigma[i]                      <- abs(z[i])/sqrt(gamma[i])
   z[i]                          ~ dnorm(0, inv.B.squared)
   gamma[i]                      ~ dgamma(0.5, 0.5)
   inv.sigma.squared[i] <- 1/pow(sigma[i], 2)
}
inv.B.squared                    <- 1/pow(B, 2)
B                                ~ dunif(0, 100)
mu.theta[1:2]                    ~ dmnorm(m[], T[,])
inv.Omega[1:2, 1:2]              ~ dwish(R[,], k)
Omega[1:2, 1:2]                  <- inverse(inv.Omega[,])
```

The advantage of the log-normal population distribution is that it facilitates modelling the residual variation as a function of covariates. The heavy tails of the half-Cauchy distribution, on the other hand, allow for outlying individuals with relatively large residuals, which arise frequently in these types of model due to it being difficult to fit every individual's data equally well. Note also that the half-Cauchy accommodates the possibility of variances very close to zero for some units. The Cauchy's heavy tails result in less shrinkage to the population mean, as demonstrated in Figure 10.6.

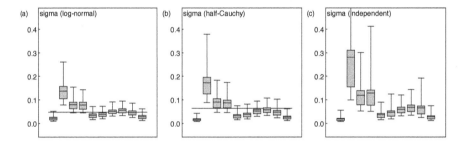

FIGURE 10.6
Posterior summaries for each σ_i, $i = 1, ..., 10$, from analysis of cadralazine data assuming: (a) log-normal exchangeability, (b) half-Cauchy exchangeability, (c) independence. In (a) and (b) the horizontal line is the posterior-mean population-median residual-standard-deviation.

10.5   Redundant parameterisations

Various authors (e.g., Gelman and Hill (2007), Chapter 19) have suggested deliberately overparameterising hierarchical models to improve convergence/mixing. Supposing $\theta_i \sim \text{Normal}(\mu, \omega^2)$, $i = 1, ..., N$, say, which can be rewritten as $\theta_i = \mu + \alpha_i$, $\alpha_i \sim \text{Normal}(0, \omega^2)$, $i = 1, ..., N$, then the basic idea is to replace each α_i with $\xi\eta_i$, $\eta_i \sim \text{Normal}(0, \omega_\eta^2)$. The η_is and ξ are then not identifiable, although their products, α_i, $i = 1, ..., N$, are. Moreover, mixing for the identifiable parameters is improved, as illustrated in the following example, which demonstrates only a mild effect; in some cases the effect may be quite dramatic.

Example 10.5.1. *Bristol (continued): overparameterisation*

Returning to the surgery data of Example 10.1.1, we wish to replace the specification $\logit \theta_i \sim \text{Normal}(\mu, \omega^2)$ with $\logit \theta_i = \mu + \xi \eta_i$, $\eta_i \sim \text{Normal}(0, \omega_\eta^2)$. A natural prior for ξ would be $\xi \sim \text{Normal}(0, v_\xi)$. However, since ω_η, the scale of the η_is, is unknown, we can without loss of generality set $v_\xi = 1$. Note that $\omega_\eta = \omega/|\xi|$.

```
for (i in 1:11) {
  y[i]                    ~ dbin(theta[i], n[i])
  logit(theta[i])         <- mu + eta[i]*xi
  eta[i]                  ~ dnorm(0, inv.omega.eta.squared)
}
inv.omega.eta.squared <- 1/pow(omega.eta, 2)
omega.eta               <- omega/abs(xi)
xi                       ~ dnorm(0, 1)
omega                    ~ dunif(0, 100)
mu                       ~ dunif(-100, 100)
```

History plots for θ_1 over the first 4000 iterations are shown in Figure 10.7 and show considerably better mixing in the overparameterised version.

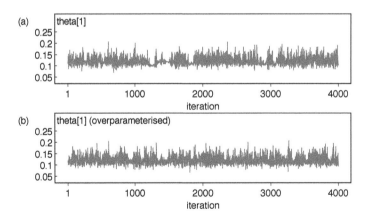

FIGURE 10.7

History plots for θ_1 from analysis of surgery data using: (a) standard model as described in Example 10.1.1, (b) overparameterised model.

10.6   More general formulations

Throughout this chapter we have dealt with hierarchical models comprising three levels: (i) a likelihood conditional on unit-specific parameters, (ii) exchangeability assumptions (population distributions) for the unit-specific parameters (or their departures from some covariate model), and (iii) prior distributions for the population parameters, also referred to as *hyperparameters*. There is, of course, no reason to stop us considering models with more levels. The appropriate number of levels is governed by the structure of the data under consideration (and is usually obvious). For example, we might be interested in modelling the outcome of children's exams at school, with observations on different children, within different schools, in different geographical areas, in different countries, with repeated observations on each child. After controlling for important covariates, we may wish to make the following subjective judgements, thus formulating a six level model: that exam results for the same child are correlated; that the performances of children within the same school are correlated; that schools within the same geographical area are correlated; that geographical areas within the same country are correlated; that different countries are all similar in some sense; and that the parameters of the "population distribution" from which different countries arise are unknown with appropriate priors.

An m-level hierarchical model is typically formulated by assuming observations or unknown parameters at one level are conditionally independent, given the parameters at the next level. If there is no next level, then a prior distribution for the parameters of the current level is specified. In general, given data y, we define a likelihood $p(y|\theta)$ and then specify a prior distribution for θ via a series of structural judgements:

$$p(\theta) = \int \int ... \int p(\theta|\phi_2)p(\phi_2|\phi_3)...p(\phi_{m-1})\, d\phi_2 d\phi_3...d\phi_{m-1},$$

where ϕ_k, $k = 2, ..., m-1$, are referred to as the "hyperparameters of level k" (or just "hyperparameters" when $k = m - 1$). The structural judgements are usually exchangeability assumptions, but other types of assumption are also possible, such as spatial correlation — see §11.3.

10.7   Checking of hierarchical models

We have seen in Chapter 8 that model checking is ideally carried out in a true predictive, or cross-validatory, framework where parts of the data, say y_i, are removed and predicted from the remainder. This is represented for hierarchical models in Figure 10.8, which shows that in order to predict a replicate y_i^{rep} it is necessary to also generate a replicate random effect θ_i^{rep}: this is known as "mixed replication."

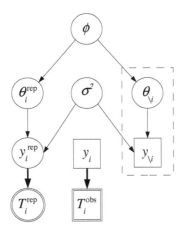

FIGURE 10.8

Cross-validatory "mixed" replication of random effects θ_i and data y_i in a hierarchical model, where ϕ are the parameters of the random effects distribution and σ^2 is the observation-level variance. Data point y_i is left out and predicted using a model fitted to the remaining data $y_{\setminus i}$. The bold black arrow indicates a logical function. The double-edged nodes represent quantities to be compared in order to assess divergence.

Example 10.7.1. *Bristol (continued): cross-validatory check of random effects model*

We consider the Bristol data, including all 12 hospitals, as listed in Example 8.3.1, but adopting the random effects model used previously (Example 10.1.1) on the 11 hospitals excluding Bristol.

```
for (i in 2:N) {
  y[i]               ~ dbin(theta[i], n[i])
  logit(theta[i]) <- logit.theta[i]
  logit.theta[i]     ~ dnorm(mu, inv.omega.squared)
}
inv.omega.squared <- 1/pow(omega, 2)
omega                  ~ dunif(0, 10)
mu                     ~ dunif(-10, 10)
# Mixed predictions of centre 1:
```

```
# generate replicate log-odds:
logit.theta1.cv    ~ dnorm(mu, inv.omega.squared)
logit(theta1.cv)   <- logit.theta1.cv
# generate replicate deaths:
y1.cv              ~ dbin(theta1.cv, n[1])
# use mid p-value:
P.mixed            <- step(y1.cv - y[1] - 0.00001)
                    + 0.5*equals(y1.cv, y[1])
```

node	mean	sd	MC error	2.5%	median	97.5%	start	sample
P.mixed	0.00165	0.03965	4.267E-4	0.0	0.0	0.0	1001	10000
omega	0.1996	0.1132	0.004138	0.01784	0.1857	0.4557	1001	10000
y1.cv	16.77	5.38	0.06203	7.0	16.0	28.0	1001	10000

The cross-validatory predictive distribution for the number of deaths in Bristol has median 16 and 95% interval 7 to 28, compared to the observed value of 41. Bristol is therefore a clear outlier, with a one-sided p-value of around 0.002. The between-hospital SD ω (excluding Bristol) has posterior mean 0.20.

Cross-validation is time consuming and, as described in Chapter 8, is generally replaced by generating replicate data from the posterior distribution of the parameters, taking into account *all* the observed data. As for nonhierarchical models, we need to be careful that the checking function assesses quantities that are not directly estimated by the probability model, otherwise it will be unduly influenced by the data which it is intended to be critiquing, inducing conservatism in the procedure and losing power to detect discrepancies. Within a hierarchical framework this means that we need to be very clear about what a replicate involves. For example, suppose we simply generate a new Y_i^{rep} conditional on the θ_i being simulated within the MCMC run and compare the replicate with the observed y_i — this is likely to be deeply conservative, as y_i will have been extremely influential in the simulated value for θ_i.

The recommended procedure is therefore to essentially repeat the cross-validation procedure but without leaving y_i out of the analysis, thus generating $\theta_i^{\text{rep}}|y$ followed by Y_i^{rep}. The generation of a new θ_i^{rep} can be viewed as "ghosting": for each unit in turn, a ghost unit is created in a parallel universe, and observations are generated. Some conservatism will be introduced, but this may be only moderate, as y_i^{obs} only influences θ_i^{rep} through ϕ (Figure 10.8).

Example 10.7.2. *Bristol (continued): approximate cross-validatory check of random effects model*
We again analyse the surgery mortality data, as in Example 10.7.1, but this time re-predicting the Bristol data based on fitting the random effects model to all 12 hospitals, including Bristol.

node	mean	sd	MC error	2.5%	median	97.5%	start	sample
P.mixed	0.02225	0.1457	0.001438	0.0	0.0	0.0	1001	10000
omega	0.4248	0.1297	0.001985	0.23	0.4059	0.7387	1001	10000
y1.cv	19.35	8.917	0.08326	6.0	18.0	40.0	1001	10000

The estimated between-hospital SD is now substantially larger due to the inclusion of Bristol, and hence the predictive distribution for the number of deaths in Bristol now has median 18 and 95% interval 6 to 40. This is considerably wider than the cross-validatory predictive distribution, but the p-value of 0.02 is still reasonably small. The lack of cross-validation has therefore introduced conservatism but not completely masked the outlier.

The issue remains of selecting a discrepancy measure T_i for a vector y_i. If T_i is chosen to be the full data y_i, then the individual observations y_{ij} each contribute to the overall measure of divergence. However, the reference distribution is then a convolution of the *likelihood* $p(y_{ij}|\theta_i)$ with the *prior* $p(\theta_i|\phi)$, and so loses power if our interest is solely in checking for divergent θ_i and we are willing to assume the likelihood is correct. This is clearly illustrated if sufficient statistics s_i exist, since by definition the likelihood factorises $p(y_i|\theta_i)$ into $p(y_i|s_i)p(s_i|\theta_i)$. The first term contains no information about θ_i and hence its inclusion in a reference distribution for a discrepancy measure can only add noise to the procedure. Thus it will be more efficient to use s_i as a discrepancy measure or, more generally, if closed-form estimators $\hat{\theta}_i$ exist, to set $T_i = \hat{\theta}_i(y_i)$ and then compare $\hat{\theta}_i$ with $p(\hat{\theta}_i^{\text{rep}}|y_{\backslash i})$.

Example 10.7.3. *Rats: checking a random effects growth model for outliers*
Recall the data introduced in Example 6.1.1 on the growth of a single rat. We now consider the growth of all 30 rats in the experiment, whose weight was measured at $T = 5$ time points: age 8, 15, 22, 29 and 36 days. A null model H_0 assumes normal errors and random-coefficient linear growth curves with time x measured in days minus 22. Intercept and gradient may be given a bivariate normal prior, so that

$$y_{ij} \sim \mathrm{N}(\mu_{ij}, \sigma^2), \quad \mu_{ij} = \alpha_{i1} + \alpha_{i2}(x_j - \bar{x}), \quad \alpha_i \sim \mathrm{MVN}_2(\mu_\alpha, \Omega)$$

where $\mu_\alpha, \Omega, \sigma^2$ are given proper but very diffuse prior distributions.
We first consider identifying divergent *rats*. In this simple linear model we can obtain closed-form sufficient statistics for the intercept and gradient parameters α_i, specifically $\hat{\alpha}_{i1} = \bar{y}_i, \hat{\alpha}_{i2} = \sum_j y_{ij}(x_j - \bar{x})/\sum_j(x_j - \bar{x})^2$. A mixed replicate $\hat{\alpha}_i^{\text{rep}}$ can then be obtained at each iteration by simulating α_i^{rep} and then using the known sampling theory $\hat{\alpha}_{i1} \sim \mathrm{N}(\alpha_{i1}, \sigma^2/T), \hat{\alpha}_{i2} \sim \mathrm{N}(\alpha_{i2}, \sigma^2/\sum_j(x_j - \bar{x})^2)$ to generate $\hat{\alpha}_i^{\text{rep}}$. A standard p-value comparison can then be made between $\hat{\alpha}_i^{\text{rep}}$ and $\hat{\alpha}_i$. Of course this procedure relies on closed-form sufficient statistics being available, and the next example will show how we can manage when this is not the case.

Divergent individual *observations* may also be of interest. This could be checked
by summaries of residuals. Specifically, if we calculate $X_i^2 = \sum_j (y_{ij} - \mu_{ij})^2/\sigma^2$,
then we know that a replicate version has a χ_T^2 distribution. A p-value can be
therefore be obtained by comparing X_i^2 with a χ_T^2 observation at each iteration.

```
for (i in 1:N) {
  for (j in 1:T) {
    Y[i,j]                ~ dnorm(mu[i,j], tau)
    mu[i,j]               <- alpha[i,1] + alpha[i,2]
                             *(x[j] - mean(x[]))
    ## cross-product statistics
    XY[i,j]               <-  Y[i,j]*(x[j] - mean(x[]))
    XX[i,j]               <- (x[j] - mean(x[]))*(x[j] - mean(x[]))
    ## posterior predictive
    ## N(0,1) under null
    resid[i,j]            <- (Y[i,j] - mu[i,j])*sqrt(tau)
    resid2[i,j]           <- resid[i,j]*resid[i,j]
  }
  X2[i]                   <- sum(resid2[i,])  # chisq(T) under null
  chi.sqr[i]              ~ dchisqr(T)    # comparison under null
  P.resid[i]              <- step(X2[i] - chi.sqr[i])
  # prior for intercept and gradient
  alpha[i,1:2]            ~ dmnorm(mu.alpha[],R[,])
  # replicated intercept and gradient
  alpha.pred[i,1:2]       ~ dmnorm(mu.alpha[],R[,])
  # summary statistics for intercept and gradient
  alpha.est[i,1]          <- mean(Y[i, ])
  alpha.est[i,2]          <- sum(XY[i,])/sum(XX[i,])
  ## precision of intercept estimates
  alpha.est.prec[i,1] <- tau*T
  ## precision of gradient estimates
  alpha.est.prec[i,2] <- tau*sum(XX[i,])
  alpha.est.pred[i,1]   ~ dnorm(alpha.pred[i,1],
                              alpha.est.prec[i,1])
  alpha.est.pred[i,2]   ~ dnorm(alpha.pred[i,2],
                              alpha.est.prec[i,2])
  P.alpha[i,1]            <- step(alpha.est[i,1] -
                              alpha.est.pred[i,1])
  P.alpha[i,2]            <- step(alpha.est[i,2] -
                              alpha.est.pred[i,2])
}
mu.alpha[1]               ~ dunif(-1000,1000)
mu.alpha[2]               ~ dunif(-1000,1000)
R[1:2,1:2]                ~ dwish(Omega[,],2)
Omega[1,1]                <- 1;
```

```
Omega[1,2]              <- 0;
Omega[2,1]              <- 0;
Omega[2,2]              <- 1
tau                     ~ dgamma(0.001,0.001)
sigma                   <- 1/tau
```

Table 10.1 selects rats with at least one p-value being less than 0.05 or greater than 0.95, and the trajectories for these rats are highlighted among the 30 rats shown in Figure 10.9. The clearly erratic behaviour of rat 3 is identified by the aggregate residual measure. The intercepts of rats 9, 14, and 29 are highlighted, as well as the gradients of rats 2, 4, and 9. However, we are performing three hypothesis tests for each of 30 rats; therefore we are more likely to find low p-values merely by chance. If we applied Bonferroni's correction, we would only highlight p-values less than $0.05/(3 \times 30) = 0.00056$ or greater than 0.9994. Therefore, even the intercept of rat 9 is not as extreme as the p-value of 0.997 might suggest. This is apparent from Figure 10.9, which also shows that rats 14 and 29 do not appear to have intercepts which diverge from the rest of the population.

TABLE 10.1
p-values for rat growth model checking which are less than 0.05 or greater than 0.95. Diagnostics for divergent rats are based on mixed-parameter replication of estimated intercepts and gradients. Divergent observations are examined by a χ^2 statistic based on standardised residuals.

| | Divergent rats | | Divergent observations |
Rat	Intercept	Gradient	within rats
2		0.959	
3			0.999
4		0.044	
9	0.997	0.961	0.951
14	0.963		
29	0.043		

We note that we have not carried out full cross-validation, but the influence of individual rats on the random-effects distribution will be limited and so the conservatism will not be great. If desired, individual rats could be removed and the analysis repeated.

An alternative way of detecting outlying observations or units in a hierarchical model is by treating the issue as one of potential conflict between the posterior inferences $p_R(\theta_i|y_i)$ that would arise from the data y_i alone assuming a reference prior for θ_i, which we consider as "fixed-effects" estimates,

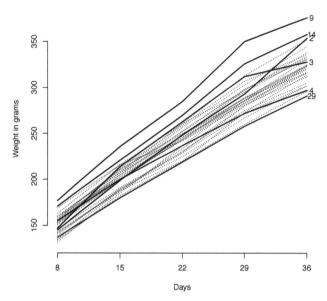

FIGURE 10.9
Growth curves for 30 rats, highlighting "outlying" rats with p-values less than 0.05 or greater than 0.95.

and the predictive prior distribution $p(\theta_i|y_{\setminus i})$ that arises from the remaining data. Contrasting these two sources results in "conflict p-values" (Marshall and Spiegelhalter, 2007). See also §8.10 on prior-data conflict.

We consider the rats example using this method, but without full cross-validation, so that the predictive prior $p(\theta_i|y_{\setminus i})$ is approximated by $p(\theta_i|y)$. This procedure requires duplication of the data in order to contrast fixed-effect estimates with posterior predictive distributions, but the nuisance parameter $\sigma^2 = 1/\tau$ should only be estimated from one set of data. We therefore make use of the "cut" function (see §9.4) so that information about τ can feed through to the fixed-effect estimates without "double counting."

Example 10.7.4. *Rats: checking a longitudinal growth model using conflict p-values*
We assess the rats' growth model using replications of both intercept and gradient, in which (approximate) predictive replicates generated conditionally on all observed data are contrasted with fixed-effects replicates generated conditionally only on the data for that rat. Uniform priors are assumed for α_{i1} and α_{i2}, and σ^2 needs to be estimated from the entire data.

```
for (i in 1:N) {
  for (j in 1:T) {
```

```
   . . .
   Y.fix[i,j]      <- Y[i,j]     # duplicate data
   Y.fix[i,j]       ~ dnorm(mu.fix[i,j], tau.fixed)
   mu.fix[i,j]     <- alpha.fix[i,1] +
                      alpha.fix[i,2]*(x[j] - mean(x[]))
   }
   ## priors for fixed effects parameters
   alpha.fix[i,1]   ~ dunif(-1000,1000)
   alpha.fix[i,2]   ~ dunif(-1000,1000)
   . . .
   P.alpha.fix[i,1] <- step(alpha.fix[i,1] - alpha.pred[i,1])
   P.alpha.fix[i,2] <- step(alpha.fix[i,2] - alpha.pred[i,2])
   }
   . . .
   ## prevent learning about tau from duplicate data
   tau.fixed <- cut(tau)
```

The resulting p-values are almost identical to those shown in the final column of Table 10.1. We note that this method did not require any knowledge of closed-form estimates and their sampling distributions.

10.8 Comparison of hierarchical models

As already illustrated in Example 10.3.3, model comparison in hierarchical models can be carried out using the tools developed in Chapter 8. Indeed the concepts of the "effective number of parameters" p_D and the Deviance Information Criterion (DIC) were developed for random effects models, where it is possible to apparently have more parameters than observations and simple counting of parameters is inappropriate. Essentially the hierarchical model can be thought of as an estimated common prior distribution for the random effect parameters, which are no longer independent and hence do not remove a full degree of freedom in the fitting procedure.

Example 10.8.1. *Hepatitis B (continued): measurement error*
Returning to the hepatitis B data from Example 10.3.2, we replace the model with the baseline measurement y_{i0} as a covariate,

$$\psi_{ij} = \alpha_i + \beta_i(t_{ij} - \bar{t}) + \gamma(y_{0i} - \overline{y_0}),$$

with a model that includes classical measurement error on y_{0i}. The predictor y_{0i} is replaced by its "true" value μ_{0i}, which is given a hierarchical prior.

$$\psi_{ij} = \alpha_i + \beta_i(t_{ij} - \bar{t}) + \gamma(\mu_{0i} - \overline{y_0}),$$
$$y_{0i} \sim \mathrm{N}(\mu_{0i}, \sigma^2), \quad \mu_{0i} \sim \mathrm{N}(\mu_\epsilon, \sigma_\epsilon^2).$$

```
psi[i,j] <- alpha[i] + beta[i]*(t[i,j] - tbar)
            + gamma*(mu0[i] - y0bar)
```

We obtain the following summary statistics and DIC statistics from WinBUGS.

node	mean	sd	MC error	2.5%	median	97.5%	start	sample
gamma	1.004	0.1597	0.004629	0.7242	0.9923	1.353	10001	50000
mu.alpha	6.134	0.1619	0.00212	5.819	6.135	6.453	10001	50000
mu.beta	-1.064	0.1406	0.005251	-1.336	-1.065	-0.789	10001	50000

	Dbar	Dhat	pD	DIC
y	815.571	719.083	96.488	912.059
y0	303.517	250.644	52.873	356.389
total	1119.090	969.727	149.361	1268.450

With y_{0i} as a direct covariate, but with the same model for y_{0i} in terms of μ_{0i}, we obtain the following:

node	mean	sd	MC error	2.5%	median	97.5%	start	sample
gamma	0.675	0.08526	0.001505	0.5055	0.6752	0.8403	10001	50000
mu.alpha	6.137	0.1518	7.699E-4	5.839	6.136	6.434	10001	50000
mu.beta	-1.056	0.1466	0.006404	-1.339	-1.056	-0.7769	10001	50000

	Dbar	Dhat	pD	DIC
y	814.684	716.752	97.932	912.616
y0	300.487	226.607	73.879	374.366
total	1115.170	943.359	171.811	1286.980

The DIC for the data y alone is very similar between the two models. However, the measurement error model gives improved explanation for the baseline data y_{i0} — in this model we learn about μ_{i0} both directly from y_{i0} and indirectly from the rest of the data, via the inclusion of μ_{i0} in the regression model for $y_{ij}: j > 0$. Also the posterior estimate of $\gamma \approx 1$ from the measurement error model is more consistent with scientific belief (Coursaget et al., 1991).

10.8.1 "Focus": The crucial element of model comparison in hierarchical models

Consider the basic hierarchical model outlined in Figure 10.1 (left), which expresses the assumption that

$$p(y, \theta, \phi) = p(y|\theta)p(\theta|\phi)p(\phi)$$

There are three broad approaches that could be taken to comparing alternative models with this structure. These are compared below according to their construction and their computational ease, and then we consider their interpretation.

- Deviance Information Criterion

$$\text{DIC} = D(\bar{\theta}) + 2p_D$$

  which is based on $p(y|\theta)$ and is trivial to compute.

- Akaike Information Criterion

$$\text{AIC} = -2\log p(y|\hat{\phi}) + 2p_\phi$$

  where p_ϕ is the number of hyperparameters. This relies on being able to integrate out the θ's to give

$$p(y|\phi) = \int_\Theta p(y|\theta)p(\theta|\phi)d\theta.$$

- Bayesian Information Criterion

$$\text{BIC} = -2\log p(y|\hat{\phi}) + p_\phi \log n$$

  which is an approximation to $-2\log p(y)$, where

$$p(y) = \int_\phi p(y|\phi)p(\phi)d\phi.$$

  This depends on a proper prior formulation and is generally computationally difficult to evaluate (see §8.7).

The crucial observation is that the "likelihood" is not well defined in a hierarchical model. Is it $p(y|\theta)$, $p(y|\phi)$, or $p(y)$? The three criteria use different definitions, according to what can be thought of as the "focus" of the analysis: either θ, ϕ, or the model structure without any unknown parameters. So it is not a matter of which model comparison criterion is "correct," but which is appropriate for the purposes intended.

One way to identify the focus is to think of the prediction problem of interest, since prediction is not well defined in a hierarchical model without stating the focus, which is essentially what remains fixed when making predictions.

For example, suppose the three levels of our model concerned classes within schools within a country. Then:

- If we were interested in predicting results of future *classes* in those actual schools, then θ is the focus and deviance-based methods such as DIC are appropriate. This would be the case if we wanted, for example, to create "league tables" of schools according to their likely results on future classes.

- If we were interested in predicting results of future *schools* in that country, then ϕ is the focus and marginal-likelihood methods such as AIC are appropriate; this would be the relevant procedure if we wanted to make statements about the education in the whole country.

- If we were interested in predicting results for a new *country*, then no parameters are in focus, and Bayes factors are appropriate to compare models — this would be the case if we wanted to make very general statements about education in the whole world, outside the specific country being studied.

This strongly suggests that Bayes factors may in almost all circumstances be inappropriate measures by which to compare hierarchical models.

10.9   Further resources

Gelman and Hill (2007) is a practically focused book on regression and hierarchical modelling, which discusses model development, checking, and presentation, often in a Bayesian framework. Causal inference is a topic covered there which we have not mentioned here. Hierarchical models are also discussed in the more general book on Bayesian analysis by Gelman et al. (2004). Congdon (2010) describes a variety of hierarchical Bayesian models and includes WinBUGS implementations of many examples.

MLwiN (Rasbash et al., 2009) is a robust and well-featured software package for multilevel/hierarchical modelling with as long a history as BUGS. It implements many more complex models than those discussed in this chapter, including models for non-nested or cross-classified data structures (for example, children in the same school who live in different neighbourhoods) and data where lower-level units belong to more than one higher-level unit (for example, a child who changes schools). Many of these are implemented from a Bayesian perspective using MCMC sampling (Browne, 2009). The underlying principles behind a wide variety of models such as these are covered in detail by Goldstein (2010), with many applications in social and medical sciences.

11

Specialised models

Perhaps the greatest strength of BUGS is its flexibility — the language can represent statistical models and data structures of arbitrary complexity. We now have the tools and skills to tackle a theoretically limitless range of models, knowing the basics of Bayesian analysis and MCMC computation (Chapters 1–5), the Bayesian view of the standard linear and generalised linear models pervasive in applied statistics (Chapters 6–7), how to assess and compare models (Chapter 8), and how BUGS can deal with the common complications of real data analysis (Chapter 9). Chapter 10 discussed hierarchical models, which are naturally suited to the graphical modelling principles of BUGS, and now we present an overview of many other specialised applications. Of course, the flexibility of BUGS comes at the cost of learning the many idiosyncrasies of the language and its practical limitations, and we hope to illustrate these issues in the examples here.

11.1 Time-to-event data

Data representing times $t_1, \ldots, t_n$ until the occurrence of an event are often known as *survival* data (Kalbfleisch and Prentice, 2002). They are characterised by *censored* observations — event times which are only known to be within a certain range. For example, if patient i was alive at the end of the study follow-up, their time of death t_i is known to be greater than their last follow-up time c_i — this is called *right-censoring*.

In WinBUGS, censored survival data are usually modelled by using the I() construct (see §9.6) to indicate that a censored event time t[i] is drawn from the same model as the observed data, but it is unobserved and known only to be greater than c[i]. For example, consider n observations of exponentially distributed survival times t[i]:

```
for (i in 1:n) {
  t[i] ~ dexp(mu)I(c[i], )
}
```

Censored times are considered to be unknowns, in the same way as model parameters. If t[i] is observed, the observed value is supplied in the data for t[i], and the variable c[i] is set to 0, representing no constraint on the exponential distribution of t[i].

```
list(n=7,
     t=c(10, 13, 6, 14, 1, NA, NA, NA),
     cens=c(0, 0, 0, 0, 0, 15, 15, 15))
```

If the event time is censored, then t[i] is set to NA in the data, and c[i] contains the time of censoring. The I() construct indicates that the true t[i] could be any value greater than c[i].

In OpenBUGS the C() function is preferred to I() (see §12.5.1) and JAGS uses a different syntax (see §12.6.2).

11.1.1 Parametric survival regression

Standard survival models include a "location" parameter which can be expressed as a linear function of covariates. Usually this defines a *proportional hazards* (PH) model, where the *hazard*, or instantaneous risk of the event, $h(t) = f(t)/F(t)$, is proportional to covariate values: $h(t) = h_0(t)e^{\beta' x_i}$, or an *accelerated failure time* (AFT) model where the survival time T_i is such that $T_i \exp(\beta' x_i)$ has a fixed distribution. For Weibull-distributed survival times, $t_i \sim \text{Weibull}(r, \mu_i)$, the model $\log(\mu_i) = \alpha + \beta x_i$ is both PH and AFT.

Frailty models can be applied to grouped survival data, such as times to related events on the same individual, or groups j of related individuals. These include a random effect b_j representing any heterogeneity in survival between groups: $\log(\mu_{ij}) = \alpha + \beta x_i + b_j$, and can be expressed in BUGS just as any other hierarchical model (Chapter 10), for example,

```
log(mu[i]) <- a + b.age*age[i] + b.sex*sex[i] + b[group[i]]
for (j in 1:Ngroups) {
  b[j]        ~ dnorm(0.0, inv.omega.squared)
}
```

Semiparametric survival models are often used as a better-fitting alternative to simple parametric models such as the Weibull. The Cox regression model has hazard $h(t) = h_0(t)e^{\beta' x_i}$ proportional to covariates, but the baseline hazard $h_0(t)$ is nonparametric. A Bayesian analogue of this model can be implemented in BUGS, as described in §11.7.3.

Example 11.1.1. *Icelandic volcano eruptions: predicting event times*
In April 2010, the volcano Eyjafjallajökull in Iceland erupted. The resulting ash cloud was blown towards Western Europe and caused severe disruption to air travel for the following few weeks. A report into the eruption and its impact (UCL Institute for Risk and Disaster Reduction, 2010) reviewed how well the risk had been managed. One question was whether potentially more devastating eruptions from the larger neighbouring volcano Katla can be predicted from a recent eruption of

Eyjafjallajökull. The report provides the dates of all 18 eruptions of Katla since the year 1177, with a corresponding indicator of whether Eyjafjallajökull had also erupted within the previous year.

(a) Simple prediction. In this first model, we consider only the data on Katla. We fit a Weibull distribution to the times t[i] from one Katla eruption to the next, given the dates supplied as D[i]. Note we reparameterise the Weibull distribution in terms of $\sigma = (\mu)^{-1/r}$, so we can place an approximate Jeffreys prior on σ (Sun and Berger, 1994). We predict the time of the next Katla eruption, given the dates of all previous eruptions and the knowledge that it has not erupted from 1918 to 2010. In other words, we predict the next eruption interval t[19] given that it is censored at 92 years. The I() construct, as above, is used to indicate censored observations.

We also estimate the probability that Katla will erupt in the next t years, given that it hasn't erupted in the 92 years since 1918, for $t = 1, 5, 10$ or 50 years. This is $1 - S(t + 92)/S(92)$, where $S()$ is the survivor function, or 1 minus the cumulative density of the Weibull distribution.

```
for (i in 2:(Nk-1)) {
   t[i]       <- D[i] - D[i-1]
}
for (i in 2:Nk) {
   t[i]        ~ dweib(r, mu)I(cens[i],)
}
r               ~ dunif(0, 10)
sigma          ~ dgamma(0.001, 0.001)      # Jeffreys prior
mu             <- 1/pow(sigma, r)
median         <- sigma*pow(log(2), 1/r)   # median time to event
p.erupt.1  <- 1 - exp(pow(92/sigma,r) - pow((92+1)/sigma,r))
p.erupt.5  <- 1 - exp(pow(92/sigma,r) - pow((92+5)/sigma,r))
p.erupt.10 <- 1 - exp(pow(92/sigma,r) - pow((92+10)/sigma,r))
p.erupt.50 <- 1 - exp(pow(92/sigma,r) - pow((92+50)/sigma,r))

list(Nk=19,
D = c(1177, 1262, 1311, 1357, 1416, 1440, 1450, 1500, 1550,
      1580, 1612, 1625, 1660, 1721, 1755, 1823, 1860, 1918, NA),
cens = c(0,0,0,0,0,0,0,0,0,0,0,0,0,0,0,0,0,0,92))
```

A burn-in of 1000 and 5000 subsequent iterations gives the following summary statistics. There is a 95% credible interval of 92–148 for t[19], or 2010–2066 for the actual next eruption year. We estimate a substantial probability (0.03–0.15) that Katla will erupt within the next year following 2010 and a high likelihood of eruption (0.79–0.99) within the next 50 years.

node	mean	sd	MC error	2.5%	median	97.5%	start	sample
t[19]	106.2	16.22	0.3188	92.33	101.3	148.4	1000	5001

```
median       45.23   6.158    0.1093   33.42    45.17   57.87   1000  5001
p.erupt.1   0.07364 0.03261 5.453E-4 0.02759  0.06752 0.1525  1000  5001
p.erupt.5   0.3161  0.1157   0.002027 0.132    0.3009  0.5782  1000  5001
p.erupt.10  0.5286  0.1539   0.002845 0.2502   0.5212  0.8376  1000  5001
p.erupt.50  0.9647  0.05829  0.00123  0.7927   0.9892  1.0     1000  5001
```

(b) Prediction using a time-dependent covariate. In the second model, we try to improve prediction by including eruption of Eyjafjallajökull as a time-dependent covariate for the time t_i until the next Katla eruption. Again, we model t_i as Weibull, but this time we assume the hazard changes in periods when Eyjafjallajökull is erupting. We approximate this by assuming that there are instantaneous spikes in the hazard of Katla eruption at the recorded eruption times of its neighbour.

The likelihood for parameters $\boldsymbol{\theta}$ is a product of contributions from the survivor function from the 19 periods when Katla is inactive and contributions from the hazard at 18 eruption times. The 19th period ends in the year 2010, so that the 19th eruption time is censored. The survivor function for the Weibull distribution is $S(t) = \exp(-\mu t^r)$, and the instantaneous hazard is $h(t) = \mu(t)rt^{r-1}$, where $\mu(t) = \mu e^{\beta}$ at the instant of an Eyjafjallajökull eruption, and $\mu(t) = \mu$ otherwise; therefore the likelihood is

$$l(\boldsymbol{\theta}|\mathbf{t}) = \prod_{i=1}^{19} S(t_i|\boldsymbol{\theta}) \prod_{i=1}^{18} h(t_i|\boldsymbol{\theta}) = \prod_{i=1}^{19} \exp(-\mu t_i^r) \prod_{i=1}^{18} \mu \exp(\beta x_i) rt_i^{r-1}$$

This is written in the BUGS language using the "ones trick":

```
for (i in 2:19) {
   t[i]           <- D[i] - D[i-1]
   inactive[i]    <- 1
   erupt[i]       <- 1
}
# Likelihood contributions from inactive periods
for (i in 2:19) {
   inactive[i]     ~ dbern(p.inactive[i])
   p.inactive[i] <- exp(-mu*pow(t[i], r))
}
# Likelihood contributions at eruption times
for (i in 2:18) {
   erupt[i]        ~ dbern(q[i])
   q[i]          <- mu*r*pow(t[i], r-1)*exp(beta*eyja[i])
}
beta              ~ dnorm(0, 0.7) # weakly informative
rel.risk        <- exp(beta)
r                 ~ dunif(0, 10)
sigma             ~ dgamma(0.001, 0.001)
mu              <- 1/pow(sigma, r)
```

```
p.erupt.1      <- 1 - exp(mu*rel.risk*
                        (pow(92, r) - pow(92+1, r)))
p.erupt.50     <- 1 - exp(mu*rel.risk*
                        (pow(92, r) - pow(92+1, r)))*
                    exp(mu*(pow(92+1, r) - pow(92+50, r)))

list(
D= c(1177, 1262, 1311, 1357, 1416, 1440, 1450, 1500, 1550,
     1580, 1612, 1625,1660, 1721, 1755, 1823, 1860, 1918, 2010),
eyja = c(0,0,0,0,0,0,0,0,0,0,0,1,0,0,0,0,1,0,0,1))
```

Notes:

- The prior precision of $\tau = 0.7$ on the log relative risk β represents the belief that the risk is unlikely to be raised by more than a factor of 10, so that the prior standard deviation is about $2/\sqrt{\tau} = \log(10)$.

- As before, we wish to estimate the probability that Katla will erupt within 1 year, or within 50 years, from now. In calculating this, we assume that the hazard ratio e^β applies only in the year after the eruption of Eyjafjallajökull.

Even with a fairly conservative prior distribution on β, we obtain a posterior median of 17 (1.7, 181) for the relative hazard e^β of Katla eruption while the neighbouring volcano is active. The predictions of the next Katla eruption time are more pessimistic now this extra information has been included — we now estimate a 70% (10%–100%) probability of eruption within the next year.[*]

11.2 Time series models

There are many forms of model for time series data. They typically relate the response at time t, or its mean, to the preceding values of those (or related) quantities. For example, an *autoregressive* (AR) model relates successive responses y_t, $t = 1, ..., n$, via:

$$y_t = c + \sum_{i=1}^{p} \theta_i y_{t-i} + \epsilon_t, \quad t = p+1, ..., n, \tag{11.1}$$

---

[*]At the time this book was completed (May 2012) Katla still hadn't erupted again, though intense seismic activity was detected there in late 2011. The Grimsvotn volcano, in a different part of Iceland, erupted in May 2011.

with $\epsilon_t \sim \text{Normal}(0, \sigma^2)$, say. Here p is the order of the AR process, which is thus denoted AR(p). Note that we cannot specify the model in the form of (11.1) in BUGS, since data cannot be assigned a logical value. Instead we specify $y_t \sim \text{Normal}(m_t, \sigma^2)$, with $m_t = c + \sum_{i=1}^{p} \theta_i y_{t-i}$. Note, however, that m_t is undefined for $t = 1, ..., p$. We can circumvent this problem by specifying appropriate priors for the undefined m_ts, or for the corresponding residuals. In the latter case we may then define $m_t = y_t - \epsilon_t$.

A related model is the *moving average* (MA) model, which relates responses to previous values of the residuals, e.g.,

$$y_t = c + \sum_{i=1}^{q} \phi_i \epsilon_{t-i} + \epsilon_t, \quad t = q+1, ..., n.$$

This is denoted MA(q), and the early time points are handled in the same way as for the AR model. AR(p) and MA(q) may be combined to form an ARMA(p, q) model:

$$y_t = c + \sum_{i=1}^{p} \theta_i y_{t-i} + \sum_{i=1}^{q} \phi_i \epsilon_{t-i} + \epsilon_t,$$

the use of which is illustrated in the following example, after first fitting an AR(1) process.

Example 11.2.1. *Sunspots*
Our data for this example count the number of sunspots observed each year from 1770 to 1869 (Anderson, 1971, p. 660). We begin by fitting an AR(1) process to the observed data:

```
# AR(1):
for (t in 1:n) {
   y[t]      ~ dnorm(m[t], tau)
   yr[t]   <- 1769 + t
}
for (t in 2:n) {
   m[t]    <- c + theta*y[t-1]
   eps[t] <- y[t] - m[t]
}
m[1]       <- y[1] - eps[1]
eps[1]      ~ dnorm(0, 0.0001)
theta       ~ dnorm(0, 0.0001)
c           ~ dnorm(0, 0.0001)
tau        <- 1/pow(sigma, 2)
sigma       ~ dunif(0, 100)
```

The resulting model fit is shown in Figure 11.1(a), and the main parameter estimates are given in the BUGS output below. The model fit is quite good, considering there are only four parameters, but it can be improved by including more terms in the model. For example, an ARMA(2,1) model (code below) gives the fit shown in Figure 11.1(b); parameter estimates are also given in the BUGS output. Note the substantial reduction in sigma, the residual standard deviation, and that the DIC is reduced from 903 to 832 when fitting the ARMA(2,1) model.

```
# ARMA(2,1):
for (t in 1:n) {
   y[t]       ~ dnorm(m[t], tau)
   yr[t]      <- 1769 + t
}
for (t in 3:n) {
   m[t]       <- c + theta[1]*y[t-1] + theta[2]*y[t-2]
                 + phi*eps[t-1]
   eps[t]     <- y[t] - m[t]
}
m[1]          <- y[1] - eps[1]
m[2]          <- y[2] - eps[2]
eps[1]        ~ dnorm(0, 0.0001)
eps[2]        ~ dnorm(0, 0.0001)
for (i in 1:2) {
   theta[i]   ~ dnorm(0, 0.0001)
}
phi           ~ dnorm(0, 0.0001)
c             ~ dnorm(0, 0.0001)
tau           <- 1/pow(sigma, 2)
sigma         ~ dunif(0, 100)
```

AR(1) parameter estimates:

node	mean	sd	MC error	2.5%	median	97.5%	start	sample
c	8.539	3.506	0.03699	1.436	8.567	15.35	501	9500
sigma	21.81	1.597	0.01637	18.97	21.72	25.26	501	9500
theta	0.8113	0.05886	5.718E-4	0.6976	0.8103	0.9295	501	9500

ARMA(2,1) parameter estimates:

node	mean	sd	MC error	2.5%	median	97.5%	start	sample
c	15.89	3.473	0.04195	9.321	15.85	22.83	4001	6000
sigma	15.11	1.135	0.01632	13.12	15.03	17.51	4001	6000
theta[1]	1.204	0.1189	0.003421	0.9601	1.207	1.429	4001	6000
theta[2]	-0.5408	0.1139	0.003274	-0.7561	-0.5444	-0.3036	4001	6000
phi	0.4013	0.1329	0.005149	0.1205	0.41	0.6383	4001	6000

Another type of time series model is the *hidden Markov model* (HMM), which relates the *mean*, say, of the response to preceding values of the mean. We illustrate the use of such a model in the example below. We emphasise that while the BUGS language can accommodate many time series models, a Gibbs

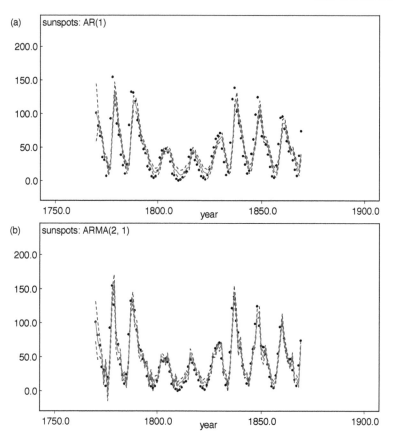

FIGURE 11.1
Model fits for sunspot data: (a) AR(1) model, (b) ARMA(2,1) model. Solid
line: posterior mean of `m[t]`; dashed lines: 95% interval; dots: observed data.

sampling scheme is often not the best approach to parameter estimation,
due to high posterior correlations induced by the nature of the model. With
HMMs in particular, convergence can be improved by specifying the model in
a certain way. For example, suppose we wish to assume $y_t \sim \text{Normal}(m_t, \sigma^2)$,
$t = 1, ..., n$, with

$$m_t = f(m_{t-1}) + u_t, \quad t = 2, ..., n,$$

and $u_t \sim \text{Normal}(0, \omega^2)$, say, along with appropriate priors for σ and ω. An
alternative, but equivalent, specification is $m_t \sim \text{Normal}(f(m_{t-1}), \omega^2)$, which
typically leads to much better mixing in the simulated Markov chains. Such
so-called *hierarchical centering* works in many settings (Gelfand et al., 1995),
not just in time series analysis. With HMMs, it may be the only option anyway,

due to one of BUGS' many idiosyncracies. We do not fully understand why, but in cases where the function $f(.)$ contains more than one instance of m_{t-1}, e.g., $f(m_{t-1}) = \beta m_{t-1}/(1 - m_{t-1})$, WinBUGS will "hang," when building the internal representation of the model, with anything more than a few data $(n > 20, \text{ say})$.

Example 11.2.2. *Tuna*

Meyer and Millar (1999) consider tuna (albacore) stocks in the South Atlantic between 1967 and 1989. Denoting the natural logarithm of the "catch per unit effort" (kg per 100 hooks) by y_t, the authors propose the following model:

$$y_t \sim \text{Normal}(\log(qKP_t), \sigma^2), \quad t = 1, ..., n = 23,$$

where q, K, and P_t denote the "catchability parameter," "carrying capacity" of the environment, and biomass in year t expressed as a proportion of the carrying capacity, respectively. In addition, the biomass dynamics are given by

$$\log P_t = f(P_{t-1}) + u_t, \quad f(P_{t-1}) = \log\left[P_{t-1} + rP_{t-1}(1 - P_{t-1}) - \frac{C_{t-1}}{K}\right],$$

where $u_t \sim \text{Normal}(0, \omega^2)$, r is the "intrinsic growth rate," and C_{t-1} denotes the total catch, in kilotonnes, during year $t - 1$. To avoid the issues discussed above, we will express this, instead, as

$$\log P_t \sim \text{Normal}(f(P_{t-1}), \omega^2), \quad t = 1, ..., n.$$

BUGS code incorporating Meyer and Millar's informative priors for q, K, r, σ^{-2}, and ω^{-2}, and the assumption $f(P_0) = 0$, is given below; the model fit is shown in Figure 11.2.

```
for (t in 1:n) {
    y[t]               ~ dnorm(mu[t], inv.sigma.squared)
    yr[t]              <- t + 1966
    mu[t]              <- log(q*K*P[t])
    log(P[t])          <- LP[t]
    LP[t]              ~ dnorm(f[t], inv.omega.squared)
}
f[1]                   <- 0
for (t in 2:n) {
    f[t]               <- log(P[t-1] + r*P[t-1]*(1 - P[t-1])
                          - C[t-1]/K)
}
q                      <- 1/inv.q
log(K)                 <- log.K
log(r)                 <- log.r
sigma                  <- 1/sqrt(inv.sigma.squared)
```

```
omega                <- 1/sqrt(inv.omega.squared)
inv.q                ~ dgamma(0.001, 0.001)I(0.5, 100)
log.K                ~ dnorm(5.043, 3.760)I(2.303, 6.908)
log.r                ~ dnorm(-1.380, 3.845)I(-4.605, 0.1823)
inv.sigma.squared ~ dgamma(1.709, 0.008614)
inv.omega.squared ~ dgamma(3.786, 0.01022)
```

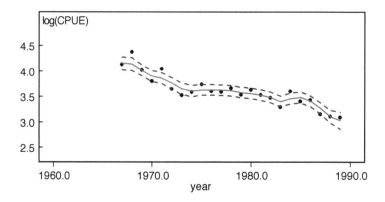

FIGURE 11.2

Model fit for tuna catch data showing log catch per unit effort (CPUE) against time. Solid and dashed lines show posterior mean and 95% interval, respectively, for mu[t]; dots show observed data y[t].

node	mean	sd	MC error	2.5%	median	97.5%	start	sample
K	275.5	60.24	2.227	183.6	266.1	418.8	4001	100000
omega	0.05383	0.01472	2.334E-4	0.03371	0.05106	0.09079	4001	100000
q	0.2425	0.05435	0.002133	0.1475	0.239	0.3611	4001	100000
r	0.2976	0.08411	0.00295	0.1483	0.2932	0.4757	4001	100000
sigma	0.1086	0.01939	1.755E-4	0.07575	0.1067	0.152	4001	100000

11.3   Spatial models

While time series models express the correlations between quantities measured at different points in time, *spatial* models are used for quantities which vary over two dimensions. These are used extensively in the ecological and environmental sciences (Ripley, 2004; Bivand et al., 2008) and in medical image analysis (Penny et al., 2006). Epidemiologists also study how the risk of disease varies systematically over areas, or with spatially varying predictors such as socio-economic characteristics or environmental exposures (Elliott et al., 2000; Lawson et al., 2003).

The most flexible way of modelling spatial dependence is via a spatially structured random effects distribution in a hierarchical model, with observations assumed conditionally independent given the random effects. Some spatial models can also be used directly as sampling distributions. Let $S = (S_1, S_2, \ldots, S_n)$ be a vector of random variables associated with locations $i = 1, \ldots, n$. WinBUGS and OpenBUGS contain a substantial set of features for implementing spatial models for S and for mapping posterior estimates, designated "GeoBUGS" and collected in the `Map` menu in the Windows interface.[†] The particular choice of model to use depends in part on whether the locations are areas (lattice data) or spatial coordinates (point data), whether the goal of the analysis is estimation or prediction, and whether we wish to model dependence via a jointly specified distribution $p(S)$ or a set of conditionally specified distributions $p(S_i | S_{\setminus i})$.

11.3.1 Intrinsic conditionally autoregressive (CAR) models

Conditionally autoregressive (CAR) models specify how each S_i is related to the S_j at all other locations via a set of univariate conditional distributions. One of the most commonly used formulations (Besag et al., 1991) is

$$
S_i | S_{\setminus i} \sim \text{Normal}\left(\sum_{j \neq i} \frac{w_{ij} S_j}{w_{i+}}, \frac{\omega^2}{w_{i+}} \right) \tag{11.2}
$$

where w_{ij} are weights used to express spatial dependence between locations i and j, with $w_{ij} = w_{ji}$, $w_{ii} = 0$ and $w_{i+} = \sum_j w_{ij}$. Thus the conditional mean (and mode) of S_i is a weighted average of the other S_js. This model is available in BUGS as

```
S[1:n] ~ car.normal(adj[], weights[], num[], inv.omega.squared)
```

(see §11.3.2 for explanation of `adj`, `weights`, and `num`). A heavier-tailed double exponential model is available as `car.l1`, which leads to the conditional mode of S_i being a weighted median of other S_js. A simple and widely used choice of weights for areal (lattice) data is $w_{ij} = 1$ if i and j are adjacent areas, and $w_{ij} = 0$ otherwise, which is usually sufficient for regions with regularly spaced, similarly sized areas. This is an example of a *Markov random field* in which

---

[†]All references to WinBUGS in this section also include OpenBUGS, though none of the spatial modelling or mapping facilities described here are currently available in JAGS.

each S_i is independent of others conditionally on the S_j of its neighbours. However, any pair of (S_i, S_j) will be marginally correlated due to the chain of dependencies on the paths linking them.

Model (11.2) and its double exponential equivalent are termed *intrinsic* CAR models, where the joint distribution of S is not explicitly defined. The implied joint distribution is in fact improper with an undefined mean and so cannot be used as a sampling distribution for data. However, if used as a prior, these models will still lead to a proper posterior distribution for S. A trick to remove the impropriety is to constrain the S_i $(i = 1, \ldots, n)$ to sum to zero (automatically imposed in the car.normal and car.l1 distributions). Whilst this still does not constitute a useful model for data (constraining the sampling distribution of a set of observations to have zero mean is not meaningful in general), it does allow car.normal or car.l1 to be used as zero-mean random effects distributions in a hierarchical regression model with a separate intercept term, or in combination with other sets of random effects. If an improper uniform prior (denoted dflat in BUGS) is specified for the intercept term, then the joint prior distribution for the intercept and constrained CAR random effects is equivalent to an intrinsic CAR prior on the unconstrained random effects.

Note that the elements of S form an *undirected* acyclic graph; however, their posterior distribution can be sampled within the BUGS graphical modelling framework (§4.2.2) by considering S as a multivariate node within the overall directed acyclic graph, with elements constrained to sum to zero.

The intrinsic CAR models can also be used for temporal smoothing of time series data, as a random walk prior for latent parameters treated as if defined on a single collapsed spatial dimension (see the GeoBUGS manual for an example).

11.3.2 Supplying map polygon data to WinBUGS and creating adjacency matrices

Information about spatial locations for areal data is typically provided as a polygon file giving geographical coordinates of the boundaries of each area. Various polygon map file formats (labelled "Splus," "ArcInfo," and "Epimap") can be imported directly into WinBUGS using the Import option in the Map menu (see GeoBUGS manual). The R package maptools (Bivand et al., 2008) also enables ArcView shapefiles (ESRI, Ltd.) to be imported and can manipulate map data exported from WinBUGS. The Map->Mapping Tool dialog box in WinBUGS can then be used to draw maps of data or posterior summary statistics from a fitted model (as in Figure 11.3).

For the car.normal and car.l1 distributions, spatial *adjacency* information should be supplied with the rest of the data required by the model. The vector adj[] is a collapsed vector for each area i indicating its neighbours, that is, which areas j have non-zero weights in the CAR model (Equation 11.2). num[] is a vector of length n indicating the number of neighbours for each area and is used to determine which entries of `adj[]` correspond to which area. For example,

```
adj = c(19, 9, 5, 10, 7, ...), num = c(3, 2, ...)
```

indicates area 1 has 3 neighbours (areas 19, 9, and 5), area 2 has 2 neighbours (areas 10 and 7), and so on. The vector `weights[]` is the same length as `adj[]` and provides the weights w_{ij} in model (11.2) that are associated with each pair of areas. For the simple CAR model described in §11.3.1, this will be a vector of 1's. Other choices of weights are possible, but the user must ensure that the specified weights are symmetric (i.e., $w_{ij} = w_{ji}$). Given polygon data supplied to WinBUGS in map format, `num[]` and `adj[]` may be created by the `adj matrix` option of the `Map->Adjacency Tool` dialog box in WinBUGS.

Example 11.3.1. *Mapping lip cancer in Scotland*

"Disease mapping" is the production of area-level summaries of disease outcomes. A simple measure of area-level disease risk is the standardised morbidity or mortality ratio (SMR). This is defined as Y_i/E_i, where Y_i is the observed number of disease cases and E_i is the expected number of cases given the age-specific population distribution in the area. However this is unhelpful for small areas and rare diseases, when sampling variability will obscure any systematic differences in disease risk between areas. A common method of distinguishing true risk variations from random noise is to "smooth" the data using a spatial hierarchical model.

Here we use data from Clayton and Kaldor (1987) on the numbers of cases `Y[i]` of lip cancer registered in each of $i = 1, \ldots, 56$ counties of Scotland during the 6 years from 1975 to 1980. The aim is to estimate the underlying relative risk `RR[i]` of lip cancer in each area i compared to the expected risk, after accounting for random variability. The expected count `E[i]` in area i, calculated from the age-specific population distribution in area i and overall age-specific lip cancer rates in Scotland, is included as a constant offset on the log scale. The covariate `X[i]` represents the percentage of the population occupied in agriculture, fishing, or forestry in county i and is used as a proxy for the population average exposure to sunlight — a known risk factor for lip cancer. An area-level random effect `S[i]` with an intrinsic normal CAR prior is included to spatially smooth the relative risks. `S[i]` is often interpreted as capturing the effects of unobserved spatially structured latent covariates.

The posterior variance of the random effects can be sensitive to the prior for ω^{-2}, as explained in §10.2.3. A Gamma(ϵ,ϵ) prior with ϵ small, for example, would be inappropriately biased away from zero in situations where spatial dependence is negligible. Instead, following Kelsall and Wakefield (1999), we use a Gamma(0.5, 0.0005) prior, equivalent to the belief that the random effect standard deviation is centered around 0.05 with a 1% prior probability of being smaller than 0.01 or larger than 2.5.

```
for (i in 1:n) {
```

```
   Y[i]                ~ dpois(mu[i])
   log(mu[i])         <- log(E[i]) + alpha0 + alpha1*X[i]/10 + S[i]
   RR[i]              <- exp(alpha0 + alpha1*X[i]/10 + S[i])
}
S[1:n]                 ~ car.normal(adj[], weights[], num[],
                                    inv.omega.squared)
for(k in 1:sumNumNeigh) { # sumNumNeigh = length of adj[],
   weights[k]        <- 1     # obtained from Map->Adjacency Tool
}
alpha0                 ~ dflat()
alpha1                 ~ dnorm(0.0, 1.0E-5)
rr.x                  <- exp(alpha1)
inv.omega.squared ~ dgamma(0.5, 0.0005)
```

Care is needed when specifying initial values for S []. These cannot be generated from the prior since it is improper, so if the gen inits option is selected in WinBUGS or OpenBUGS, initial values of zero will be assigned to all elements of S []. It is therefore preferable to provide user-specified initial values for S [] if multiple chains are being run to assess convergence. Note also that three of the regions in Scotland consist of islands that have no "neighbours," and so the conditional distributions (11.2) are not defined for these areas. BUGS assigns a constant value of zero to any element of S [] that has no neighbours and does not update these during the MCMC simulation. Such elements must be assigned a value NA in the initial values file.

Figure 11.3 maps the observed SMRs (left) and the posterior means of the spatially smoothed estimates of the relative risk of lip cancer (middle), which includes effects due to the occupational factors that were adjusted for. The geographical variation in the latter is smoother, reflecting the assumption of spatial dependence imposed by the random effects and separation of the between-area (random effects) variation from the random Poisson variation.

A feature of the intrinsic CAR model is that spatial structure is imposed a priori and there is no Bayesian *learning* about the strength of spatial dependence in the data. Besag et al. (1991) propose an alternative formulation (commonly referred to as the convolution or "BYM" model, after its authors) that includes a second unstructured random effect H[i] to capture effects of unobserved unstructured latent covariates. In practice, one or the other of S[i] or H[i] will often dominate the other, but which one will not usually be known in advance. Bayesian learning about the extent of spatial versus unstructured variation in the data can then be based on the posterior proportion of the total between-area variation in log relative risks explained by each random effect (denoted pS and pH, respectively, in the code below). We also estimate the proportion of variance of the log relative risks which is explained by the population age structure (pE), and by the covariate (pX).

Another useful summary of the between-area variation in relative risks is the quantile ratio (QR), which quantifies the spread of the empirical distribution of

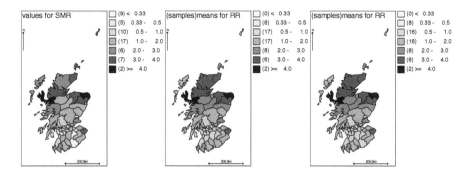

FIGURE 11.3
Left: Standardised mortality ratios for lip cancer in Scottish counties. Middle: smoothed using spatial random effects. Right: smoothed using spatial and exchangeable effects.

the random effects. For example, the 80% QR is calculated by ranking the random effects across areas and taking the exponentiated difference between the random effects in areas ranked at the 90% and 10% quantiles. For the current example, the 80% QR may be calculated based on the quantiles of the total relative risk, including the effects of observed covariates (QR80), and/or the residual relative risk after adjusting for observed covariates (resQR80). In BUGS this is calculated by using ranked() to find the 6th and 51st out of the 56 areas in order of (residual) relative risk, corresponding to the 10% and 90% quantiles.

```
for (i in 1:n) {
   ...
   log(mu[i])      <- log(E[i]) + alpha0 + alpha1*X[i]/10
                      + S[i] + H[i]
   RR[i]           <- exp(alpha0 + alpha1*X[i]/10 + S[i] + H[i])
   residRR[i]      <- exp(S[i] + H[i]) # residual RR
   H[i]             ~ dnorm(0, inv.omega.sq.h)
   X.pred[i]       <- alpha1*X[i]/10
   lE[i]           <- log(E[i])
}
...
inv.omega.sq.h   ~ dgamma(0.5, 0.0005)
sdS              <- sd(S[])
sdH              <- sd(H[])
sdX              <- sd(X.pred[])
sdE              <- sd(lE[])
sumvar           <- sdS*sdS + sdH*sdH + sdX*sdX + sdE*sdE
pS               <- sdS*sdS/sumvar
pH               <- sdH*sdH/sumvar
pX               <- sdX*sdX/sumvar
```

```
pE                  <- sdE*sdE/sumvar
QR80                <- ranked(RR[], 51)/ranked(RR[], 6)
resQR80             <- ranked(residRR[], 51)/ranked(residRR[], 6)
```

node	mean	sd	MC error	2.5%	median	97.5%	start	sample
QR80	7.555	1.297	0.01185	5.412	7.424	10.46	1001	40000
pE	0.6238	0.03673	5.298E-4	0.549	0.6246	0.6931	1001	40000
pH	0.0182	0.03086	0.001449	1.703E-4	0.003441	0.1125	1001	40000
pS	0.2718	0.05497	0.001634	0.1559	0.2733	0.3773	1001	40000
pX	0.08619	0.04071	9.25E-4	0.01754	0.0829	0.1749	1001	40000
resQR80	4.934	0.9109	0.01352	3.481	4.818	7.016	1001	40000
rr.x	1.587	0.1907	0.004348	1.236	1.58	1.986	1001	40000

The resulting mapped relative risks (Figure 11.3, right) are very similar to those from the previous model, suggesting that there is negligible extra non-spatial variation. From the posterior means under this model, 62% of the variation in log disease rates is associated with the age structure of the population and 9% with the occupational characteristics. The estimated relative risk of lip cancer is 1.59 (1.24, 1.99) for an increase of 10% in the proportion of the population working in agriculture, fishing, or forestry. The remainder of the variation is associated mainly with unobserved spatially dependent factors (27%), with less than 2% of the variation attributable to unobserved non-spatial factors. The quantile ratio suggests large heterogeneity (7.6-fold variation across the central 80% of areas) in the risk of lip cancer across Scotland. Occupational characteristics explain some of this heterogeneity, but nearly 5-fold variation across areas remains unexplained according to the posterior mean of resQR80.

Adjustments to deal with potential biases are often desirable in geographical epidemiological studies of this kind. As discussed by Best et al. (2001) and Richardson and Best (2003), the Bayesian graphical modelling approach allows, for example, measurement error in potential predictors to be accounted for, along with *ecological bias* caused by ignoring within-area variations in the predictor (see Example 11.4.1). Each model elaboration is typically implemented with a few extra lines of BUGS code.

11.3.3 Multivariate CAR models

The intrinsic CAR model can be used to define spatially correlated random effects for any form of response model. The Gaussian version of this model generalises straightforwardly to multivariate random effects. A p-dimensional set of spatially correlated effects can be modelled by replacing the univariate conditional distribution (Equation 11.2) with a multivariate conditional distribution. The multivariate normal intrinsic CAR model is implemented as

```
S[1:p,1:n] ~ mv.car(adj[], weights[], num[], inv.Omega[,])
```

where `inv.Omega[]` is the precision matrix, the multivariate equivalent of ω^{-2}. As in the univariate case, spatial dependence *between* areas is modelled via the adjacency weights and is assumed to be the same for all p sets of random effects. The *within*-area conditional correlation of the vector of p random effects for area i given the neighbouring effects is modelled via the off-diagonal terms in the inverse precision matrix.

11.3.4 Proper CAR model

The *proper* CAR model introduces an additional parameter γ, often termed an autocorrelation coefficient, into the conditional specification (Model 11.2). Provided γ satisfies certain constraints, this leads to a joint distribution for S that is uniquely defined:

$$S \sim MVN_n(\mu, \sigma^2(I - \gamma C)^{-1}M),$$

where C is an $n \times n$ normalised weight matrix with elements $C_{ij} = w_{ij}/w_{i+}$ (with w defined as in §11.3.1), reflecting the spatial dependence between areas i and j, M is an $n \times n$ diagonal matrix with elements $M_{ii} = 1/w_{i+}$, and γ controls the overall strength of dependence. This model is implemented in WinBUGS as

```
S[1:n] ~ car.proper(mu[], C[], adj[], num[], M[], tau, gamma)
```

To define a proper joint distribution, `gamma` must be bounded by the inverse of the maximum and minimum eigenvalues of $M^{-\frac{1}{2}}CM^{\frac{1}{2}}$, which can be calculated using the `min.bound` and `max.bound` functions in BUGS (see the GeoBUGS manual). In practice, `gamma` often needs to be very close to its upper bound in order to represent even moderate spatial dependence and so the BYM model (see Example 11.3.1) is often preferred as a random effects prior that allows learning about the strength of spatial dependence. Unlike the intrinsic CAR, however, `car.proper` can be used as a distribution for continuous spatially correlated data, as well as a prior for random effects.

11.3.5 Poisson-gamma moving average models

An alternative to the CAR model (Ickstadt and Wolpert, 1998; Best et al., 2000a) expresses spatial correlation among a set of Poisson counts $Y_i \sim$ Poisson(μ_i) as a *convolution* or *moving average*. This assumes a set of independent gamma-distributed random effects γ_j representing the underlying risks in a (possibly different) partition of the study region indexed by j. The risk in area i is modelled as

$$\mu_i = A_i \sum_j k_{ij}\gamma_j,$$

where A_i is a scaling factor for area i (usually related to its size or population) and k_{ij} are a set of kernel weights which govern the contribution of the area

j random effect γ_j to the area i risk. The k_{ij} typically decrease as the distance between the areas increases. For example, Best et al. (2000b) employed a Gaussian kernel with $k_{ij} = 1/(2\pi\rho^2)\exp(-d_{ij}^2/(2\rho^2))$, where d_{ij} is the distance between the centroids of the areas. Best et al. (2000b) also discussed an advantage of this model over the log-linear CAR model discussed above — if the Poisson mean μ_i is modelled as a linear function of covariates, then the covariate effect is consistent under spatial aggregation, avoiding *ecological bias* (§11.4.1).

This convolution model is implemented in WinBUGS as

```
S ~ dpois.conv(mu[])
```

though the user must provide definitions of the k_{ij} and γ_j which the elements of `mu[]` depend on. See the GeoBUGS manual.

11.3.6   Geostatistical models

While CAR models are typically used to model areal or lattice data, point-referenced spatial data are more often modelled using a geostatistical model. Such models have their origins in the kriging literature and can be used as a sampling distribution for continuous spatial data or as a spatial random effects distribution on an appropriately transformed scale (Diggle et al., 1998). In either case, let $S = (S_1, S_2, \ldots, S_n)$ be a vector of random variables associated with point locations $(x_i, y_i), i = 1, \ldots, n$. Then S can be modelled with a multivariate normal distribution $S \sim MVN_n(\mu, \Sigma)$ whose covariance matrix Σ is specified as a function of the distance d_{ij} between points i and j. This covariance matrix must be symmetric and positive-definite, and only certain parameterisations will guarantee this. BUGS implements one of the most common spatial covariance matrix parameterisations, the *powered exponential* function $\Sigma_{ij} = (1/\tau)\exp(-(\phi d_{ij})^\kappa)$, as

```
S[1:n] ~ spatial.exp(mu[], x[], y[], tau, phi, kappa)
```

where `x[]` and `y[]` contain the x and y coordinates of each point, which are used by BUGS to compute the distances d_{ij}. An alternative covariance function, `spatial.disc`, with correlation decreasing approximately linearly to zero at a specified distance, is described in the GeoBUGS manual. Spatial interpolation or prediction from these models (*kriging*) can be carried out in WinBUGS using the functions `spatial.pred` and `spatial.unipred`. The former jointly predicts all target locations simultaneously, whereas the latter carries out prediction site by site and so ignores correlation between the predictions at different locations. This tends to yield slightly wider prediction intervals (the predicted means will be the same for both versions) but is faster.

Note that none of the spatial models in BUGS allows a complete covariance matrix to be defined explicitly; only specific parameterisations which implicitly guarantee a symmetric positive definite covariance matrix are available.

Example 11.3.2. *Estimating radioactivity levels on Rongelap Island*

Rongelap Island in the Pacific Ocean experienced contamination due to fall-out from the Bikini Atoll nuclear weapons testing programme in the 1950s. Its former inhabitants have lived in self-imposed exile on another island since 1985. Diggle et al. (1998) used Bayesian kriging to estimate radioactivity levels on the island using data from a radiological survey conducted to establish whether Rongelap could be safely re-settled. We analyse a subset of the data covering the western half of the island, comprising photon emission counts Y[] attributable to radioactive caesium measured at each of 73 locations given by coordinate vectors x[] and y[]. A Poisson model is specified for Y[] with offset log(t[]), the (log) time over which the counts were recorded. The log emission rate is modelled as a latent Gaussian spatial field with exponential correlation function. The fitted model is used to predict emission rates at a further 93 locations with coordinate vectors x.pred[] and y.pred[], in order to estimate the location and level of maximum contamination on the island.

Weakly informative priors are recommended for the parameters of the spatial covariance matrix, since there is often little information in the data to estimate these. The choice may be guided by examination of the empirical variogram and by plotting correlation-distance decay curves simulated from the prior. Figure 11.4 (top left) shows a plot of $\exp(-(\phi d_{ij})^{\kappa})$ versus d_{ij} for values of $\phi \sim \text{Uniform}(0, 120)$ and $\kappa \sim \text{Uniform}(0.1, 0.95)$, the priors used by Diggle et al. (1998). We also use the same weakly informative uniform prior as Diggle et al. for the overall mean beta. However, in order to overcome convergence difficulties with this analysis, we specify an informative half-standard-normal prior for the standard deviation (sigma) of the spatial process.

```
for(i in 1:73) {
   Y[i]                  ~ dpois(lambda[i])
   log(lambda[i])        <- log(t[i]) + beta + S[i]
}
# spatial field representing true log contamination intensity
S[1:73]                  ~ spatial.exp(mu[], x[], y[],
                                        tau, phi, kappa)
for(i in 1:73) {
   mu[i]                 <- 0
}
# mean log contamination intensity
beta                     ~ dunif(-3, 7)
# priors on parameters of spatial covariance matrix
phi                      ~ dunif(0, 120)
kappa                    ~ dunif(0.1, 1.95)
sigma                    ~ dnorm(0, 1)I(0,)
tau                      <- 1/pow(sigma, 2)
for(j in 1:93) {                              # prediction
```

```
    T[j]                      ~ spatial.unipred(mu.pred[j], x.pred[j],
                                               y.pred[j], S[])
    exp.T[j]                  <- exp(T[j] + beta) # predicted intensity
    mu.pred[j]                <- 0
  }
  # combine observed and predicted locations
  for(i in 1:73)    {
    pred[i]                   <- exp(S[i])
  }
  for(i in 74:166) {
    pred[i]                   <- exp.T[i-73]
  }
  # max value of contamination
  max.level                   <- ranked(pred[], 166)
  for(i in 1:166) {                              # location of maximum
    pred.rank[i]              <- rank(pred[], i)
    prob.max[i]               <- equals(pred.rank[i], 166)
  }
  # prob that count/sec > 15 at location i
  for(i in 1:166) {
    prob.exceeds.15[i] <- step(pred[i] - 15)
  }
```

node	mean	sd	MC error	2.5%	median	97.5%	start	sample
beta	1.841	0.3141	0.014	1.228	1.856	2.491	10001	250000
kappa	0.9327	0.2239	0.002256	0.4973	0.9317	1.368	10001	250000
max.level	32.72	12.44	0.08155	18.93	29.82	63.6	10001	250000
phi	8.547	6.106	0.1143	1.008	7.58	21.0	10001	250000
sigma	0.8377	0.2211	0.004254	0.5677	0.787	1.413	10001	250000

As noted for some of the time series models, a Gibbs sampling updating scheme is often not the best approach for models such as this, due to high posterior correlations between parameters. Convergence of phi, sigma, kappa, and (in particular) beta is poor for this example, and long run lengths are needed. Posterior medians for these parameters differ somewhat from those obtained by Diggle et al. (1998) for the whole island, but posterior estimates of the maximum contamination are similar. Figure 11.4 maps the predicted photon emission rates for the western half of Rongelap (top right), the locations where there is at least a 5% posterior probability that the rate exceeds 15 counts per unit time (bottom left), and the posterior probability that each location is the maximum (bottom right). The latter two maps highlight areas with high levels of radioactivity that are of specific concern when considering the habitability of the island.

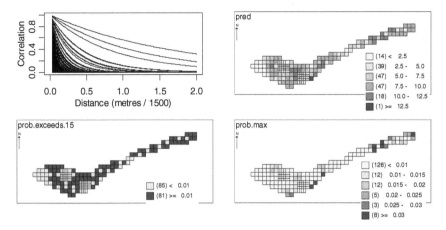

FIGURE 11.4
Top left: Prior simulations of correlation versus distance for the Rongelap example. Top right: Predicted photon emission rates (counts per unit time). Bottom left: Posterior probability that rate exceeds 15 counts per unit time. Bottom right: Posterior probability that each location is the maximum.

11.4 Evidence synthesis

"Evidence synthesis" can be broadly described as making inference on a quantity from more than one dataset at the same time. *Meta-analysis* is a typical example, where estimates of the same quantity obtained from different studies are combined. A BUGS graphical model forms a natural framework for evidence synthesis. If there are N datasets, each assumed to be generated by a different model, but with a parameter θ in common, then all the datasets simultaneously provide information about θ.

11.4.1 Meta-analysis

In random-effects meta-analysis, for example, suppose there are eight trials indexed by k, rt[k] deaths out of nt[k] patients in the treatment group and rc[k] deaths out of nc[k] patients in the control group. A different binomial model is assumed to generate each trial dataset; however, the models are linked hierarchically using an exchangeable normal model on the control log-odds phi[k] and log-odds ratios of treatment theta[k]. All the data simultaneously inform the parameters of interest — the "pooled" mean treatment effect mu and baseline mu.base.

```
for (k in 1:8) {
```

```
    rt[k]           ~ dbin(pt[k], nt[k])
    rc[k]           ~ dbin(pc[k], nc[k])
    logit(pt[k]) <- phi[k] + theta[k]
    logit(pc[k]) <- phi[k]
    theta[k]        ~ dnorm(mu, inv.omega.sq)
    phi[k]          ~ dnorm(mu.base, inv.omega.sq.base)
}
```

As the random effects variance 1/inv.omega.sq tends to zero, this tends to a "fixed effects" analysis where each study is assumed to estimate a common pooled effect theta. This common effect would be approximately the same if the theta[k] were estimated *independently* from each study k and the estimates were then pooled using a weighted average, as in a classical fixed-effects meta-analysis. In a random-effects meta-analysis, the advantage of a Bayesian approach is that uncertainty about the random effects variance is acknowledged in the pooled effect, unlike a classical analysis where the pooled estimate uses just a point estimate of the variance (DerSimonian and Laird, 1986). See Spiegelhalter et al. (2004) for more discussion.

This can be extended to *mixed treatment comparisons* involving several treatments, where different pairs of treatments may appear in different studies, but any pair can be connected by a chain of comparisons. The aim is to estimate relative effects between every pair of treatments using the whole network of studies, not just those which directly compare each pair. See, for example, Caldwell et al. (2005), and Lu and Ades (2004) for examples of WinBUGS code to implement these.

11.4.2   Generalised evidence synthesis

Spiegelhalter et al. (2004) discuss further generalisations of meta-analysis, including adjusting for bias when combining observational and randomised studies. Studies of a common design or with some other common aspect may be assumed to be exchangeable (§ 10.1), though prior assumptions may be important if there are few studies of a particular type.

Economic evaluations, such as those conducted by the National Institute for Health and Clinical Excellence in the U.K., are increasingly important to health policy making. These usually depend on complex combinations of evidence about the treatments and patients. The treatment effect is often estimated from a meta-analysis, while long-term costs and health effects under different policies are projected using a Markov or similar multi-state transition model (as in Example 7.2.1). Spiegelhalter and Best (2003) describe such a simultaneous evidence synthesis and cost-effectiveness analysis, with associated programs available in the hips example for WinBUGS 1.4. Demiris and Sharples (2006) synthesised survival data from a population register and a randomised trial to inform a health policy evaluation, providing WinBUGS code.

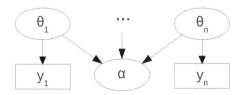

FIGURE 11.5

Generalised evidence synthesis. Evidence is required on a parameter α which is a function of more fundamental parameters $\theta_1, \ldots, \theta_n$, each of which generates observed data $y_1, \ldots, y_n$.

Typically in complex Bayesian evidence syntheses, evidence is required on a quantity α which is a function of more fundamental parameters $\theta_1, \ldots, \theta_n$. Each of the fundamental parameters θ_i defines a statistical model which generates observed data y_i, as illustrated by the directed acyclic graph in Figure 11.5. These have been used in epidemiology. Goubar et al. (2008) and Presanis et al. (2008), for example, estimate the prevalence of human immunodeficiency virus (HIV) infection in England and Wales. Multiple survey datasets are used to simultaneously estimate over 30 parameters or functions of parameters in WinBUGS, such as the underlying prevalence and probability of HIV diagnosis in selected samples with particular risk profiles. Conflicts or inconsistencies, where the model fitted well to some datasets but not others, were identified using the posterior mean deviance and deviance information criterion.

Example 11.4.1. *Limiting long-term illness: combining individual and aggregate data*

Here we give an example of combining evidence from observational data in different forms from different sources. Determining individual-level risk factors from data aggregated over areas is fraught with problems known as *ecological bias*. The effect of an area-level average predictor on the area-level average outcome is not necessarily the same as the effect of the individual predictor for an individual's outcome. Jackson et al. (2006) studied the benefit of combining aggregate and individual data to alleviate this bias and improve power. A model was fitted for the risk of self-reported limiting long-term illness (LLTI) among men aged from 45 to 59 years in London. Potential risk factors included non-white ethnicity, household income (continuous, on the log scale), and area-level deprivation as measured by the Carstairs index. Aggregate data (mainly from the census) were available from ngr = 255 electoral wards in London, supplemented by samples of survey data from n = 412 individuals.

To estimate the risk factors from a synthesis of *both datasets*, we simply assume they are generated by models with the same underlying individual-level parameters.

Individual-level data First we present the portion of the model representing the individual-level data llti. This is simply a logistic regression, which includes a random area-specific effect mu.base[ward[i]] to account for heterogeneity between areas due to unobserved factors.

```
for (i in 1:n) {
   llti[i]         ~ dbern(pzy[i])
   logit(pzy[i]) <- mu.base[ward[i]] + mu.scale
                    + a.carstairs*carstairs[ward[i]]
                    + a.nonwhite*nonwhite[i]
                    + a.lincome*lincome[i]
}
mu.scale          <- 0.7
```

A danger of combining evidence from different sources is potential inconsistency between the sources — and this was encountered here. There is known to be a higher rate of reporting limiting long-term illness in surveys compared to population censuses. A fixed "fudge factor" mu.scale was therefore added to the individual-level model, based on external data on this discrepancy. This could easily be extended with a prior distribution to represent uncertainty about mu.scale.

Aggregate data The second half of the model represents the number of individuals per ward N.llti reporting limiting long-term illness. This needs careful modelling to avoid ecological bias. Naively using a binomial logistic regression, with the area-level proportion of non-white residents p.nonwhite and mean log income mean.lincome as covariates, would yield the effects of the area-level average predictors on area-level outcomes, whereas we are interested in inferring effects of *individual-level* risk factors. Therefore the area-level binomial probability p_i is obtained by integrating the individual level model over the within-area distribution of the covariates. Firstly, we integrate over the binary covariate, non-white ethnicity,

$$p_i = q_{0i}(1 - \phi_i) + q_{1i}\phi_i$$

where ϕ_i is the proportion of non-white residents in area i (p.nonwhite[i]) and q_{0i}, q_{1i} are the probabilities of LLTI for white and non-white individuals, respectively (pw[i] and pnw[i]). q_{0i}, q_{1i} are obtained by integrating over the within-area distribution $g_i(x)$ of the continuous variable, log income, assumed normal with standard deviation s_i. Salway and Wakefield (2005) show that this integral is approximated by

$$q_{0i} = \int \text{expit}(\mu_i + \beta x)g_i(x)dx \approx \text{expit}\left\{(1 + c^2\beta^2 s_i^2)^{-1/2}(\mu_i + \beta m_i)\right\}$$
$$q_{1i} = \int \text{expit}(\mu_i + \alpha + \beta x)g_i(x)dx \approx \text{expit}\left\{(1 + c^2\beta^2 s_i^2)^{-1/2}(\mu_i + \alpha + \beta m_i)\right\}$$

where $c = 16\sqrt{3}/(15\pi)$, α, β are the log odds ratios for non-white and log income, respectively, and μ_i is the baseline log odds for area i, modelled as a normal random effect with a regression on the area's Carstairs deprivation index. Further details are given in Jackson et al. (2006).

```
for (i in 1:ngr) {
  N.llti[i]      ~ dbin(p[i], N.pop[i])
  p[i]          <- p.nonwhite[i]*pnw[i]
                 + (1 - p.nonwhite[i])*pw[i]
  logit(pw[i])  <- (mu.base[i] + a.carstairs*carstairs[i]
                 + a.lincome*mean.lincome[i])
                 / sqrt(1 + pow(c*a.lincome, 2)/tau.x[i])
  logit(pnw[i]) <- (mu.base[i] + a.carstairs*carstairs[i]
                 + a.nonwhite + a.lincome*mean.lincome[i])
                 / sqrt(1 + pow(c*a.lincome, 2)/tau.x[i])
  tau.x[i]      <- 1/pow(sd.lincome[i], 2)
  mu.base[i]     ~ dnorm(base.mu, base.tau)
}
c               <- 16*sqrt(3)/(15*pi)
pi              <- 3.141592654
```

Priors for shared parameters The "weakly informative" normal priors with mean 0 and precision 1.48 for the log odds ratios for income and ethnicity represent a 95% prior belief that the odds ratio is between 1/5 and 5. The logistic prior on the baseline log odds base.mu is equivalent to a uniform prior on the probability scale (§5.2.5).

```
a.carstairs    ~ dnorm(0, 0.1)
a.nonwhite     ~ dnorm(0, 1.48)
a.lincome      ~ dnorm(0, 1.48)
or.nonwhite   <- exp(a.nonwhite)
or.lincome    <- exp(a.lincome)
or.carstairs  <- exp(a.carstairs)
base.mu        ~ dlogis(0, 1)
base.tau       ~ dgamma(1, 0.01)
base.sig      <- 1/sqrt(base.tau)
```

After 30,000 iterations from two parallel chains, with the first 4000 from each discarded, we obtain the following summary statistics. Non-white residents are estimated to have a 43% increased odds of limiting long-term illness, and low household income is also a significant risk factor. Living in a deprived area, as measured by the Carstairs index, also seems to confer an additional risk of LLTI.

node	mean	sd	MC error	2.5%	median	97.5%
base.mu	-2.108	0.03027	0.001257	-2.171	-2.107	-2.052
base.sig	0.1843	0.01172	1.438E-4	0.1622	0.184	0.2082
or.carstairs	1.071	0.004904	1.836E-4	1.062	1.072	1.081
or.lincome	0.5617	0.04406	0.00158	0.4785	0.5608	0.6513
or.nonwhite	1.432	0.1649	0.007305	1.133	1.424	1.763

Using the individual-level data alone, there is insufficient power to detect significant associations with ethnicity. Using the aggregate data alone gives estimates which are very similar to those from the combined data — because in this case the individual data sample size is small. However, simulation studies (Jackson et al., 2006) show that individual data can improve inference from area-level data when the area-level data contain little between-area variation in average covariate values and thus little information about the relationship between covariate and outcome.

Riley et al. (2008) describe similar methods for combining individual and aggregate data in the context of meta-analysis, using shared parameters in WinBUGS.

11.5 Differential equation and pharmacokinetic models

In nonlinear regression, within the field of pharmacokinetics in particular, it is sometimes more convenient to express the regression function via differential equations. Consider, for example, the cadralazine regression function used in Example 10.4.1. For the jth plasma cadralazine concentration taken from individual i (at time t_{ij}):

$$\psi_{ij} = \frac{D}{V_i} \exp\left(-\frac{CL_i}{V_i} t_{ij}\right).$$

This is the *unique* solution to the following differential equation and initial condition, at time $t = t_{ij}$:

$$\frac{dC(t)}{dt} = -\frac{CL_i}{V_i} C(t), \quad C(t=0) = \frac{D}{V_i}, \tag{11.3}$$

where $C(t)$ denotes cadralazine concentration at time t. Hence we might specify the model in terms of (11.3) instead, as illustrated below. Why might we want to do this? In pharmacokinetics, the regression function is typically derived from a compartmental model, which represents the body as a number of compartments between which drug transfer occurs at various rates. The movement of drug between these compartments is naturally expressed in terms of differential equations, but, in general, we may not know (or want to derive) the analytic solution. Compartmental models are used in many areas, but there are other fields with alternative motives for expressing models in terms of differential equations. WinBUGS and OpenBUGS provide an interface, therefore, to facilitate such specification.

Example 11.5.1. *Cadralazine (continued): differential equations*
With *WinBUGS Differential Equation Interface* (WBDiff) installed,[‡] the likelihood
for our cadralazine example (10.4.1) can be specified in WinBUGS as follows:

```
for (i in 1:10) {
  for (j in offset[i]:(offset[i+1]-1)) {
    y[j]           ~ dnorm(psi[j, 1], inv.sigma.squared[i])
  }
  psi[offset[i]:(offset[i+1]-1), 1:dim]
              <- ode(init[i, 1:dim],
                     time[offset[i]:(offset[i+1]-1)],
                     D(C[i, 1:dim], t), origin, tol)
  D(C[i, 1], t) <- -CL[i]*C[i, 1]/V[i]
  init[i, 1]    <- D/V[i]
  ...
}
```

The ode(.) syntax represents a matrix-valued function that returns the numerical
solution to the differential equation(s) at the times specified, in this case the
elements of time[] between offset[i] and offset[i+1]-1. In OpenBUGS,
this facility is available without needing to install an add-on package.

Here there is only one differential equation, for a single quantity C (the underlying system is a one compartment model). More generally, however, there may be
dim equations for dim quantities. Each column of the output from ode(.) (psi[]
in this case) corresponds to one of the quantities being modelled, whereas rows
correspond to the times at which the solution is required. The various arguments
of ode(.) are described as follows: (i) a vector of dim initial conditions, one for
each of the quantities being modelled; (ii) the set of times at which the solution is
to be evaluated; (iii) the differential equations themselves, specified elsewhere in
the code via the D(, t) notation; (iv) the "origin" to which the initial conditions
relate; and (v) the level of numerical accuracy required in the solution, typically
10^{-6}. There is a slight inconsistency in that the same "dummy variable" t is
used to specify differential equations for all individuals. In OpenBUGS, a different
dummy variable must be used for each individual.

WBDiff also allows users to package their differential equation system as new
BUGS syntax. The above regression function could be re-expressed as follows, for
example,

```
psi[offset[i]:(offset[i+1]-1), 1:dim]
            <- one.comp(init[i, 1:dim],
                        time[offset[i]:(offset[i+1]-1)],
                        theta[i, 1:numpar], origin, tol)
```

[‡]Available from http://www.winbugs-development.org.uk.

Here one.comp(.) is a new function, written in Component Pascal by the user, in which the differential equations are specified and linked to a numerical solver (the reader is referred to the WBDiff documentation for details). In terms of arguments, the function is virtually identical with the more general ode(.) function above. The only difference is that instead of supplying the equations, we provide the parameters required to define the equations internally, theta[i,]. In the above specification, the elements of theta[i,] may correspond to CL_i and V_i, say — it depends on how one.comp(.) has been defined internally.

The main advantage of "hard-wiring" differential equation systems into BUGS, as in the latter example above, is that as compiled code they can be computed much more rapidly, especially as the complexity and/or the number of equations increases.

Most of the compartmental systems used in basic pharmacokinetics have known solutions. Many of these are available via the PKBugs software, which is compatible only with Version 1.3 of WinBUGS, or via the *Pharmaco* interface in WinBUGS 1.4.[§] In both cases the likelihood for our cadralazine example can be specified as

```
for (i in 1:10) {
  for (j in offset[i]:(offset[i+1]-1)) {
    y[j]      ~ dnorm(psi[j], inv.sigma.squared[i])
    psi[j] <- pkIVbol1(theta[i, 1:2], time[j], D)
  }
}
```

where pkIVbol1(.) is a hard-wired function representing single compartment kinetics following an intravenous bolus dose. In this case the elements of theta[i,] must correspond to $\log CL_i$ and $\log V_i$ (in that order).

11.6   Finite mixture and latent class models

Discrete mixtures of distributions arise in a number of contexts. The prior distribution may be one of a set of alternatives, such as a mixture of two components, described in §5.4. A similar analysis arises when assuming that all the data come from one of a set of competing models (§8.7).

A *mixture model*, however, usually means that each individual data point y_i is assumed drawn from one of a list of possible distributions. This can be

---

[§]Both PKBugs and Pharmaco are available from http://www.winbugs-development.org. uk.

considered as a clustering of the points into groups $G_j, j = 1, .., J$, where y_i is a member of group T_i and has a distribution parameterised by θ_{T_i}. We may write this general model as

$$y_i \sim p(y_i|\theta_{T_i}, G_{T_i}), \qquad T_i \sim \text{Categorical}(p) \qquad (11.4)$$

so that the probability that y_i is in the jth group G_j is $\Pr(T_i = j) = p_j$.

Particular difficulties can arise when using MCMC to analyse such models using this formulation. Each iteration involves generating a possible clustering of the points, and if a group j is left empty at a particular iteration, then there will be no likelihood term to contribute to the next simulation of the group's parameters θ_j, and this can lead to convergence problems. Similarly, the model is invariant to permutations of the labels of the groups, so that careful constraints are required to minimise the effect of points switching between different groups at different MCMC iterations (Richardson and Green, 1997; Stephens, 2000).

The JAGS implementation of BUGS includes a specialised module `mix` for finite mixtures of normal distributions. This provides a distribution for a normal mixture

```
y[i] ~ dnormmix(mu, tau, pi)
```

where the component-specific means `mu`, precisions `tau`, and membership probabilities `pi` are vectors with length equal to the number of alternative distributions. A specialised random-walk Metropolis–Hastings sampler, with tempered transitions to jump between multiple modes of the posterior density (Neal, 1996; Celeux et al., 2000), is also used.

Mixture models with unknown numbers of components may be implemented in BUGS using Bayesian nonparametric methods, as discussed in §11.8.

Example 11.6.1. *Eyes*

Bowmaker et al. (1985) analyse data on the peak sensitivity wavelengths for individual microspectrophotometric records on the eyes of monkeys. A scaled histogram of the data from one monkey ($S14$ in the paper) is shown in Figure 11.6 with a kernel density estimate suggesting a two-component mixture.

Part of the analysis involves fitting a mixture of two normal distributions with common variance to this distribution, so that each observation y_i is assumed drawn from one of two groups. Let $T_i = 1, 2$ be the true group of the ith observation, where group j has a normal distribution with mean λ_j and precision τ. We assume an unknown fraction p_1 of observations are in group 1, $p_2 = 1 - p_1$ in group 2.

The model is thus

$$y_i \sim \text{Normal}(\lambda_{T_i}, \tau), \qquad T_i \sim \text{Categorical}(p).$$

This formulation easily generalises to additional components in the mixture, although for identifiability, an order constraint is put onto the group means. This

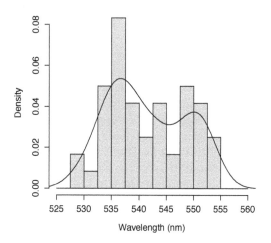

FIGURE 11.6

Monkey eye tracking measurements with a kernel density estimate.

avoids "label switching" between MCMC iterations. However, in general, particularly with greater numbers of groups, inferences about the groups may be sensitive to the constraint chosen to identify them, and Stephens (2000) discusses specialised algorithms to alleviate this problem.

Diebolt and Robert (1994) pointed out that there is a danger using this model that at some iteration all the data will go into one component of the mixture, and this state will be difficult to escape from — this matches our experience. They suggested a re-parameterisation, a simplified version of which is to assume

$$\lambda_2 = \lambda_1 + \theta, \quad \theta > 0.$$

$\lambda_1, \theta, \tau, p$ are given independent vague priors.

```
for (i in 1:N) {
   y[i]      ~ dnorm(mu[i], tau)
   mu[i]    <- lambda[T[i]]
   T[i]      ~ dcat(p[])
}
p[1:2]       ~ ddirch(alpha[])
alpha[1]  <- 1
alpha[2]  <- 1
theta        ~ dunif(0, 1000)
lambda[2] <- lambda[1] + theta
lambda[1]  ~ dunif(-1000, 1000)
```

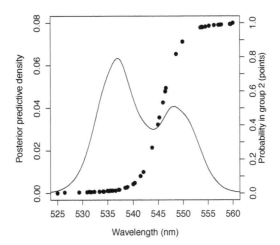

FIGURE 11.7
Posterior predictive distribution for an eye tracking observation. The probabilities that each data point is in mixture group 2 are overlaid.

```
sigma      ~ dunif(0, 100)
tau        <- 1 / pow(sigma, 2)
```

Posterior medians and 95% CIs for the two group means are 537 (535, 539) and 549 (545, 551) nm, respectively, with common standard deviation 3.7 (3.0, 6.2). About 60% (41%, 77%) of observations are estimated to be in group 1.

Figure 11.7 illustrates how the probability that each observation is in group 2 increases with the size of the observation. Overlaid is the posterior predictive distribution for a new observation, which approximately follows the kernel density estimate for the data.

Since there are only two categories, we could also have used a Bernoulli instead of a categorical distribution for T[i]. See Example 11.6.3.

11.6.1 Mixture models using an explicit likelihood

The finite mixture model can be reformulated by summing over the possible memberships T_i to give the the following distribution, from which every observation is assumed to arise:

$$p'(y_i|\theta) = \sum_j p_j p(y_i|\theta_j, G_j).$$

This sampling distribution can be employed using the zeros trick (§9.5). This allows a multimodal density to be fitted to the data, but cannot allocate observations T_i to distinct groups. An advantage of this formulation is that the DIC for model comparison can be calculated. In §11.4 the model involves a discrete parameter T_i for which the posterior mean, and therefore DIC, is undefined (see §8.6.4), whereas here this parameter is integrated over. Though again, as discussed by Celeux et al. (2006), the DIC may be sensitive to the constraint chosen to avoid label switching.

Example 11.6.2. *Eyes (continued): DIC with the zeros trick*
In this mixture formulation, we assume each observation is drawn from a density

$$p(y_i|) = p \times \text{Normal}(\lambda_1, \sigma^2) + (1 - p) \times \text{Normal}(\lambda_2, \sigma^2).$$

implemented using the zeros trick.

```
for (i in 1:N) {
  Zero[i]  <- 0
  Zero[i]   ~ dpois(phi[i])
  phi[i]   <- -log((p[1]*exp(-0.5*tau*pow(y[i] - lambda[1], 2))
                  + p[2]*exp(-0.5*tau*pow(y[i] - lambda[2], 2)))
                  * sqrt(tau/(2*3.14159))) + C
}
C           <- 100
p[2]        <- 1 - p[1]
p[1]         ~ dunif(0, 1)
theta        ~ dunif(0, 1000)
lam          ~ dunif(0, 1000)
lambda[2]   <- lam + theta/2
lambda[1]   <- lam - theta/2
sigma        ~ dunif(0, 100)
tau         <- 1/pow(sigma, 2)
```

The mixing is considerably poorer than under the previous formulation, requiring about ten times as many iterations for a similar degree of accuracy. Indeed, a further alteration to the parameterisation was required for satisfactory convergence, replacing the group means λ_1, $\lambda_2 = \lambda_1 + \theta$ by a "centred" parameterisation $\lambda_1 = \lambda - \theta/2$, $\lambda_2 = \lambda + \theta/2$.

Note also that in this example, the component membership probabilities p cannot be explicitly given a Dirichlet prior in WinBUGS, where the ddirch distribution can only be employed as a conjugate prior for the parameter of a categorical or multinomial distribution, or for forward sampling. The equivalent p[1] ~ dunif(0,1) is therefore specified. In OpenBUGS or JAGS, a Dirichlet prior would be allowed.

To compare the DIC for a model implemented with the zeros trick with the DIC of a model defined using a built-in distribution, for example, a simple normal model

```
for (i in 1:N) {
   y[i]   ~ dnorm(lam, tau)
}
lam       ~ dunif(0, 1000)
sigma     ~ dunif(0, 100)
tau       <- 1/pow(sigma, 2)
```

an adjustment must be made. The DIC which WinBUGS calculates when the zeros trick is used (D_{zero}, say) refers to the Poisson model for the dummy Zero data $Z_i = 0$, $i = 1, \ldots, n$. This is on a different scale to the DIC (D, say) which would be reported from a direct model for y. The sampling distribution for a single Poisson-distributed Z_i is $p(Z_i | \phi_i) = e^{-\phi_i} \phi_i^{Z_i} / Z_i! = e^{-\phi_i}$. Since $\phi_i = C - \log(g(y_i | \lambda_1, \lambda_2, \sigma))$, where $g(y_i | .)$ is the sampling distribution of the normal mixture model for y_i, the Poisson sampling distribution is $e^{-C} g(y_i | .)$. Taking $-2 \log()$ and summing over the n observations, we conclude that D_{zero} will differ from the "natural" DIC D by a constant

$$D_{zero} = D + 2nC$$

where C is the constant applied to ensure that the Poisson rate is positive, and n is the number of observations (100 and 48, respectively, in this example).

WinBUGS reports a DIC of 9918 for the two-component mixture model implemented using the zeros trick. For the simple normal model implemented using dnorm, the DIC reported is 326. To compare these, a constant $2nC = 9600$ is added to 326, obtaining 9926. Since this is greater than 9918, the two-component mixture model with an effective additional two parameters is therefore preferred to the simple normal model. Indeed, if the simple normal model were implemented using the zeros trick, WinBUGS would report a DIC of 9926.

Example 11.6.3. *Zero-inflated Poisson*
In Example 8.4.6 we used a focussed checking function to identify an excess of zeros in a set of 35 observations assumed to come from a Poisson distribution. A natural extension is a "zero-inflated Poisson" (ZIP) model , which is a two-component mixture model in which an unknown proportion $1 - p$ of observations is constrained to be exactly 0, while the remainder p are drawn from a standard Poisson distribution (some of these observations may, by chance, also be 0). There are a number of possible ways of expressing such a model in BUGS — perhaps the simplest is to assume each observation y_i comes from one of two groups so that $g_i = 0$ or $g_i = 1$: those in group 0 are Poisson with mean 0 (and so are all 0), while those from group 1 are Poisson with mean μ. This can be expressed as $y_i \sim \text{Pois}(m_i)$, where $m_i = g_i \mu$, and $g_i \sim \text{Bern}(p)$. BUGS code is as follows:

```
for (i in 1:N) {
   y[i]      ~ dpois(m[i])
```

```
    m[i]      <- group[i]*mu           # mean is 0 if group is 0
    group[i]  ~ dbern(p)
  }
  # proportion of claims that could be positive
  p                 ~ dunif(0,1)
  mu                ~ dgamma(0.5, 0.0001) # approximate Jeffreys prior
```

We estimate p to be 0.81 (95% interval 0.63 to 0.97) and μ to be 2.07 (1.50 to 2.73).

11.7 Piecewise parametric models

11.7.1 Change-point models

Arbitrarily flexible models can be built in BUGS by piecing together simpler parametric forms. A simple example is a *change point* model for a regression of y on x consisting of two straight lines, which meet at a certain point $x = \theta$.

$$y_i \sim \text{Normal}(\mu_i, \sigma^2), \qquad \mu_i = \alpha + \beta x + \beta_2 (x - \theta)_+ \qquad (11.5)$$

Here $(x - \theta)_+$ denotes the positive part of $(x - \theta)$, which is implemented by the step() function in BUGS. The regression slope is then β for $x \leq \theta$ and $\beta + \beta_2$ for $x > \theta$ (as in Figure 11.8).

Example 11.7.1. *Stagnant water: change point model*
Carlin et al. (1992) analyse data on water stagnation and flow rates. y_i is the log flow rate down an inclined channel, and x_i is the log height of stagnant surface layers for different surfactants i. The rate of decline in flow rate seems to suddenly increase around $x = 0$ (Figure 11.8). The linear change-point model (11.5) can be expressed simply in BUGS; however, sensible parameterisation and prior choice are necessary. The change point θ is given a continuous uniform prior over the region of the observed x_i. Constraining this point to lie at one of the $x_i = x_c$, with a *discrete* uniform prior on c, would produce a poorly converging chain due to the strong correlation between c and α ($\alpha = E[y]$ at $x = 0$).

```
  for (i in 1:N) {
    y[i]      ~ dnorm(mu[i], tau)
    mu[i]     <- alpha + beta[1]*x[i] + beta[2]*(x[i] - theta)
              * step(x[i] - theta)
  }
  tau              ~ dgamma(0.001, 0.001)
```

```
alpha       ~ dnorm(0.0, 1.0E-6)
for (j in 1:2) {
  beta[j] ~ dnorm(0.0, 1.0E-6)
}
sigma       <- 1/sqrt(tau)
theta       ~ dunif(-1.3, 1.1)
```

Ten thousand iterations after a burn-in of 500 produces a posterior mean of 0.026 (−0.040, 0.087) for the change point. Posterior means and 95% pointwise credible intervals for the regression means mu[i] are illustrated in Figure 11.8.

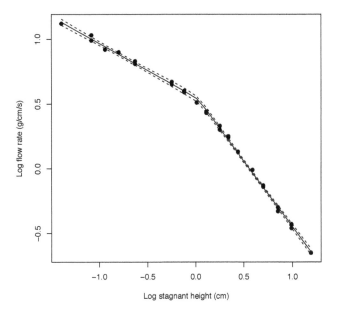

FIGURE 11.8
Change-point model for stagnant water data. Solid and dashed lines show posterior means and 95% intervals, respectively, for the fitted values mu[i]; dots show observed data y[i].

11.7.2   Splines

The idea of the change-point model generalises to *spline* models consisting of piecewise parametric functions, commonly polynomials. These are defined to change at K *knots* $\theta_1, \ldots, \theta_K$, and may optionally include smoothness constraints at the knots. An example of a cubic spline is

$$\mu_i = \alpha + \sum_{j=1}^{3} \beta_j x_i^j + \sum_{k=1}^{K} \beta_{k+3}(x_i - \theta_k)_+^3$$

The location of the knots may be estimated as part of the model. Ordering constraints on the locations (§9.7.2) are necessary with more than one knot. The required BUGS model specification becomes quite verbose for splines more complex than a quadratic polynomial with a couple of knots. Some examples are given by Congdon (2005). However, the Jump add-on to WinBUGS[¶] contains functions to automatically construct a variety of polynomial splines. Even the number of knots can be considered as an unknown parameter, since the posterior distribution is estimated using *reversible jump* MCMC (§8.8.2). This facility is under development for OpenBUGS.

11.7.3   Semiparametric survival models

Piecewise parametric models have an interesting application to flexible survival analysis. In the Cox proportional hazards regression model, the hazard is modelled as a function of covariates x_i as $h(t) = h_0(t)e^{\beta x_i}$. In the usual classical implementation by partial likelihood, $h_0(t)$ is left unspecified. The Bayesian framework, however, requires a full probability model; therefore we specify $h_0(t)$ as a piecewise-constant function.

The BUGS implementation relies on a *counting process* formulation (Andersen et al., 1993). For individual i, $N_i(t)$ counts the number of deaths up to time t. The increment of the counting process $dN_i(t)$ over the time interval $[t, t + dt]$ takes the value 1 if individual i dies in this interval and 0 otherwise. In allowing multiple "deaths" for each individual, the counting process enables more general event history models. The hazard function of the process is $\lambda_i(t) = lim_{dt \to 0} E(dN_i(t)|\mathcal{F}_t)/dt$, where $\mathcal{F}_t$ is the history of the process up to t. Let $t_j : j = 1, 2, \ldots$ represent all unique death or censoring times ($t_0 = 0$). If the hazard $\lambda_i(t)$ is assumed to be piecewise-constant, the likelihood for these data can be computed by multiplying probabilities of survival or death over intervals up to each individual's observed death time T_i or censoring time T_i^*:

$$\prod_{i} \prod_{j:t_j <= min(T_i, T_i^*)} \lambda_i(t_j)^{dN_i(t_j)} \exp(-\lambda_i(t_j)(t_j - t_{j-1})), \qquad (11.6)$$

[¶]Available from `http://www.winbugs-development.org.uk`.

equivalent to a Poisson model for $dN_i(t_j)$ with mean $\lambda_i(t_j)(t_j - t_{j-1})$. The most flexible possible model allows $\lambda_i(t)$ to change at every observed death time. But a more efficient model can be built by adapting the change points to the data, using more frequent hazard changes in regions where there are more observed deaths. See Demiris and Sharples (2006) and Jackson et al. (2010b) for further discussion of practical issues with these models.

Example 11.7.2. *Leukaemia: survival models with piecewise-constant hazards*
Times in remission were recorded for 42 leukaemia patients from a trial of the drug 6-mercaptopurine versus placebo (Gehan, 1965). A piecewise-constant hazard model for these data is built. The time in remission obs.t[i] in weeks, indicator ind[i] (1 if they relapsed at the end of the remission time or 0 if censored), and a covariate Z (0.5 representing control, and −0.5 treated) are supplied as usual (§11.1) for each individual i. To set up the counting process, a vector t[] of all unique times in remission is also given.

```
list(N=42, T=23, eps=1.0E-10,
obs.t=c(1,1,2,2,3,4,4,5,5,8,8,8,8,11,11,12,12,15,17,22,23,
6,6,6,6,7,9,10,10,11,13,16,17,19,20,22,23,25,32,32,34,35),
ind=c(1,1,1,1,1,1,1,1,1,1,1,1,1,1,1,1,1,1,1,1,1,
1,1,1,1,1,0,1,0,1,0,0,1,1,0,0,0,1,1,0,0,0,0,0),
Z=c(0.5,0.5,0.5,0.5,0.5,0.5,0.5,0.5,0.5,0.5,0.5,
    0.5,0.5,0.5,0.5,0.5,0.5,0.5,0.5,0.5,0.5,
    -0.5,-0.5,-0.5,-0.5,-0.5,-0.5,-0.5,-0.5,-0.5,-0.5,
    -0.5,-0.5,-0.5,-0.5,-0.5,-0.5,-0.5,-0.5,-0.5,-0.5,-0.5),
t=c(1,2,3,4,5,6,7,8,9,10,11,12,13,15,16,17,19,20,22,23,25,32,34,35),
period=c(1,2,3,4,5,6,7,8,8,9,10,11,12,13,14,15,15,15,16,17,17,17,17,17),
period4=c(1,1,1,1,1,1,2,2,2,2,2,2,3,3,3,3,3,3,4,4,4,4,4,4),
ndtimes=17)
```

First, the counting process increments dN are calculated for each individual i and unique relapse or censoring time j in terms of the data. Y[i,j] are indicators for obs.t >= t, in other words, whether individual i is under observation at time t[j]. The increment dN[i,j] is 1 if individual i is observed to relapse at t[j], and zero otherwise.

```
for (i in 1:N) {
    for (j in 1:T) {
        Y[i,j]  <- step(obs.t[i] - t[j] + eps)
        dN[i,j] <- Y[i,j]*step(t[j+1] - obs.t[i] - eps)*ind[i]
    }
}
```

The Poisson model (11.6) is then applied to the dN[i,j], with rate Idt[i,j] equal to the hazard multiplied by the length of the time interval ending in t[j], or zero if individual i is not at risk. The hazard is assumed proportional to the treatment covariates Z, with a baseline hazard lam which changes at every distinct relapse time. Periods with identical hazard are indicated by period in the data —

note the hazard does not change at times when censoring occurs but no relapses. Diffuse priors are specified for each distinct hazard and for the covariate effect. Finally, the survivor function $S(t)$ is then calculated in terms of the cumulative hazard $H(t)$ as $S(t) = \exp(-H(t))$ for each treatment group.

```
dt[1]                   <- t[1]
for (j in 2:(T+1)) {
   dt[j]                <- t[j] - t[j-1]
}
for (j in 1:T) {
   for (i in 1:N) {
      dN[i,j]            ~ dpois(Idt[i,j])
      Idt[i,j]          <- Y[i,j]*exp(beta*Z[i])
                         * lam[period[j]]*dt[j]
   }
}
cumhaz.treat[1]     <- 0
cumhaz.placebo[1]   <- 0
for (j in 2:(T+1)) {
   cumhaz.treat[j]   <- cumhaz.treat[j-1] + lam[period[j]]
                      * dt[j]*exp(beta*-0.5)
   cumhaz.placebo[j] <- cumhaz.placebo[j-1] + lam[period[j]]
                      * dt[j]*exp(beta*0.5)
   S.treat[j]        <- exp(-cumhaz.treat[j])
   S.placebo[j]      <- exp(-cumhaz.placebo[j])
}
for (j in 1:ndtimes) {
   lam[j]             ~ dgamma(0.001, 0.001)
}
beta                   ~ dnorm(0.0, 0.000001)
```

The posterior mean relapse-free survival curves are superimposed on standard Kaplan–Meier estimates in Figure 11.9. In a second model, the unique observed relapse times j are partitioned into four periods according to the sample quartiles, so that period[j] is replaced by period4[j], which takes the values 1, 2, 3, 4. This model is more efficient than the model with a hazard change per relapse time, as judged by its DIC of 221, compared to 246, and the treatment effect β is unchanged.

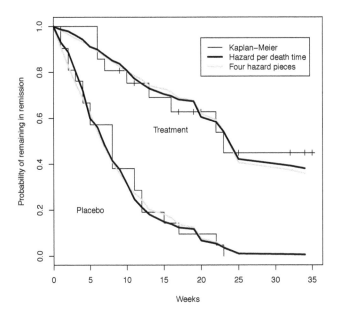

FIGURE 11.9
Relapse-free survival curves for leukaemia trial patients.

11.8   Bayesian nonparametric models

There is growing interest in the use of Bayesian nonparametric methods based
on distributions over *spaces of distributions* (Hjort et al., 2010). A typical
motivation is to account for *model uncertainty* (§ 8.8) about the choice of a
parametric distribution. For instance, normal distributions are often used un-
critically for random effects in hierarchical models or as the error distribution
in a regression. But sampling distributions for data, or priors for parameters,
often do not follow any standard parametric shape. Unlike typical classical
nonparametric methods (such as rank and permutation tests, or Kaplan–Meier
survival estimators), Bayesian nonparametric methods can provide full probability models for the data-generating process. Thus they may be used to
estimate a predictive distribution for any point outside or inside the sample
data, which accounts for uncertainty about distributional shape.

The most common models of this type are based on the *Dirichlet process*

(DP), which is a distribution over distributions

$$G() \sim \mathrm{DP}(G_0(), \alpha).$$

$G_0()$ represents the central or "mean" distribution, for instance Normal$(0,1)$ or Normal(μ, σ^2), while the precision or *concentration* parameter α governs how close realisations of $G()$ are to $G_0()$. A distribution is essentially defined by the set of all probabilities for any partition of the sample space into any number of intervals k, which follow

$$(p_1, \ldots, p_k) \sim \mathrm{Dirichlet}(\alpha p_{01}, \ldots, \alpha p_{0k}),$$

where $(p_{01}, \ldots, p_{0k})$ are the probabilities under $G_0()$.

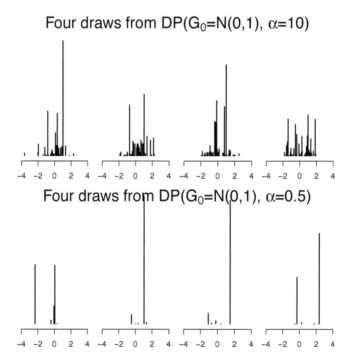

FIGURE 11.10
Samples of distributions generated from the Dirichlet process.

A realisation of the Dirichlet process is a discrete distribution, defined as an infinite collection of point masses at random locations $\theta_1, \theta_2, \ldots$ in the sample space, with weights $p_1, p_2, \ldots : \sum_1^\infty p_k = 1$. Smaller values of the precision parameter α result in a few large p_k dominating the distribution (Figure 11.10).

11.8.1 Dirichlet process mixtures

To define a continuous distribution, the DP can be used as the basis of a mixture model, for example, a mixture of Normal(θ_k, σ_k) with mixing proportions defined by the p_k. There are a theoretically infinite number of mixture components $k = 1, \ldots, \infty$, giving an arbitrarily flexible choice of distributional shapes. Multimodal or heavy-tailed distributions are naturally modelled as mixtures (§11.6). Although a finite number of mixture components may only be necessary in practice, Dirichlet process models acknowledge uncertainty about this number. Smaller values of α result in a smaller number of mixture components.

11.8.2 Stick-breaking implementation

As shown by Congdon (2006)(§6.7), Dirichlet-process-based models can be implemented in the BUGS language using a *constructive* definition, which represents the distribution (as illustrated in Figure 11.10) as a stick of length 1 which has been successively broken into smaller and smaller pieces. For $q_1, q_2, q_3 \ldots \sim \text{Beta}(1, \alpha)$,

- Define $p_1 = q_1$ (break proportion q_1 of a stick)

- $p_2 = (1 - q_1)q_2$ (break proportion q_2 of remainder)

- $p_3 = (1 - q_1)(1 - q_2)q_3$ (break proportion q_3 of remainder)

and so on, with the recursive relation $p_j = (1 - q_{j-1})q_j p_{j-1}/q_{j-1}$. The stick-breaking definition is *truncated* so that the stick is only broken C times, which assumes that the data can be represented by a maximum of C possible clusters. Finally, the sticks are placed at locations $x_1, x_2, \ldots$ sampled from the baseline distribution G_0.

This method was used by Ohlssen et al. (2007) to fit flexible random effects models for identifying health-care providers with outlying or unexpected performance. The R package `DPpackage` (Jara et al., 2011) also implements MCMC posterior inference for the DP and related models using exact and more efficient algorithms; however, the BUGS language allows a freer choice of component-specific and prior distributions, and allows DP models to be embedded within a larger graphical model.

Example 11.8.1. *Galaxy clustering: Dirichlet process mixture models*
Roeder (1990) and Escobar and West (1995) studied the velocities, relative to our own galaxy, of 82 galaxies drawn from six well-separated conic sections of the Corona Borealis region. The distribution of velocities (Figure 11.11) appears multimodal, suggesting that galaxies are clustered, as predicted by the Big Bang theory. To estimate a posterior density for the velocity distribution, we fit a Dirichlet process mixture of normals. This may also provide a posterior distribution for

the number of clusters, to distinguish between genuine clusters and artefacts of sampling.

First, the mixture model is defined, as in §11.6. The n galaxy velocities y_i are assumed to arise from a maximum of $C = 20$ normal distributions with different means and precisions. The stick-breaking definition then generates the probabilities p_j that galaxy i belongs to cluster $G_i = j : j = 1, \ldots, C$.

$$y_i \sim \text{Normal}(\mu_{G_i}, \sigma^2_{G_i})$$

$$\mu_j \sim \text{Normal}(a_\mu, \sigma^2_j/b_\mu), \qquad 1/\sigma^2_j \sim \text{Gamma}(a_\tau, b_\tau)$$

The prior distributions are derived from Escobar and West (1995). A diffuse normal prior is used for the mean a_μ of the normal cluster means, and the between-cluster variance of μ_j is a scaled version of the within-cluster variance $\sigma^2_j = 1/\tau_j$. But with diffuse priors on b_μ, a_τ, b_τ and the DP precision parameter α, the chain fails to converge. The within-cluster variances σ^2_j need more careful constraints for identifiability therefore scientific belief is used to define an inverse gamma prior with fixed shape $a_\tau = 2$ and uncertain scale $b_\tau \sim \text{Gamma}(2, 1)$, and also $b_\mu \sim \text{Gamma}(0.5, 50)$. The DP precision parameter, which governs the expected number of clusters, is initially fixed at $\alpha = 1$, which implies about 90% prior weight on between three and seven clusters to generate a dataset of $n = 82$. See Ohlssen et al. (2007) and Jara et al. (2011) for more discussion of prior specification in this class of models.

```
for (i in 1:n) {
   velocity[i]   ~ dnorm(mu[i], tau[i])
   mu[i]         <- mu.mix[group[i]]
   tau[i]        <- tau.mix[group[i]]
   group[i]      ~ dcat(pi[])
   for (j in 1:C) {
      gind[i,j] <- equals(j, group[i])
   }
}
p[1]             <- q[1]
for (j in 2:C) {
   p[j]          <- q[j]*(1 - q[j-1])*p[j-1]/q[j-1]
}
for (j in 1:C) {
   q[j]          ~ dbeta(1, alpha)
   pi[j]         <- p[j]/sum(p[])
   mu.mix[j]     ~ dnorm(amu, mu.prec[j])
   mu.prec[j]    <- bmu*tau.mix[j]
   tau.mix[j]    ~ dgamma(aprec, bprec)
}
alpha            <- 1
amu              ~ dnorm(0, 0.001)
```

```
bmu              ~ dgamma(0.5, 50)
aprec            <- 2
bprec            ~ dgamma(2, 1)
K                <- sum(cl[])
for (j in 1:C) {
  sumind[j]      <- sum(gind[,j])
  cl[j]          <- step(sumind[j]-1+0.001) # cluster j used in
                                            # this iteration
}
for (j in 1:ndens) {
  for (i in 1:C) {
    dens.cpt[i,j] <- pi[i]*
                    sqrt(tau.mix[i] / (2*3.141592654))*
                    exp(-0.5*tau.mix[i]*(mu.mix[i] - dens.x[j])
                                      *(mu.mix[i] - dens.x[j]))
  }
  dens[j]          <- sum(dens.cpt[,j])
}
```

Twenty thousand iterations are required for sufficiently precise estimation of the number of clusters K after a burn-in of 1000, giving a posterior median of 7 (4–10) clusters. The posterior mean and pointwise 95% posterior credible limits for the density function of galaxy velocities (dens in the code above, evaluated at ndens=271 points at equal increments of 0.1 between 8 and 35) are illustrated in Figure 11.11.

Ideally, α would be estimated from the data. In this example, the constant $\alpha = 1$ was replaced with a Gamma(2, 4) prior. The resulting chain mixes fairly poorly, but supports values close to $\alpha = 1$. The uncertainty around the number of clusters is slightly increased, with a posterior median of $K = 6$ (3–11). There is sensitivity to this prior choice, however — under an alternative Uniform(0.3, 10) prior, the posterior median is $K = 13$ (6–18). However, the qualitative shape of the posterior density under this alternative prior is unchanged (Figure 11.11, top row).

We can get some idea of the distribution of clusters by plotting the posterior means of the probabilities pi[i] assigned to each cluster [i] (Figure 11.11, bottom row) — note that under the uniform prior, there are more components with smaller estimated probabilities. We cannot obtain meaningful posteriors describing particular clusters, however. Just as for standard mixture models (§ 11.6), the integer labels of the mixture components stored in group[] do not retain their substantive meaning in terms of data clusters between MCMC iterations. However, under the stick-breaking implementation, the lower labels will tend to be applied to the most common clusters with the highest probabilities pi[j], particularly if α is low.

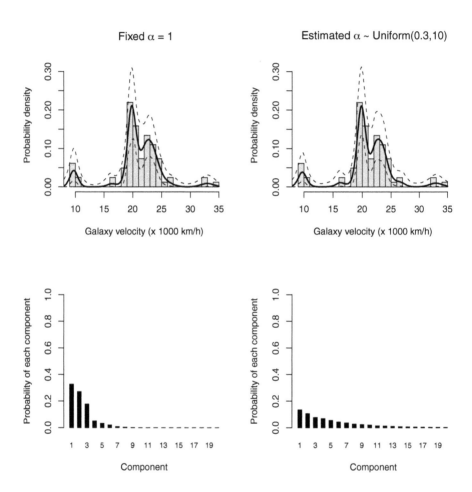

FIGURE 11.11
Galaxy velocities with posterior density from the Dirichlet process mixture model
and probabilities of mixture components, under two prior specifications. Note that
due to label switching the components may represent different clusters at each
iteration.

12

Different implementations of BUGS

Through the years, a variety of software has been developed to perform MCMC sampling for models specified in the BUGS language. This chapter describes in detail the alternative implementations and interfaces to BUGS which are currently available and will hopefully clear up some confusion. We give brief examples of how to perform a typical analysis under each interface. All examples elsewhere in this book use WinBUGS 1.4.3, though where appropriate, important differences between BUGS implementations for particular examples have been highlighted.

12.1 Introduction — BUGS engines and interfaces

We distinguish between different BUGS *engines* and BUGS *interfaces*. The engine is the core of the software — the underlying code which interprets the model and performs MCMC simulation. The interface is the means of controlling the engine — the actual software application which users open before starting their analysis, and the commands or buttons used to perform the analysis.

Currently there are three BUGS engines in widespread use:

- WinBUGS

- OpenBUGS

- JAGS

The older "Classic" BUGS is virtually obsolete. Different engines have slightly different scientific capabilities, in terms of what functions, distributions, and MCMC sampling algorithms can be used in models. There are also some minor differences in BUGS language syntax.

Each engine can be controlled though different interfaces. For a model run under the same engine, different interfaces will give identical results. For example, WinBUGS can be run by clicking in its menus, by using a "script" to run a batch of menu commands at once, or through other software (such as R, via the R2WinBUGS package). Engines also differ in the computer operating systems and hardware platforms that they can run on. For example,

TABLE 12.1

Sampling method hierarchy used by WinBUGS. Each method is only used
if no previous method in the hierarchy is appropriate.

Target full conditional	Sampling method
Discrete	Inversion of cumulative distribution function
Closed form (conjugate)	Direct sampling using standard algorithms
Log-concave	Derivative-free adaptive rejection sampling (Gilks and Wild, 1992)
Restricted range	Slice sampling (Neal, 2003)
Unrestricted range	Metropolis–Hastings (Metropolis et al., 1953; Hastings, 1970)

WinBUGS is restricted to Windows or "emulators" of Windows on other operating systems. OpenBUGS may also run natively on Linux on PC hardware,
and JAGS is even more portable.

12.2 Expert systems and MCMC methods

Each BUGS engine implements an "expert system" to decide which of the
available MCMC algorithms should/could be used for a given problem. This
is one area where the various engines differ substantially. All of the expert systems begin with classification, whereby the full conditional distribution of each
node, or block of nodes, within the specified graphical model is classified. If
the full conditional can be derived in closed form, then the appropriate classification is the corresponding density type, e.g., normal, gamma. Alternatively,
if a closed form is not available, the expert system may still be able to identify useful properties that can be exploited by particular sampling algorithms,
such as log-concavity, which enables adaptive rejection sampling (Gilks, 1992;
Gilks and Wild, 1992). In WinBUGS, there is a one-to-one mapping between
classifications and MCMC methods; in other words, the classification uniquely
determines the sampling algorithm. A simplified representation of this mapping is given in Table 12.1.

By contrast, OpenBUGS uses a more flexible system whereby sampling
methods "choose" nodes (using classification as a guide) as opposed to nodes
"choosing" methods. First, all available methods are arranged in order of increasing generality. Then, starting with the more specialised methods, such as
conjugate samplers, the software works through each available method in turn,
allocating that method to any node, or block of nodes, for which it is applicable. By focusing on sampling methods one by one, the software can go beyond
simple classifications and probe the graphical model for further, contextual

information that will facilitate the composition of better tailored sampling schemes. This approach also facilitates expansion of the software's capabilities, since new sampling methods can be added without requiring new classifications. In addition, an interface exists within OpenBUGS to disable various algorithms temporarily, so different schemes can be experimented with. Currently OpenBUGS includes many of the sampling methods used by WinBUGS along with various other "semi-general" methods, such as multivariate Metropolis, various forms of adaptive Metropolis, hybrid sampling (Hanson, 2001), and differential evolution (Ter Braak, 2006). JAGS uses largely the same approach as OpenBUGS but with a reduced set of methods, currently restricted to conjugate samplers, slice samplers, and specialised samplers for mixture models and generalised linear models.

In light of the above discussion of sampling methods, the name BUGS might be considered somewhat of a misnomer, since, strictly speaking, the software does not generally perform Gibbs sampling. However, the sampling scheme still has the flavour of Gibbs sampling since it traverses nodes (or blocks) in the graphical model and considers their full conditionals as target distributions even if they cannot be sampled exactly.

12.3   Classic BUGS

The BUGS project began in the late 1980s and grew from several strands of research on artificial intelligence, graphical modelling, and Bayesian computation. Several years' development led to the first stable release of the software in 1995, then simply called BUGS version 0.5, now referred to as "Classic" BUGS. This was written in the Modula-2 programming language, had a purely text-based, command line interface, and ran on MS-DOS and several varieties of Unix. The BUGS language was largely the same as in future versions, although the software employed a more limited range of MCMC algorithms — Gibbs sampling using conjugacy, inversion, or adaptive rejection sampling for log-concave full conditionals. Development became focused on WinBUGS around the mid-1990s. The new developments included an interactive graphical user interface, more sophisticated Metropolis–Hastings sampling algorithms, and a new, highly modular, "component-oriented" infrastructure for the program source code. This early history of the BUGS project is related in more detail by Lunn et al. (2009b).

12.4   WinBUGS

WinBUGS is an application for Microsoft Windows containing an engine and a graphical user interface for BUGS. It is freely available from http://www.mrc-bsu.cam.ac.uk/bugs. WinBUGS was written in the Component Pascal programming language, using the Blackbox Component Builder development environment and libraries.

Example 12.4.1. *Example of using WinBUGS: Seeds*
This example will be used to illustrate how to run a typical analysis in WinBUGS and other BUGS implementations. We observe seeds on 21 plates, arranged, according to a 2 by 2 factorial layout, by seed and type of root extract (Crowder, 1978; Breslow and Clayton, 1993). The number and proportion of seeds that germinated on each plate are shown below. We study how this varies by seed type (*O. aegyptiaco* 73 or 75) and root extract (bean or cucumber). r_i and n_i are the number of germinated and the total number of seeds on the ith plate, $i = 1, \ldots, N$, respectively.

| *O. aegyptiaco* 75 | | | | | | *O. aegyptiaco* 73 | | | | | |
| Bean | | | Cucumber | | | Bean | | | Cucumber | | |
r_i	n_i	r_i/n_i	r_i	n_i	r_i/n_i	r_i	n_i	r_i/n_i	r_i	n_i	r_i/n_i
10	39	0.26	5	6	0.83	8	16	0.50	3	12	0.25
23	62	0.37	53	74	0.72	10	30	0.33	22	41	0.54
23	81	0.28	55	72	0.76	8	28	0.29	15	30	0.50
26	51	0.51	32	51	0.63	23	45	0.51	32	51	0.63
17	39	0.44	46	79	0.58	0	4	0.00	3	7	0.43
			10	13	0.77						

The model is a random effects logistic regression, allowing for over-dispersion:

$$r_i \sim \text{Binomial}(p_i, n_i)$$
$$\text{logit}(p_i) = a_0 + a_1 x_{1i} + a_2 x_{2i} + a_{12} x_{1i} x_{2i} + b_i$$
$$b_i \sim \text{Normal}(0, 1/\tau)$$

where p_i is the probability of germination on the ith plate, x_{1i} and x_{2i} are the seed type and root extract of the ith plate, and an interaction term $a_{12} x_{1i} x_{2i}$ is included. The associated BUGS model code (the same for all BUGS implementations) is

```
for (i in 1:N) {
   r[i]            ~ dbin(p[i], n[i])
   b[i]            ~ dnorm(0, tau)
   logit(p[i]) <- alpha0 + alpha1*x1[i] + alpha2*x2[i]
```

```
                        + alpha12*x1[i]*x2[i] + b[i]
}
alpha0          ~ dnorm(0, 1.0E-6)
alpha1          ~ dnorm(0, 1.0E-6)
alpha2          ~ dnorm(0, 1.0E-6)
alpha12         ~ dnorm(0, 1.0E-6)
tau             ~ dgamma(0.001, 0.001)
sigma           <- 1/sqrt(tau)
```

12.4.1 Using WinBUGS: compound documents

WinBUGS uses Blackbox's *compound document* file format, with the file extension .odc. Graphics, tables, and formulae may be mixed alongside text of different fonts and colours in the same compound document; therefore Win-BUGS can be used as a simple word processor. On opening the WinBUGS application, the user is presented with a menu interface. To create a new compound document, click on the File menu, then New. To open an existing document or text file, click on File, then Open.

There are two alternative ways of organising the model and data for an analysis in WinBUGS.

1. Model, data, and initial values are kept next to each other in the same "compound document." Results such as summary statistics and posterior density plots may also be presented alongside the model inputs. This is tidier and clearer for simpler, self-contained examples and online tutorials. Using the *fold* facility (Tools->Create Fold) large chunks of text or graphics, such as data files or results, can be hidden within a pair of arrows and revealed again with a single click.

2. Each model, dataset, and set of initial values is kept in a separate file. This strategy is preferable for more complex analyses with several alternative models and datasets, and is required for analyses which are controlled by scripts or from other software. Each file may be either in .odc format or in plain text (.txt), though plain text is recommended for compatibility with other software. In the Seeds example we use separate plain text files.

12.4.2 Formatting data

Data for WinBUGS can be formatted in two alternative ways.

R/S-Plus format In most of the examples in the WinBUGS manual and this book, the "R/S-Plus" format is used. This begins with the word list, and the syntax is the same as that used in R or S-Plus to create a list of vectors of possibly different lengths. Observations on the same variable, separated by commas, are collected together with the c() (collection or concatenation) operator. The exact formatting of white space (spaces and line breaks) doesn't matter. For the Seeds example, the data are formatted as follows:

```
list(
r = c(10, 23, 23, 26, 17, 5, 53, 55, 32, 46, 10,
        8, 10, 8, 23, 0, 3, 22, 15, 32, 3),
n = c(39, 62, 81, 51, 39, 6, 74, 72, 51, 79, 13,
        16, 30, 28, 45, 4, 12, 41, 30, 51, 7),
x1 = c(0, 0, 0, 0, 0, 0, 0, 0, 0, 0, 0,
        1, 1, 1, 1, 1, 1, 1, 1, 1, 1),
x2 = c(0, 0, 0, 0, 0, 1, 1, 1, 1, 1, 1,
        0, 0, 0, 0, 0, 1, 1, 1, 1, 1),
N = 21)
```

Arrays can also be defined in this format using the **structure** operator as in the following example (a different dataset). The .Data component contains the array data strung out as a single vector, while the .Dim component defines the dimensions of the array. In this case Y is a matrix with 30 rows and 5 columns (only the first and last 2 rows are shown below, to save space for illustration). The .Data component is filled up by reading across the first row, then across the second row, and so on.

```
list(
    xbar = 22, N = 30, T = 5,
    x = c(8.0, 15.0, 22.0, 29.0, 36.0),
    Y = structure(
    .Data = c(
        151, 199, 246, 283, 320,
        145, 199, 249, 293, 354,
        [intermediate rows omitted here...]
        137, 180, 219, 258, 291,
        153, 200, 244, 286, 324),
    .Dim = c(30, 5)
    )
)
```

For arrays of general dimension, the *final index* increases fastest, followed by the penultimate index, and so on. But be warned that R and S-Plus fill arrays by reading the left-most index first, i.e., filling matrix columns before rows. The **dput** function in S-Plus produces the above format but with matrix columns filled first, so that, for example,

`.Dim=c(5,30)` must be changed to `.Dim=c(30,5)` for use in WinBUGS. In R, a convenient alternative is to use the **BRugs** or **R2WinBUGS** packages (§12.4.7, §12.5.3). A list of variables (of possibly different lengths and dimensions) stored in the R list object **dat** can be written out to the data file **data.txt** using the R command

```
BRugs:::write.datafile(dat, "data.txt")
```

In **BRugs** (though not **R2WinBUGS**, currently), categorical variables ("factors" in R) are automatically converted to numeric format by this function. Be careful to check that the resulting numeric codes represent the desired ordering of categories.

Rectangular format The alternative "rectangular" format is more familiar to users of spreadsheets or standard statistical software, where variables are supplied in different columns, headed by the variable name, and observations are in different rows. The first two and last two rows of the Seeds data can be formatted in this way as follows:

```
r[] n[] x1[] x2[]
10  39  0    0
23  62  0    0
[intermediate rows omitted for illustration...]
32  51  1    1
3   7   1    1
END
```

The file must end with the key word **END** followed by a line break. This format can be exported from other software, typically as tab-separated text, with a little extra work to define the variable names and end of file. The restriction of rectangular format is that variables of different lengths cannot be mixed. Note that each individual column of an array must be labelled when using rectangular format. For example, the columns of the 30 × 5 matrix from the previous example would be labelled

```
Y[,1]  Y[,2]  Y[,3]  Y[,4]  Y[,5]
```

Constants, such as the number of observations, should be supplied in a separate dataset. For the Seeds example, this is `list(N = 21)`. The value of a constant *node* can also be supplied as an extra variable in the model file, even if it is also given a stochastic distribution (like the variable y in Example 3.3.1) An *index* of a set, however (like N in the Seeds example), cannot be supplied in this way.

Notes on supplying data

- Initial values for unknown parameters in the model are supplied using exactly the same format as observed data.

- Data can be supplied as a single file or with different groups of variables in different files of either format. Arrays cannot be split over different files, however.

- Missing values, in either data format, are specified by NA. Prior distributions can be supplied in the model file for these values, giving an automatic procedure for multiple imputation (§9.1).

- In WinBUGS (though not JAGS or newer versions of OpenBUGS) all variables supplied in a data file (and initial values file) must be defined in the model. This is inconvenient, for example, when developing regression models and selecting covariates to include. To work around this restriction, a dummy variable can be defined in the model which is a function of elements of unused variables (x and y for instance) but unconnected to the rest of the model.

```
dummy <- x[1] + y[1]
```

12.4.3   Using the WinBUGS graphical interface

In § 2.1, we explained the basic steps involved in running a BUGS model, common to all implementations of BUGS. In WinBUGS, these steps can be performed by clicking on menus and dialog boxes.

The data, model and initial values for the Seeds example are supplied in separate plain text files in a directory called Test, which is a subdirectory of the directory where WinBUGS is installed, commonly C:\Program Files\WinBUGS14.*

1. Click on the Model menu, then Specification..., to show the Specification Tool (Figure 2.2 in Chapter 2).

2. To check the syntax of the model specification, open the model file (Seeds_mod.txt here), highlight the word model with the mouse (e.g., by double-clicking), and click on check model. If there is a syntax error in the model, an error message will appear in the status bar at the bottom of the WinBUGS window, and a cursor will highlight the position of the error. Otherwise the message model is syntactically correct will

---

*They are also supplied in the Examples directory in OpenBUGS, and with the material accompanying this book, as Seedsmodel.txt, Seedsdata.txt, Seedsinits1.txt and Seedsinits2.txt.

appear. All messages produced by WinBUGS will also appear in the log (`Info->Open Log`).

3. To load a dataset in R/S-Plus format, open the data file (`Seeds_dat.txt`), highlight the word `list`, and click on `load data`.

4. To load a dataset in rectangular format, highlight the row containing the variable names and click on `load data`. The message `data loaded` should appear, or an error message if the data formatting is incorrect.

5. Repeat until all datasets have been loaded.

6. Select the number of chains to run in `num chains` in the Specification Tool (2 in this example). Running multiple chains, with different initial values, can be useful for checking convergence (§4.4). A single chain is sufficient as a "pilot" run when developing and testing models.

7. Click on `compile`. This sets up the internal graphical model structure and chooses which MCMC updating algorithms are used for each node. Check for any error message in the status bar or a successful `model compiled` message.

8. Initial values must now be supplied for each chain. These can be arbitrary, although if extreme values are chosen, convergence can be poor, or WinBUGS may even crash with a "trap" (a window full of debugging information aimed at developers, though generally opaque to users). A different set of initial values is used for each chain. Open the file `Seeds_in.txt` containing initial values, highlight in the same way as for datasets, and click on `load inits`. A message `chain initialized but other chain(s) contain uninitialized variables` should appear.

9. Load initial values for the second chain (`Seeds_in1.txt`) in the same way.

10. It is not necessary to supply initial values for every parameter in a file. If we had not initialized all stochastic nodes in the model, we would see a message `model contains uninitialized nodes`. To get WinBUGS to generate initial values for all remaining parameters from their prior distributions, click `gen inits`. However, it is advisable to supply initial values for nodes with vague priors to avoid extreme values being sampled (see §4.3).

Once the chains have been initialised, we can start performing MCMC simulation. But first we define which variables to store sampled values for.

- Open the Sample Monitor Tool by clicking on the `Inference` menu, followed by `Samples` (Figure 2.4).

- For the Seeds example we want to monitor `alpha0`, `alpha1`, `alpha2`, `alpha12`, `sigma`. First, type `alpha0` in the box marked `node`, then click `set`. Only monitor the nodes which are of interest to avoid wasting memory and processing time.

- Repeat this for `alpha1`, `alpha2`, `alpha12`, and `sigma`.

- Instead of storing every sampled value, we could save memory by monitoring running means, standard deviations, or approximate running quantiles through the `Inference->Summary` dialog in a similar way. This is only advisable after convergence has been reached.

Then to run the simulation, open the `Update` dialog (Figure 2.3). Type the desired number of updates in the `updates` box, and click `update`.

- The number in the `refresh` box tells WinBUGS to refresh its display after the given number of iterations during the course of a simulation. This defaults to 100 iterations. For slow simulations, a lower value should be typed into this box (e.g., 10 or 1), so that WinBUGS does not appear to "freeze" while running. For very fast simulations, where redrawing the display takes a non-negligible amount of time compared to a simulation, higher refresh intervals (e.g., 1000, 10,000, or more) are advisable.

- The "over-relax" option is a technique (Neal, 1998) which can be used for some updating algorithms to improve mixing of chains. This generates multiple samples at each iteration and then selects one that is negatively correlated with the current value. The time per iteration will be increased, but the within-chain correlations should be reduced so that fewer iterations are necessary. However, due to this trade-off, this method is not always beneficial. The over-relax option can be switched on for a particular MCMC simulation by ticking the `over relax` check box in the `Update` dialog. Default settings can be modified via `Options->Update`.

- The `adapting` box is automatically ticked if an adaptive Metropolis or slice sampling algorithm is being used and the parameters of the algorithm are being "tuned" to improve the acceptance rate. WinBUGS prevents the user from calculating summary statistics when the sampler is adapting, since the chain has not converged. The user cannot un-tick this box, but the length of the adaptive phase can be controlled via `Options->Update`.

- Run a sufficient number of iterations until convergence has been reached (§ 4.4). Convergence is typically examined informally from a complete history of the sampled values (Figure 12.1, top). To produce this, return to the Sample Monitor Tool (`Inference->Samples`). Type the the name of a parameter into `node`, or the character `*` to refer to all monitored parameters, and click `history`. A "burn-in" of at least around 1000 is required for the Seeds example.

- Run a sufficient number of iterations after convergence to produce posterior summary statistics with the desired accuracy (§4.5). Ten thousand updates are sufficient for the Seeds example. To produce posterior summary statistics, use the `stats` button. The boxes marked `beg` and `end` specify the iterations you want to use for posterior summaries — increase the number in `beg` to discard the "burn-in." Here we select `beg=4001` to discard the first 4000 out of a total of 14,000 iterations. There are two parallel chains, so the summaries are based on a pooled sample of 20,000.

```
node     mean     sd      MC error 2.5%     median   97.5%    start sample
alpha0   -0.5533  0.1941  0.005749 -0.9431  -0.554   -0.162   4001  20000
alpha1   0.08043  0.3144  0.009024 -0.5799  0.09154  0.6722   4001  20000
alpha12  -0.8256  0.4274  0.01186  -1.686   -0.8206  0.01917  4001  20000
alpha2   1.358    0.2726  0.0078   0.8281   1.35     1.928    4001  20000
sigma    0.2818   0.1436  0.00511  0.04597  0.2726   0.5901   4001  20000
```

- To output the *current values* of all stochastic parameters, click `Model->Save state`. These are formatted just as initial values; therefore this facility can be used to stop a simulation, leave WinBUGS, and restart it later at the same point from these initial values. However, the new chain will be different from a continuation of the original run, since the seed used for generating random numbers will be different and an adapting phase may have been interrupted.

- To export the *entire monitored sample* for a selected variable (or all variables, using `*`) use the `coda` button in the Sample Monitor Tool. This produces one or more windows containing the samples for each parallel chain and an index window listing which rows of the output correspond to which variables. After saving these windows to text files, they can be imported, for example, by the R or S-Plus `coda` package for MCMC convergence diagnostics and output analysis.

- Many other graphical and numerical posterior summaries are available through the Sample Monitor Tool and elsewhere in the `Inference` menu. For example, the shape of the the marginal posterior distribution for a node can be displayed with a kernel density estimate (Figure 12.1, bottom). See the WinBUGS manual for more details on these. Graphics properties such as axis titles and limits can be changed by right-clicking in the plot region and selecting `Properties....` The resulting dialog commonly contains a button `Special...` which leads to settings specific to the particular type of graphic being drawn. The graphics facilities of WinBUGS are limited, however, and publication-quality figures are usually better drawn in software such as R, after exporting appropriate MCMC samples or summaries.

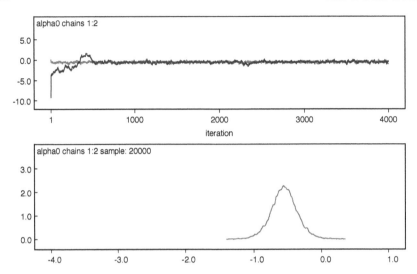

FIGURE 12.1
Sample history (top) and posterior density (bottom) for a single node in the
Seeds example.

12.4.4   Doodles

As described in §2.2, models in WinBUGS can also be specified by drawing a
graph called a *Doodle*. This illustrates relationships between all model quantities as a directed acyclic graph and is exactly equivalent to BUGS model
code. The BUGS code can be generated from any given Doodle, but WinBUGS
cannot convert BUGS language to Doodles. Doodles can help to explain the
principles of graphical modelling, but they are cumbersome to construct for
all but the simplest models.

12.4.5   Scripting

A typical BUGS session consists of a long sequence of actions: load the model
specification and data, compile and initialise a certain number of chains, monitor certain nodes, update the chain for many iterations, check convergence,
discard the first few iterations as a burn-in, and compute summary statistics.
WinBUGS 1.4 contains a *scripting* facility which enables the entire session
to be run using a single action, instead of using the mouse to perform the
relevant menu choice for every action. The sequence of actions is given as a
list of text commands in a plain text or `.odc` format script file. Each command is exactly equivalent to a menu choice, so that both scripts and mouse
clicks may be used in the same session. To run the script, open the file and
select `Script` in the `Model` menu. For example, the following script runs the

WinBUGS session described above for the Seeds example. Change the paths to the model, data, and initial value files according to where they are stored on your system, e.g., check('c:/path/to/Seeds_mod.txt').

```
display('log')
check('Test/Seeds_mod.txt')
data('Test/Seeds_dat.txt')
compile(2)
inits(1, 'Test/Seeds_in.txt')
inits(2, 'Test/Seeds_in1.txt')
gen.inits()
update(4000)
set(alpha0)
set(alpha1)
set(alpha12)
set(alpha2)
set(sigma)
update(10000)
history(*)
density(*)
stats(*)
coda(*,seeds)
save('Test/seedsLog')
```

Two chains are compiled and initialised. After a burn-in of 4000 iterations, the parameters alpha0, alpha1, alpha12, alpha2, sigma are monitored. A sample of 10,000 is drawn and a full sample trace, density plot, and summary statistics are produced for the monitored variables. All the outputs are directed to the WinBUGS log window, which is saved at the end of the script. See the user manual for a full description of all possible scripting commands.

Scripting has several advantages compared to running analyses entirely by point-and-click. Indeed, any statistical software driven by text commands shares these advantages.

Convenience The cycle of developing models, criticising their results, revising and comparing models, and correcting the inevitable errors, usually needs tens to hundreds of model "runs." Scripting makes each individual run more convenient by replacing a sequence of (usually tens of) mouse clicks with just one click.

Reproducibility All scientific analyses should be reproducible. Analyses controlled by point-and-click can in principle be reproducible, but the exact sequence of clicks needs to be recorded. Given a BUGS model, a seed for random number generation, the length of the burn-in, and the number of iterations to keep, any WinBUGS analysis can be replicated. When using the scripting facility, these steps are always recorded and saved in the script file.

Automated analyses We often want to run a batch of similar analyses in succession. For example, a novel statistical model or method is commonly assessed by a simulation study. An analysis is run repeatedly on different datasets simulated from a known model. Finally, the results are aggregated to assess whether point estimates are biased or interval estimates have the appropriate coverage. Running more than about ten simulation replicates would not be practical if WinBUGS were purely driven through its menus, but it can be accomplished by scripting. Automated analyses should, however, be used with care — in particular, when running several models in succession, convergence should be assessed for each model separately.

12.4.6   Interfaces with other software

A WinBUGS analysis with a given script saved in the file `script.txt` can be executed from the Windows command line interface as

`"C:\Program Files\WinBUGS14\WinBUGS14.exe" /PAR "C:\script.txt"`

assuming that WinBUGS is installed in `C:\Program Files\WinBUGS14` and the script is located in the root directory of the `C:` drive.

An entire WinBUGS analysis can then be triggered by external software, without the users opening WinBUGS themselves. The following procedure allows repeated analyses to be automated with a slightly modified model or dataset every time.

1. Generate the model and data; write them to files.

2. Run WinBUGS using the above Windows command and a script which always includes the following:

   - `display('log')` to ensure that results are written to the log file instead of separate windows,

   - the file names of the current model and data,

   - summary statistics are produced using `stats()`, or full MCMC samples saved to separate text files (with names beginning with 'output') using `coda(*,output)`

   - finally, `save('log.txt')` to save the entire log in a text file.

3. Parse the log file `'log.txt'`, saving the summary statistics.

4. Repeat the above steps with different data (or model file) if necessary.

The external software must be able to run the above Windows command and should handle the burden of converting the data from its native format to

WinBUGS format, parsing the log file, and saving the statistics. This usually needs a specialised package for that software to be written. A full list of these interfaces that we currently know of is maintained at `http://www.mrc-bsu.cam.ac.uk/bugs/winbugs/remote14.shtml`, including packages for R, Excel (Woodward, 2011), Stata, SAS, and MATLAB®.

12.4.7 R2WinBUGS

The external interface to WinBUGS we are most familiar with is the R2WinBUGS package for R (Sturtz et al., 2005) available from the CRAN repository of contributed R packages (`http://cran.r-project.org`). This is based around an R function `bugs()`, which runs an entire WinBUGS session and returns MCMC samples, summary statistics, and related information.

The model is provided in a text file. Again for illustration we use the Seeds example with the model in `Seeds_mod.txt`. Begin by calling `setwd()` in R to change to the directory where this is stored. R2WinBUGS requires data as an R `list` object. Since the data in `Seeds_dat.txt` are a text representation of such an object, we can load it into the R workspace using `dget`.[†]

```
> library(R2WinBUGS)
# change the following to wherever the files are stored
> setwd("c:/Program Files/WinBUGS14/Test")
> model.file <- "Seeds_mod.txt"
> data <- dget("Seeds_dat.txt")
```

Note this will not work if the data contain matrices or arrays, for which the ordering of the dimensions will need to be reversed — though in practice the data are likely to have originated in some other format which is easier to import into R than data in WinBUGS format.

Initial values can be supplied as a list of lists, one list for each chain,

```
> inits <- list(
    list(alpha0 = 0, alpha1 = 0, alpha2 = 0, alpha12 = 0,
        tau = 1),
    list(alpha0 = 10, alpha1 = 10, alpha2 = 10, alpha12 = 10,
        tau = 0.1))
```

or as an R function returning a (potentially random) list, which will be called for each chain.

```
> inits <- function(){
        list(alpha0=rnorm(0, 10), alpha1=rnorm(0, 10),
            alpha2=rnorm(0, 10), alpha12=rnorm(0, 10),
            tau=runif(0, 5))}
```

[†]Any error message about an "incomplete final line" can safely be ignored.

The parameters to monitor are defined, and the `bugs()` function runs Win-
BUGS in the background with the given model, data, initial values, number of chains `n.chains`, total iterations `n.iter`, initial iterations to discard
`n.burnin`, and thinning interval `n.thin`.

```
> parameters <- c("alpha0","alpha1","alpha2","alpha12","sigma")
> seeds.sim <- bugs(data, inits, parameters, model.file,
    n.chains=2, n.iter=14000, n.burnin=4000, n.thin=1,
    bugs.directory="c:/Program Files/WinBUGS14/")
```

The directory `bugs.directory` where WinBUGS is installed may need to be
changed from the default. Other useful options include `debug=TRUE`, which
keeps WinBUGS open to allow further interactive investigation into the
progress of the MCMC chain, before returning to R.

Printing the object returned by `bugs()` shows the posterior mean and standard deviation, a set of five quantiles for the monitored parameters, and the
DIC and effective number of parameters p_D for model comparison. Also shown
are convergence diagnostics using the Brooks–Gelman–Rubin method (§4.4.2).

```
> seeds.sim
Inference for Bugs model at "Seeds_mod.txt", fit using WinBUGS,
 2 chains, each with 14000 iterations (first 4000 discarded)
 n.sims = 20000 iterations saved
              mean  sd  2.5%   25%    50%    75% 97.5% Rhat n.eff
alpha0       -0.6 0.2 -0.9  -0.7   -0.6   -0.4  -0.2    1  2000
alpha1        0.1 0.3 -0.6  -0.1    0.1    0.3   0.7    1  9500
alpha2        1.4 0.3  0.8   1.2    1.4    1.5   1.9    1  2800
alpha12      -0.8 0.4 -1.7  -1.1   -0.8   -0.6   0.0    1 20000
sigma         0.3 0.1  0.0   0.2    0.3    0.4   0.6    1  1200
deviance    102.1 7.0 90.1  96.9  101.6  107.1 116.0    1   980

For each parameter, n.eff is a crude measure of effective sample
size, and Rhat is the potential scale reduction factor (at
convergence, Rhat=1).

DIC info (using the rule, pD = Dbar-Dhat)
pD = 11.2 and DIC = 113.3
DIC is an estimate of expected predictive error (lower deviance
is better).
```

Plotting the object, `plot(seeds.sim)`, will give simple line plots of posterior credible intervals for each parameter. However, we are not restricted to the
facilities provided with R2WinBUGS — any of the data manipulation and plotting capabilities of R and its contributed packages may be exploited instantly
to perform further analyses and produce publication-quality graphs and tables. In Figure 12.2, the posterior distributions of the odds ratios $\exp(\alpha_1)$,

$\exp(\alpha_2)$, $\exp(\alpha_{12})$ of seed germination corresponding to seed type 73 (versus 75), cucumber (versus bean), and their interaction, respectively, are illustrated as *density strips* using the `denstrip` package in R. The darkness of the strip is proportional to the posterior density, fading to white at zero. The axis is plotted on the log scale, and mathematical symbols are used in the annotation. The posteriors are obtained from the MCMC samples available in `seeds.sim$sims.list`. Density strips can also be plotted directly in OpenBUGS via `Inference->Compare`. The odds of germination are about five times greater for cucumber root extract, while the seed type only affects the odds ratio for cucumber root extract.

```
install.packages(denstrip) # if necessary
library(denstrip)
plot(0, type="n", xlim=c(0.1, 6), ylim=c(0,3),
     xlab="Odds ratio", ylab="", log="x", yaxt="n",
     bty="n", cex.axis=1.5, cex.lab=1.5)
abline(v=1, col="lightgray")
for (i in 1:3) {
  sam <- exp(seeds.sim$sims.list[[i+1]])
  denstrip(sam, at=3-i, tick=quantile(sam, c(0.025, 0.975)),
           mtick = median(sam))
}
text(0.1, 2:0 + 0.2,
     c(expression(paste("Seed type 73 (vs. 75): ",
                   e^alpha[1])),
       expression(paste("Cucumber (vs. bean): ",
                   e^alpha[2])),
       expression(paste("Interaction (73 and cucumber): ",
                   e^alpha[12])),
     pos=4)
```

12.4.8 WBDev

The BUGS language can be used to perform calculations by defining new nodes as functions of other nodes, using the built-in mathematical functions available in BUGS. However, the processing and memory requirements for computing and storing nodes increase vastly as the calculations become more complex. Furthermore, the BUGS code becomes less clear as more complex calculations are incorporated into the same model. Similarly, new distributions may be defined by explicitly writing out their log-likelihoods using the zeros or ones trick (§9.5), but the BUGS language is an inefficient and cumbersome means of doing this.

The WBDev (WinBUGS development) interface (Lunn, 2003) was developed to allow users to write their own functions and statistical distributions

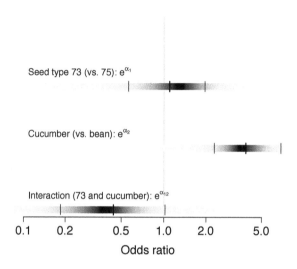

FIGURE 12.2
Posterior density strip plots for the odds ratios for seed germination using R
graphics after an R2WinBUGS analysis.

in efficient compiled code. The function or distribution is packaged into a new
modelling component which acts as a "black box" — i.e., its implementation
details are stored separately from the BUGS model in which it is employed.
This makes the BUGS model specification vastly clearer, reducing the scope
for errors. Using compiled code also offers computational improvements up to
orders of magnitude. For example, Jackson et al. (2010a) performed a health
economic analysis in WinBUGS and WBDev, where calculating expected costs
and benefits of different treatment choices involved a product of over 500 8×8
transition probability matrices from a Markov model. This would have been
practically impossible in pure BUGS code.

There are three main steps to developing a new function or distribution
using WBDev.

- Install the Blackbox Component Builder software and integrate it with
  the WinBUGS installation.

- Edit a template file, or "module" written in Component Pascal. The user
  needs only to write the minimum of code necessary to define the new
  function or the probability density of a new distribution (and optionally also the cumulative density and a random sampling procedure). All
  the facilities of Component Pascal and its libraries, used to write Win-

BUGS itself, are available to developers, providing much more scope for complex calculations.

- Add a line to a "resource file" to tell WinBUGS the name and location of the newly defined function or distribution. The new module may also be distributed for the benefit of other users.

WBDev is available to download from `http://www.winbugs-development.org.uk`. More detailed documentation and instructions are provided in the manuals distributed with WBDev. Some functions and distributions written by WinBUGS users can also be downloaded from here.

In OpenBUGS there are templates for implementing new distributions and functions in the source code in a similar manner; see § 12.5. JAGS (§ 12.6) may be extended similarly by writing modules, though this is not currently documented.

12.5   OpenBUGS

WinBUGS is stable and will remain available, but it will no longer be developed beyond the version (1.4.3) used to run the examples in this book. Development is now concentrated on the OpenBUGS project (`http://openbugs.info`). OpenBUGS began by "forking" or diverging the Component Pascal source code of WinBUGS and evolved into a completely separate engine for the BUGS language. OpenBUGS has now been validated on an extensive suite of examples to ensure that it gives the same correct results as WinBUGS, and it is now recommended for routine use.

There were several motives for the development of OpenBUGS.

Better communication with other software OpenBUGS has a Windows graphical user interface (GUI) which is largely identical to WinBUGS, but the internal engine for graphical modelling and computation has been decoupled from the user interface. OpenBUGS is distributed with a *shared library* (`.dll` on Windows, or `.so` on Linux) which provides an application programming interface (API) for graphical modelling. Any external software which can call C-based libraries can then be enhanced with the graphical modelling and MCMC capabilities of OpenBUGS. Unlike the scripting facility described in § 12.4.6, this is fully *interactive*. This means that OpenBUGS commands can be run one at a time, returning the user to the calling program after each command, instead of having to run an entire OpenBUGS session before returning. The `BRugs` package for R, described below, is the only such interactive interface we know of at the moment.

Portability The Blackbox Component Builder environment and libraries used to build WinBUGS and OpenBUGS are only available on Windows. Therefore Windows is required to modify or extend the OpenBUGS source code. However, Blackbox on Windows can be used to build a shared object (.so) in the ELF format. This is the standard format for shared libraries on Unix/Linux systems running on x86, the standard PC processor hardware. This allows OpenBUGS to run natively on a PC running Linux. Currently there is only a plain-text command-line interface and an R interface (described below) available for this library. Running the OpenBUGS GUI on Linux still requires an "emulator" such as Wine.

Open source principles The benefit of open source development in statistics has been shown by the success of R (R Development Core Team, 2011) and its huge repository of add-on packages. R has now superseded the proprietary S-Plus as the dominant statistical software in research environments. With open source software, all the algorithms used for computation may be inspected, users may modify the code for their own needs, and modifications may be shared with others. The GNU General Public License, which is applied to OpenBUGS, not only ensures that the source for OpenBUGS is available, but also that any published modifications must remain open. The potential for collaborative development on OpenBUGS, however, has been hampered in practice by the obscurity of the Component Pascal language and the restriction of the Blackbox tools to Windows. This is likely to have discouraged developers used to more widespread languages such as C++ or Java and scientific programmers accustomed to working with GNU tools in Unix/Linux (including Mac) environments.

Platform for new developments The open source infrastructure is intended to make OpenBUGS a platform for experimental developments in Bayesian modelling, which may then be disseminated for the benefit of other users. This has led, for example, to features for parallel computing (§12.5.4) and differential equation modelling (§11.5). While WinBUGS allows users to write their own functions and distributions using the WBDev facility (§12.4.8), these can be implemented (as well as new sampling algorithms) just by editing the OpenBUGS source. Templates are provided (see the OpenBUGS Developer Manual) so that the user only has to write the minimum amount of code necessary to define their function or distribution. Entirely new "subsystems," which may comprise functions, distributions, and samplers, may also be written and automatically loaded without needing to modify the existing OpenBUGS code. The `Reliability` subsystem for industrial reliability analysis (Kumar et al., 2010), is an example of this, currently distributed with OpenBUGS.

12.5.1 Differences from WinBUGS

The OpenBUGS graphical interface is largely identical to WinBUGS. Models are loaded and run, and data files are formatted, in the same way. The BUGS language used for models is also the same, though with a few additions.

- Several new functions and distributions have been added to OpenBUGS, for example, `eigen.vals`, `integral`, `gammap`, `ode`, `prod`, `p.valueM`, `solution`.

- The `I()` censoring function is still supported, but the preferred function for this purpose is now `C()` to reduce confusion between censoring and truncation (see §9.6). OpenBUGS now has a `T()` function for truncation, but it is currently only partially implemented and not well documented or tested.

- A `dloglik` distribution was added for clearer model statements when implementing the "zeros trick" (§9.5) for generic likelihoods and priors.

- The names of the scripting commands have been changed in OpenBUGS (see §12.5.3 below), and many new commands have been added.

Other new features are currently summarised on the web page `http://openbugs.info/w/OpenVsWin`. Most importantly, the choice of MCMC updating methods has been made more flexible. In WinBUGS, the updater could only be changed for all nodes of a particular type by editing a configuration file before opening WinBUGS. In OpenBUGS, before compiling, the user can select the updater to be used for each node (`Model->Updater Options`). An updater can also be disabled globally for all nodes in the session.

12.5.2 OpenBUGS on Linux

OpenBUGS on Linux is distributed with a basic command line interface. This has sometimes been called "LinBUGS," but it has the same modelling capability as any other OpenBUGS interface, since it simply calls on the OpenBUGS shared library. We therefore simply call it OpenBUGS for Linux.

After installing the package in the standard way as described on its download page, OpenBUGS is launched from the Linux command prompt by typing:

```
OpenBUGS
```

This gives an OpenBUGS command prompt:

```
OpenBUGS version 3.2.1 rev 781
type 'modelQuit()' to quit
OpenBUGS>
```

Standard scripting commands can then be entered into this prompt to load, run, and analyse the model. Perhaps more usefully, given a script file stored in the file /path/to/script.txt, a full OpenBUGS session can be run from the Linux command prompt as follows:

```
OpenBUGS /path/to/script.txt
```

or equivalently,

```
OpenBUGS < /path/to/script.txt
```

The sampled chains can be saved in CODA files using the samplesCoda() command (see below). An entire transcript of the session can be saved by redirecting the output of the Linux command to the file log.txt:

```
OpenBUGS /path/to/script.txt > log.txt
```

12.5.3　BRugs

The bugs() function from the R2OpenBUGS package in R (analogous to R2WinBUGS) may be used to run an entire OpenBUGS session in the background. The BRugs package, however, enables *fully interactive* use of Open-BUGS from R on both Windows and Linux.[‡] This makes it possible, for example, to set additional monitors, run more updates, check convergence, and so on, in the middle of a run without needing to exit and restart WinBUGS. Furthermore, the command line OpenBUGS on Linux is a very limited interface. For example, on a standard terminal, it is not possible to recall and edit previous commands. To plot traces and diagnose convergence in the middle of a model run, a user would need to save the sample history to a file using coda() and use external MCMC diagnostic software such as the coda package in R. Using BRugs, models can be controlled and queried entirely interactively from R.

All R commands provided by BRugs have the same names as their Open-BUGS scripting equivalents (though different from WinBUGS scripting commands, §12.4.5) and obey the same syntax. First type

```
library(BRugs)
setwd('c:/Program Files/OpenBUGS/OpenBUGS321/Examples')
```

in R to load the package and change the working directory to where the model files are stored. From then on, a sequence of commands like the one below can be used both under R and as a script for running from the OpenBUGS Windows program.

[‡]Partially interactive use is possible in R2OpenBUGS, which allows a running session to be saved to a file and subsequently resumed from the same point.

```
modelCheck('Seedsmodel.txt')
modelData('Seedsdata.txt')
modelCompile(2)
modelInits('Seedsinits1.txt', 1)
modelInits('Seedsinits2.txt', 2)
modelGenInits()
modelUpdate(4000)
samplesSet('alpha0')
samplesSet('alpha1')
samplesSet('alpha12')
samplesSet('alpha2')
samplesSet('sigma')
modelUpdate(10000)
samplesStats('*')
samplesHistory('*')
samplesCoda('*','Seeds')
```

The `coda` package can then be used for diagnostics and summaries. The R command

```
seeds.coda <- read.openbugs("Seeds")
```

reads the entire sample history into an R object `seeds.coda`, which is an R object of class `mcmc.list`. This is essentially a list of matrices of class `mcmc`, one for each of the two chains, with a row for each sample and a column for each variable. Most functions for summaries and plots in the `coda` package act on this type of object. For example,

```
HPDinterval(seeds.coda)
```

estimates intervals of highest posterior density for each parameter, a feature unavailable in OpenBUGS.

12.5.4 Parallel computation

For several years, the maximum clock speed of single computer processor cores has been limited to around 3–4 GHz, while the number of cores in a computer has increased. At the time of writing, a typical consumer PC contains two to four processor cores. To make the best use of these resources, different processors should be allowed to perform computations at the same time. *Parallelised* or *threaded* programming has therefore become increasingly important.

MCMC is a natural application for parallel programming, since different MCMC chains are independent conditionally on their starting values and random number seeds and do not need to communicate with each other while running. OpenBUGS now contains the infrastructure for parallel MCMC computation. An executable program called `MultiBUGS.exe` is under development and will soon be distributed with OpenBUGS.[§] This is identical to the Open-BUGS Windows interface, except that if more than one chain is compiled, then each chain is computed on a different processor. The first chain is computed using the `MultiBUGS.exe` process, which launches multiple instances of another executable called `WorkerBUGS.exe` to compute each subsequent chain. Initial values, random number seeds, and monitored samples are communicated between the master and worker processes by saving to files.

Multiple processing will not be enabled in the standard OpenBUGS program due to the overhead of launching the worker processes and communicating information. For simpler models which only take a minute of run time, there would be little benefit. But for large, computationally intensive models, substantial savings in computer time can be made. MultiBUGS is not currently available for Linux, although it is possible in principle.

Within WinBUGS, and indeed from OpenBUGS, MCMC computations can be parallelised by calling from other software, as in §12.4.6. Different instances of WinBUGS would be launched simultaneously for the same model and data, but with different initial values and random number seed specification. The results would then be combined and analysed by reading `coda` output, for example, with the `coda` package for R.

12.6   JAGS

JAGS,[¶] which stands for "Just Another Gibbs Sampler" (Plummer, 2003), is the most recent engine for the BUGS language, developed entirely independently of WinBUGS and OpenBUGS. This was written in C++, developed using the GNU compilers and packaging tools, is open source, and is freely available from `http://mcmc-jags.sourceforge.net`. JAGS may run natively on any system supported by the GNU tools, including Windows and many varieties of Unix such as Linux and Mac OS X. Currently, binary versions are available for Windows, Mac OS X, and most Linux distributions.

The current version (3.2.0) implements most of the same functionality as WinBUGS, with some minor but important differences in the language syntax (see below) and some extra features. A notable WinBUGS facility currently missing from JAGS is spatial models (§11.3), due to its restriction to directed graphs. The MCMC methods currently available in JAGS are restricted to conjugate and slice samplers, and specialised samplers for mixture models and generalised linear models, whereas OpenBUGS contains many more spe-

---

[§]The source code for an experimental version of MultiBUGS is already available with the OpenBUGS download.
[¶]Thanks to Martyn Plummer for his help in writing this section.

cialised samplers (§12.2). Since JAGS is built on a version of the numerical library (**Rmath**) used for R, many of the functions in base R for mathematical and statistical calculations are also available in the JAGS dialect of the BUGS language.

Just like OpenBUGS, JAGS includes a shared C library for BUGS language interpretation and MCMC computation. The user interfaces to JAGS all call on the same shared library. Currently there is a basic command-line interface (§12.6.4) and an R interface (§12.6.5).

12.6.1 Extensibility: modules

Like "subsystems" in OpenBUGS, users may develop *modules* for some specialised application, without needing to alter the core JAGS code. Modules may be loaded and unloaded within a single JAGS session. A single module may contain new functions, distributions, MCMC samplers, random number generators, and "monitors" (facilities to record sampled quantities for later analysis). This is intended to make JAGS a platform for new developments in Bayesian statistics. Often a new statistical model or sampling method is developed as an extension of an existing model. By working within the BUGS graphical modelling framework, researchers may implement just their specialised extension, without needing to "re-invent the wheel" by developing generic modelling functions or user interfaces. Their new module may then be distributed for the benefit of other users.

The **bugs** module, for example, contains the functions, distributions, and other facilities available in WinBUGS. The **dic** module provides functions for model comparison using deviance-based measures. In particular it implements a monitor for the model deviance, and methods for estimating the effective number of parameters (p_D, §8.6.1). These provide alternatives to the conventional DIC (§8.6.4) which were proposed by Plummer (2008) for estimating the expected predictive utility of models.

12.6.2 Language differences

The current JAGS user manual explains the differences between JAGS and WinBUGS or OpenBUGS. For example, the data format is different (see below), although formatting data is not a concern when running through the **rjags** R interface. Some specific functions and distributions are defined differently, as described in Appendices B and C — but here we explain more fundamental language differences.

Censoring and truncation Some improvements were made to clarify certain confusing aspects of the BUGS language. In particular, WinBUGS represents censoring, truncation, and prior ordering all by the I() construct, which tempts incorrect use (see § 9.6). JAGS separates these concepts.

Truncated distributions are simply given using the `T()` function, for example, `dnorm(mu, tau)T(lower,upper)` for the truncated normal. This allows truncated distributions to be used as likelihoods or priors, avoiding the danger of using `I()` wrongly in WinBUGS.

A right-censored survival time (see §11.1) is specified in WinBUGS as

```
t[i] ~ dweib(r, mu[i])I(c[i], )
```

where `t[i]` takes the value `NA` in the data if it is censored (`t[i] > c[i]`), and `c[i]=0` if `t[i]` is observed. In JAGS this becomes

```
X[i] ~ dinterval(t[i], c[i])
t[i] ~ dweib(r, mu[i])
```

where `X[i]` takes the value 1 if `t[i]` is censored and 0 otherwise. The data for `t[]` and `c[]` are defined in the same way.

`dinterval` is a distribution representing general interval-censored data. `X ~ dinterval(t, c[])` indicates that the scalar X is a coarsened version of a continuous variable t based on a vector of cut points c_m. If $t \leq c_1$ then $X = 0$, if $c_m < t \leq c_{m+1}$ then $X = m$ (for $m = 1, \ldots, M - 1$), or if $t > c_M$ then $X = M$. For simple right-censoring, as in the example above, there is just one cut point $M = 1$, then c is a scalar, and X takes the values 0 and 1.

Prior ordering of a set of parameters `a[1:n]` (see §9.7.2) can simply be implemented by the `sort` function, which sorts a vector in ascending order. The elements of `a[]` are given independent priors and replaced by `a0[1:n] <- sort(a)`.

Data as observable functions of parameters `dinterval` is a distribution because it generates a likelihood for data X. The likelihood is 1 if the unknown t is in a range consistent with the data X and zero otherwise. However, when sampling from this distribution, the outcome X has only one possible value given t and the c_m; therefore in this sense `dinterval` behaves like a deterministic function. Relations with this dual nature are called "observable functions" in JAGS.

Another observable function is `dsum`, which allows aggregate data X to be modelled as a sum of two unknown parameters a and b:

```
X ~ dsum(a, b)
```

instead of `X <- a + b`.

We would like X to contribute to the likelihood for the parameters a and b, but the latter form doesn't allow the true value of X to be supplied as data. `dsum()` enables this likelihood contribution to be obtained. `dsum` is used, for example, in ecological inference (see Wakefield et al. (2011) and Example 11.4.1), where individual-level relationships are inferred from aggregate data.

Data transformations WinBUGS allows an observed variable to be transformed and then given a model (§A.7), for example,

```
for (i in 1:N) {
  z[i] <- sqrt(y[i])
  z[i]  ~ dnorm(mu, tau)
}
```

This infringes the "declarative" principle of the BUGS language — that each variable should be defined only once in terms of its parents. To transform data in JAGS, a separate `data` block must be written:

```
data {
  for (i in 1:N) {z[i] <- sqrt(y[i])}
}
model {
  for (i in 1:N) {z[i]  ~ dnorm(mu, tau)}
  ...
}
```

This effectively defines a distinct graphical model. Stochastic variables can also be defined in the `data` block — allowing simulation studies where data are generated and analysed by different models.

Vectorised calculations In JAGS, just as in R or S-Plus, a scalar c can be added to a vector (or array) A using

```
B <- A + c
```

Any scalar function is automatically vectorised in this way when its arguments are arrays with conforming dimensions, or scalars. In WinBUGS and OpenBUGS, this more verbose form would be required:

```
for (i in 1:n) {
  B[i] <- A[i] + c
}
```

Vectorisation may slow down MCMC sampling, however. In this example, one vector node `B` is created in place of `n` scalar nodes. If only one element `A[i]` of `A` is updated, then the whole of `B` is recalculated instead of just the corresponding `B[i]`. This can create a bottleneck when updating the model, especially if `n` is large. Link functions cannot be vectorised, to avoid this problem in generalised linear models.

Deterministic nodes In JAGS, a deterministic node need not be explicitly defined to be used on the right-hand side of a stochastic relation. For example,

```
for (i in 1:n) {
   y[i] ~ dnorm(a + b*x[i], tau)
}
```

would be permissible in JAGS, but not in WinBUGS and OpenBUGS, where it must be written in a form such as

```
for (i in 1:n) {
   y[i] ~ dnorm(mu[i], tau)
   mu[i] <- a + b*x[i]
}
```

Multivariate nodes A multivariate node in JAGS must either be fully observed or fully unobserved. Therefore to model a multivariate normal vector where some variables are missing, a conditional specification is required (see §C.4).

12.6.3 Other differences from WinBUGS

Data formatting Data are supplied to JAGS in a different format to WinBUGS and OpenBUGS. For example, the Seeds data are formatted as

```
r <-
c(10, 23, 23, 26, 17, 5, 53, 55, 32, 46, 10, 8, 10, 8, 23, 0,
3, 22, 15, 32, 3)
n <-
c(39, 62, 81, 51, 39, 6, 74, 72, 51, 79, 13, 16, 30, 28, 45,
4, 12, 41, 30, 51, 7)
x1 <-
c(0, 0, 0, 0, 0, 0, 0, 0, 0, 0, 0, 1, 1, 1, 1, 1, 1, 1, 1, 1, 1)
x2 <-
c(0, 0, 0, 0, 0, 1, 1, 1, 1, 1, 1, 0, 0, 0, 0, 0, 1, 1, 1, 1, 1)
N <- 21
```

This data format is identical to the R commands required to define the corresponding variables as R objects. With this format, JAGS is consistent with R/S-Plus in filling matrices by row, instead of by column as WinBUGS expects (see §12.4.2).

Existing R objects may be saved in this format using the R command:

```
dump(c("r","n","x1","x2","N"), file="seeds-data.R")
```

The variable N does not need to be supplied in JAGS, unlike WinBUGS, since the `length` function may be used as follows to determine the length of the variable r.

```
for (i in 1:length(r)) {
  r[i] ~ dbin(p[i], n[i])
    ...
}
```

Initial values In JAGS, if initial values are not supplied for a parameter and the user requests that JAGS provides one automatically, this is not generated from the prior but instead set to a plausible central value given the prior. This value depends on the distribution but is typically the mean, median, or mode. This avoids the common problem in WinBUGS where an extreme initial value is generated, causing WinBUGS to crash.

12.6.4 Running JAGS from the command line

JAGS has a command line interface and can be invoked interactively or through a script file. The same commands are used in each case. To invoke JAGS interactively, simply type `jags` at the Unix or Windows command prompt. To invoke JAGS with a script file, say, `Seeds_jags.cmd`, type

```
jags Seeds_jags.cmd
```

A typical script file has the following commands. It is easy to see the correspondence with WinBUGS or OpenBUGS script commands.

```
model in "Seedsmodel.txt"
data in "seeds-data.R"
compile, nchains(2)
parameters in "seeds-inits1.R"
parameters in "seeds-inits2.R"
initialize
update 4000
monitor alpha0
monitor alpha1
monitor alpha12
monitor alpha2
monitor sigma
update 10000
coda *, stem(seedsCODA)
```

The last line of this script saves the monitored parameter values to files in CODA format with names beginning with `seedsCODA`. These can then be read into R using functions from the `coda` package, for example,

```
library(coda)
seeds.coda <- read.openbugs("seeds")
```

Summary statistics and diagnostics can then be produced; for example,

```
summary(seeds.coda)
plot(seeds.coda)
```

plots sample history and posterior densities.

12.6.5   Running JAGS from R

The package rjags (Plummer, 2011), available from the CRAN archive (http://cran.r-project.org), gives a fully interactive R interface to the JAGS computation engine, allowing Bayesian models to be developed entirely within the R interface. An advantage over the JAGS command-line interface is that there is no need to reformat the data and initial values, since these are simply supplied as R list objects.

```
library(rjags)
seeds <- list(r=r, n=n, x1=x1, x2=x2, N=N)
seeds.jags <- jags.model("Seedsmodel.txt", data=seeds,
                         n.chains=2)
```

The command jags.model compiles and initialises the given model, using the given data and number of chains. Initial values may also be supplied as an R list in an inits argument. For certain MCMC algorithms, an *adaptation* period is run at this stage. This may be controlled by the argument n.adapt and is 1000 iterations by default. If optimal behaviour of the sampler is not attained, a warning is given and the adaptation period can be extended by calling the adapt function.

The following command then updates the model for 4000 iterations without monitoring any parameters.

```
update(seeds.jags, 4000)
```

After this burn-in, 10,000 samples are drawn from five parameters as follows.

```
seeds.coda <- coda.samples(seeds.jags,
            c("alpha0","alpha1","alpha2","alpha12","sigma"),
            10000)
```

The coda.samples function produces an R object of the same mcmc.list class as produced by reading CODA files using read.openbugs. Thus the same functions as before may be used to produce summary statistics and plots:

```
summary(seeds.coda)
plot(seeds.coda)
```

The package `R2jags` (Su and Yajima, 2011) provides a non-interactive interface to JAGS, analogous to `R2WinBUGS`. An entire JAGS analysis can be run using a single function call from R. The package provides a facility for running multiple chains on parallel processors.

Appendix A

BUGS language syntax

A.1 Introduction

As described in Chapter 12, the BUGS language is currently implemented in WinBUGS, OpenBUGS, and JAGS, which have minor differences in their capabilities and language syntax. In this chapter we focus on BUGS as implemented in WinBUGS, but mention differences in other implementations.

The BUGS language facilitates a declarative, textual description of the probability model whereby the relationship between each node and its parents is stated explicitly; the software uses this to construct an internal, graphical representation of the model. Note that the declarative nature of the language means that the order in which the various relationships are specified is irrelevant. There are two possible types of relation:

- ~ means "is distributed as" and denotes a *stochastic* relation;

- <- means "is to be replaced by" and denotes a *logical* relation.

Nodes on the left of a ~ are stochastic nodes and those on the left of a <- are logical (or deterministic) nodes. In addition, a model may include *constant* nodes, which are fixed by the design of the study and have no parents — unless supplied as logical constants, their values must be specified as part of the data (see Chapter 2). Note that, generally speaking, each logical or stochastic node should appear once and only once on the left-hand side of a statement (although see § A.7 for exceptions to this rule). This chapter provides an overview of the syntax available for fully specifying each parent–child relationship.

A.2 Distributions

A.2.1 Standard distributions

Lists of continuous and discrete distributions available in the different implementations of BUGS, along with their full definitions and examples of their usage, are given in Appendices B and C. Note that the parameters of distributions, i.e., quantities occurring on the right-hand side of a ~ relation, must be named nodes or numerical values rather than expressions. For example, we cannot specify x ~ dnorm(a + b*z, 1), say; instead, we must use something like:

```
x    ~ dnorm(mu, 1)
mu <- a + b*z
```

A.2.2   Censoring and truncation

Suppose a stochastic quantity x has been observed to lie in the interval (a, b). In WinBUGS we can specify such *interval censoring* by following the distributional expression for x with I(a, b). That is,

```
x ~ ddist(psi)I(a, b)
```

where psi (ψ) denotes a generic set of distributional parameters. (JAGS deals with censoring differently; see §12.6.2). Note that censoring bounds must be named nodes or numerical values rather than expressions, and that leaving either censoring bound blank simply corresponds to no limit, e.g., I(a,) designates a quantity that has been observed to be greater than a, but not less than any specific amount.

It is vital to understand that use of the I(.,.) construct is *not* the same as specifying a truncated distribution. The sole effect of the I(.,.) expression is to ensure that all values sampled for x outside the specified interval are rejected. This may seem like an appropriate strategy for sampling from truncated distributions, but the full conditional distribution derived for ψ is inconsistent with truncation. To see this, note that the density of a truncated distribution, when considered as a function of its parameters (ψ, as opposed to x), is *not* simply proportional to the untruncated density. Instead, it is normalised by the integral of the untruncated density over the truncation interval, which is a complex function of ψ (and a and b). BUGS does not take account of this normalising constant when deriving the full conditional for ψ, as it is inappropriate to do so in the case of interval censoring — the correct likelihood contribution is the untruncated density.

Of course when ψ is known, for example, when $p(x|\psi)$ is a hyperprior, there is no need to sample from $p(\psi|.)$ and the effect of using I(.,.) *is* equivalent to specifying a truncated distribution for x. However, in general, it is the user's responsibility to check that appropriate sampling will result whenever the I(.,.) construct is stretched beyond its intended purpose. In WinBUGS and OpenBUGS, if truncated distributions are required in cases where ψ is unknown, then they might be handled by working out an algebraic form for the likelihood and using the techniques for implementing non-standard distributions discussed in the following subsection. In JAGS, distributions may be easily truncated in any situation, as described in §12.6.2.

Note that the I(.,.) notation is never appropriate when x is observed (the bounds will simply be ignored in this case). Note also that even when I(.,.) is used for censoring, if x, ψ, a, and b are all unknown, then a and b must not be functions of ψ.

In OpenBUGS, I(,.,) works in the same way as in WinBUGS, but the equivalent function C(.,.) is preferred to clarify that this is intended for censoring rather than truncation. A T(.,.) function is intended to be used for truncation, but currently this is only partially implemented.

A.2.3 Non-standard distributions

Distributions other than those given in Appendix C can still be implemented by making use of either the "ones trick" or the "zeros trick" (see § 9.5 for details). Alternatively, new distributions can be "hard-wired" into Win-BUGS using the WinBUGS Development Interface (WBDev (Lunn, 2003); http://www.winbugs-development.org.uk), as described in §12.4.8. Or in OpenBUGS and JAGS, the program source code may be extended directly, as discussed in Chapter 12.

A.3 Deterministic functions

A.3.1 Standard functions

Logical expressions can be built using the operators +, -, *, and / and the standard functions listed in Appendix B. Bracketing can be used to any depth. All scalar-valued parameters appearing on the right-hand side of a <- relation can be expressions as well as named nodes or numerical values, with the exception (in WinBUGS and OpenBUGS) of the "index" parameter (i) in both rank(.,.) and ranked(.,.). Note that the functions cloglog(.), log(.), logit(.), and probit(.) can be used on the left-hand side of a logical relation, as indicated in the "Usage" column (§B.1). Also note that logical nodes cannot be given data or initial values (except when using the data transformation facility described in §A.7).

A.3.2 Special functions

A.3.2.1 The "cut" function

Suppose we observe some data that we do not wish to contribute to the parameter estimation and yet we wish to consider as part of the model. This might happen, for example: (a) when we wish to make predictions on some individuals for whom we have observed partial data that we do not wish to use for parameter estimation; (b) when we want to use data to learn about some parameters but not others; or (c) when we want evidence from one part of a model to form a prior distribution for a second part of the model, but we do not want "feedback" from this second part.

The `cut(.)` function forms a kind of "valve" in the graph: prior information is allowed to flow "downwards" through the cut, but likelihood information is prevented from flowing upwards. In practical terms, the syntax `y <- cut(x)` produces a logical node `y` that is an exact copy of node `x` in the sense that it always has the same value as `x` (`x` may be stochastic or logical). However, any descendants of `y` that are introduced into the graph cannot then influence inference on `x`. The reader is referred to §9.4 for examples of appropriate usage of `cut(.)`. Note this facility is currently unavailable in JAGS.

A.3.2.2 Deviance

BUGS automatically creates a logical node called **deviance** for the specified model. This evaluates, using the current state of the model, minus twice the logarithm of the conditional likelihood. By conditional likelihood we mean the joint probability distribution of all observed and censored nodes, conditional on their stochastic parents. The **deviance** node can be monitored and contributes to the calculation of DIC (see §8.2, §8.6.4).

A.3.3 Add-on functions

There are several add-on interfaces to WinBUGS that contain libraries of specialized functions, e.g., PKBugs (Lunn et al., 2002) for pharmacokinetic modelling, and Jump (Lunn et al., 2009c) for reversible jump computation (§8.8.2) on variable selection and spline models. In addition, WBDev (Lunn, 2003) can be used to "hard-wire" one's own specialized functions into the Win-BUGS framework — see §12.4.8. This can offer massive gains in efficiency over building the relevant expression using standard BUGS syntax. In OpenBUGS and JAGS, the program source code itself can be extended to implement new functions, as discussed in Chapter 12.

A.4 Repetition

Repeated structures are specified using a "for-loop." The syntax for this is:

```
for (i in a:b) {
  list of statements to be repeated for increasing
  values of loop-variable i
}
```

Note that any depth of nesting of for-loops is permitted (as long as a different index/loop variable is used for each loop). Note also that a and b must both be integer-valued observed data or integer-valued logical expressions involving only standard operators (+, -, *, /), numerical values, observed data, and/or other for-loop indices (j, say). In particular, a and b must not be unobserved-stochastic or named-logical nodes. The step function can often be used to work around this. For example, to add up the first K out of 10 values of the vector x[], where K is random, instead of

```
for (i in 1:K) {
   xtosum[i] <- x[i]
}
s                 <- sum(xtosum[])
```

we could write

```
for (i in 1:10) {
   xtosum[i] <- x[i]*step(K - i + 0.1)
}
s                 <- sum(xtosum[])
```

As described in §12.6.2, JAGS also offers a concise R-like syntax for performing parallel scalar functions on *vectorised* arguments.

A.5   Multivariate quantities

We define a multivariate quantity as any collection of nodes, either all stochastic or all logical, that are defined simultaneously, e.g., x[] ~ dmnorm(mu[], T[,]), y[,] <- inverse(z[,]). All multivariate quantities must form contiguous elements of the array of nodes to which they belong. As we traverse contiguous elements of a given array, the fastest changing index is the final index, and so all multivariate quantities should be defined using the latter indices. For example, the following code specifies $y_i \sim \mathrm{MVN}_d(\mu, T^{-1})$ for a collection of vectors y_i, $i = 1, ..., K$, which form the rows of a $K \times d$ matrix Y:

```
for (i in 1:K) {
   Y[i, 1:d] ~ dmnorm(mu[1:d], T[1:d, 1:d])
}
```

Now suppose that each y_i has a distinct precision matrix T_i, as opposed to a common precision T, and that all of these matrices are to be stored in a three-dimensional array P[,,]. We specify a Wishart(R, k) prior for each matrix via:

```
for (i in 1:K) {
  P[i, 1:d, 1:d] ~ dwish(R[1:d, 1:d], k)
}
```

A.6 Indexing

A.6.1 Functions as indices

The four basic operators (+, -, *, and /) along with appropriate bracketing are allowed to calculate an integer function as an index, for example:

```
Y[(i + j)*k, 1]
```

On the left-hand side of a relation, an expression that always evaluates to a fixed value (given the values of any loop variables) is allowed for an index. On the right-hand side the index can be a fixed-value expression or a named node, which allows a straightforward formulation for mixture models, in which the appropriate element of an array is "picked" according to a random quantity (see §A.6.3 below). However, functions of unobserved nodes are not permitted to appear directly as index terms (named logical nodes may be introduced if such functions are required).

A.6.2 Implicit indexing

The conventions broadly follow those of S-Plus/R:

- n:m represents $n, n + 1, ..., m$;
- x[] represents all values of vector x;
- Y[,3] indicates all values in the third column of matrix Y.

Multidimensional arrays are handled internally as one-dimensional arrays with a "constructed" index. Thus functions defined on arrays must be over equally spaced nodes within the array: for example, y <- sum(i, 1:4, k).

A.6.3 Nested indexing

Nested indexing can be very effective. For example, suppose N individuals can each be in one of J groups, and g[1:N] is a vector containing the group memberships for each individual. Then group coefficients beta[j] ($j = 1, ..., J$) can be fitted using beta[g[i]] in a regression equation in which individuals are indexed by i.

In the BUGS language, nested indexing can be used for the parameters of distributions, e.g.,

```
for (i in 1:N) {
  c[i] ~ dcat(theta[1:J])
  y[i] ~ dnorm(mu[c[i]], tau)
}
```

Here $y[i]$, $i = 1, ..., N$, are realisations from a normal mixture distribution with J components, each with a distinct mean $mu[j]$, $j = 1, ..., J$, and a common precision tau. The $c[1:N]$ vector contains the indices of the components to which each observation belongs.

A.7 Data transformations

Although transformations of data can always be carried out before using BUGS, it is convenient to be able to try various transformations of dependent variables within a model description. For example, we may wish to try both y and sqrt(y) as dependent variables without creating a separate variable z = sqrt(y) in the data file.

The BUGS language therefore permits the following type of structure to occur:

```
for (i in 1:N) {
  z[i] <- sqrt(y[i])
  z[i]  ~ dnorm(mu, tau)
}
```

Strictly speaking, this goes against the declarative structure of the model specification, with the accompanying exhortation to construct a directed graph and then to make sure that each node appears once *and only once* on the left-hand side of a statement. However, a check has been built in so that when finding a logical node that also features as a stochastic node (such as z above), a stochastic node is created with the calculated values as fixed data. Note that this construction is only possible when transforming observed data (not a function of data and parameters) with no missing values.

JAGS addresses this issue using a separate **data** block, as described in §12.6.2.

A.8 Commenting

is the *comment* character, used to annotate BUGS code to help the programmer. BUGS ignores everything following # on a line.

```
model {
  Y   ~ dbin(0.5, 8)
  P2 <- step(2.5 - Y) # 1 if Y is 2 or less, 0 otherwise
}
```

Appendix B

Functions in BUGS

§B.1 lists the logical functions which are available in all current BUGS implementations, and §B.2 onwards list functions only in OpenBUGS or JAGS, with the occasional exception.

B.1 Standard functions

The standard functions listed in Table B.1 are used in the same way in WinBUGS, OpenBUGS, and JAGS, with a few exceptions listed in a footnote.

B.2 Trigonometric functions

OpenBUGS and JAGS provide a full complement of trigonometric functions and their inverse, hyperbolic, and inverse hyperbolic analogues:

```
sin arcsin sinh arcsinh
cos arccos cosh arccosh
tan arctan tanh arctanh
```

WinBUGS only includes `sin`, `cos`, and `tan`. In JAGS, the inverse functions may be called as `asin`, `acos`, etc., for consistency with R.

B.3 Matrix algebra

JAGS contains some extra functions for matrix calculations. The `t()` function transposes a matrix, the `%*%` operator multiplies two matrices of compatible dimensions, e.g., `A %*% B`, and the `mexp()` function in the `msm` module (for continuous-time Markov multi-state models) computes the matrix exponential

TABLE B.1
Functions available in all BUGS implementations.

Expression	Function	Usage	Definition
abs	absolute value	y <- abs(x)	$y = \lvert x \rvert$
cloglog	complementary log-log	y <- cloglog(x) cloglog(z) <- a + b*x	$y = \ln(-\ln(1-x)); \quad x \in (0,1)$ $\ln(-\ln(1-z)) = a + bx$
cos	cosine	y <- cos(x)	$y = \cos(x)$
equals	logical equals	y <- equals(a,b)	$y = 1$ if $a = b$, $y = 0$ otherwise
exp	exponential	y <- exp(x)	$y = e^x$
inprod	inner product	y <- inprod(a[],b[])	$y = \sum_i a_i b_i$
inverse	matrix inverse	y[1:n,1:n] <- inverse(x[,])	$Y = X^{-1}; \quad X$ & Y both $n \times n$
log	natural logarithm	y <- log(x) log(z) = a + b*x	$y = \ln(x); \quad x > 0$ $\ln(z) = a + bx$
logdet	log determinant	y <- logdet(x[,])	$y = \ln\lvert X \rvert$; X symmetric & positive definite
logfact	log factorial	y <- logfact(x)	$y = \ln(x!); \quad x = 0, 1, 2, \ldots$
loggam	log gamma function	y <- loggam(x)	$y = \ln(\Gamma(x)); \quad x > 0$
logit	logistic transform	y <- logit(x) logit(z) <- a + b*x	$y = \ln(\frac{x}{1-x}); \quad x \in (0,1)$ $\ln(\frac{z}{1-z}) = a + bx$
max	maximum	y <- max(a,b)	$y = \max(a,b)$
mean	mean	y <- mean(x[])	$y = \frac{1}{n}\sum_i x_i$
min	minimum	y <- min(a,b)	$y = \min(a,b)$

TABLE B.1
(Continued.)

Expression	Function	Usage	Definition
phi	standard normal distribution function	y <- phi(x)	$y = \Phi(x) = \int_{-\infty}^{x} \frac{1}{\sqrt{2\pi}} e^{-\frac{1}{2}t^2} dt$
pow	power	y <- pow(a,b)	$y = a^b$
probit	probit	probit(z) <- a + b*x	$\Phi^{-1}(z) = a + bx$
sqrt	square root	y <- sqrt(x)	$y = \sqrt{x}; \ x \geq 0$
rank	rank of ith element in vector	y <- rank(x[],i)	$y = \sum_j I(x_j \leq x_i)$
ranked	ith smallest value in vector	y <- ranked(x[],i)	$y = i$th smallest value in x
round	nearest integer	y <- round(x)	$y = \lfloor x + 0.5 \rfloor$
sd	standard deviation	y <- sd(x[])	$y = \sqrt{\frac{1}{n-1} \sum_i \left(x_i - \frac{1}{n} \sum_i x_i \right)^2}$
step	unit step	y <- step(x)	$y = 1$ if $x \geq 0$, $y = 0$ otherwise
sum	sum	y <- sum(x[])	$y = \sum_i x_i$
trunc	truncate towards 0	y <- trunc(x)	$y = \lfloor x + (1 - \varepsilon)I(x < 0) \rfloor$

- min(...), max(...) in JAGS accept any number of arguments and return their minimum or maximum.
- rank(v[]) in JAGS, given a vector v, returns a vector of ranks, so that the equivalent of rank(x[],i) is y <- rank(x[]); y[i]. Also ranked is not in JAGS — the equivalent of ranked(x[], i) is y <- sort(x[]); y[i].
- probit can be used on the right-hand side of a definition y <- probit(x) in JAGS, but in WinBUGS and OpenBUGS it can only be used as a link function, e.g., probit(y) <- a + b*x.

of a square matrix. The matrix exponential is a power series of matrix products $1 + A + A^2/2! + \ldots$, which is *not* the same as taking the scalar exponential of each element.

OpenBUGS has the function `eigen.vals(x)`, which returns the vector of eigenvalues of a square matrix `x`.

B.4   Distribution utilities and model checking

OpenBUGS provides the functions

```
density(s1, s2)
cumulative(s1, s2)
deviance(s1, s2)
```

to evaluate, respectively, the probability density $p(x|\theta)$, the cumulative density $\int_{-\infty}^{x} p(u|\theta)du$, and deviance $-2\log(p(x|\theta))$ of the scalar node `s1` evaluated at x defined by the current value of the scalar node `s2`, and θ defined by the current parameters of `s1`. The node `s1` must be a stochastic node.

JAGS provides a similar facility via a suite of functions with names beginning `d`,`p`,`q`, or `r`. For example,

```
dnorm(x,mu,tau)
pnorm(x,mu,tau)
qnorm(x,mu,tau)
rnorm(x,mu,tau)
```

produce the probability density, cumulative density, inverse cumulative density, and a random sample, respectively, from a normal distribution with mean `mu` and precision `tau`. Most of the distributions in Appendix C which are available in JAGS may be used in this way with analogous names — see the current JAGS user manual for further details. These functions are named like the corresponding functions in R, but the JAGS parameterisations are used.

OpenBUGS includes a set of functions intended for predictive model checking (see §8.4).

```
replicate.post(s)
replicate.prior(s)
```

generate a new node from the distribution of the scalar stochastic node `s`. In `replicate.post`, the current values of the parameters of this distribution are used, but in `replicate.prior`, these parameters are resampled in turn from their distributions. The choice depends on the desired "focus" for model assessment; see §10.7. `replicate.postM(v)` can be used if `v` is a vector. There are two corresponding functions

```
post.p.value(s)
prior.p.value(s)
```

for calculating predictive *p*-values. These return one if a sample from the distribution of s is less than the value of s, and zero otherwise. The latter function resamples any stochastic parents of s. post.p.value(s) has an equivalent p.valueM(v) for multivariate stochastic nodes v.

B.5 Functionals and differential equations

In OpenBUGS, there are some functions with arguments defined by *functions themselves*.

```
integral(F(s), s1, s2, s3)
```

returns the definite integral of function F(s) between s = s1 and s = s2 to accuracy s3, and

```
solution(F(s), s1, s2, s3)
```

returns a solution of equation F(s) = 0 lying between s = s1 and s = s2 to accuracy s3. s1 and s2 must bracket a solution.

The function argument in either case is defined by the special notation F(s). For example, an integral $y = \int_0^u \left\{ \cos(x)^2 + \sin(x)^2 \right\} dx$ with a random upper limit u can be evaluated by

```
y    <- integral(F(x), 0, u, 1.0E-6)
F(x) <- cos(x)*cos(x) + sin(x)*sin(x)
u      ~ dunif(0, 1)
```

However, since $\cos(x)^2 + \sin(x)^2 = 1$, y will have a Uniform(0,1) distribution just like u.

Ordinary differential equations can be solved in OpenBUGS using

```
ode(v1, v2, D(v3, s1), s2, s3)
```

which gives the solution at a grid of points v2, given initial values v1, at time s2, solved to accuracy s3. D(v3,s1) defines the system in terms of time s1 using the special function D(). See the Diff subdirectory of the OpenBUGS installation for detailed documentation. This is also available in WinBUGS via the add-on WBDiff package — see Example 11.5.1 for an example of defining D().

B.6   Miscellaneous

- **sort(x)** in OpenBUGS and JAGS sorts a vector **x** into ascending order.

- JAGS and OpenBUGS have the inverse logit function **ilogit(x)** defined as $\exp(x)/(1+\exp(x))$, and the inverse complementary log-log function **icloglog(x)** defined as $1 - \exp(-\exp(x))$.

- OpenBUGS has an incomplete gamma function **gammap(a, x)**, defined as the cumulative density of the gamma distribution with scale parameter 1.

$$\int_0^x \frac{t^{a-1}\exp(-t)}{\Gamma(a)}\,dt$$

For example, y and z will be the same in the following code.

```
w   ~ dgamma(3, 1)
y <- cumulative(w, 2)
z <- gammap(3, 2)
```

- **interp.lin(e, x, y)** (OpenBUGS and JAGS)

  Suppose we have a vector of function values $y_i = f(x_i)$ for $i = 1 \ldots n$ and we wish to predict the value $f(e)$ at a new point e. This function estimates this by a simple linear interpolation between the pair of points corresponding to the closest x_i above and below e:

  $$f(e) = y_p + (y_{p+1} - y_p)q, \quad q = (e - x_p)/(x_{p+1} - x_p), \quad x_p \le e \le x_{p+1}$$

  The elements of **x** must be in ascending order, **e** must be a scalar, and **x** and **y** must be vectors of the same length.

Appendix C

Distributions in BUGS

In a Bayesian analysis, as well as choosing sampling distributions for observable quantities, we must choose prior distributions to characterise uncertainty about parameters. Therefore a thorough understanding of different probability distributions and their properties is important to Bayesian work. Here we list the distributions available in current implementations of the BUGS language, with their common uses and properties. §9.5, §12.4.8 explain how to implement distributions not in this list.

Means and variances of all distributions are given where these exist, but note that the variance may not be a sensible measure to base a prior on when the distribution is very skewed. In that case, the typical rule of thumb (based on the normal distribution) that a 95% credible interval is about "mean ±2 standard deviations" will be inaccurate. A pair of quantiles will then be a better measure of spread.

A table of prior, posterior, and predictive distributions for conjugate Bayesian analyses is given in Chapter 3, Table 3.1.

C.1 Continuous univariate, unrestricted range

Normal   x ~ dnorm(mu,tau)

Density $p(x|\mu,\tau) = \sqrt{\frac{\tau}{2\pi}} e^{-\frac{\tau}{2}(x-\mu)^2}$ for $-\infty < x < \infty$, $\tau > 0$

Mean μ **Variance** $1/\tau$

The normal distribution is ubiquitous as a model (or prior) for continuously distributed quantities, due to its convenient properties and familiarity. It is fundamental to statistics as the sampling distribution for an empirical mean, from the central limit theorem. But remember that in BUGS there is rarely any algebraic or computational need to assume normality. Another distribution may be more realistic or fit the data better, perhaps due to skewness or heavier tails, and this is usually no more computationally difficult.

Since the very first version of BUGS, the normal has been parameterised by the *precision* $\tau = 1/\sigma^2$ rather than the commoner variance σ^2 or standard deviation σ. When used as a prior, for example, smaller precisions give vaguer priors. In retrospect this was an unwise decision — although using τ gives tidier expressions for the posterior distributions of the parameters under conjugate priors, it has caused a lot of confusion. However, changing the parameterisation at this stage would be likely to redouble the confusion for existing users!

Logistic x $\sim$ dlogis(mu,tau)

Density $p(x|\mu,\tau) = \tau e^{\tau(x-\mu)}/(1 + e^{\tau(x-\mu)})^2$ for $-\infty < x < \infty, \tau > 0$

Mean μ **Variance** $\frac{\pi^2}{3\tau^2}$

A standard logistic random variable with mean $\mu = 0$ and precision parameter $\tau = 1$ is the logit of a Uniform(0,1) random variable, hence its application as a prior for the intercept in a logistic regression model, equivalent to a uniform prior on the probability scale (§4.1.1, §5.2.5). Qualitatively it is very similar to the normal, though with a slightly heavier tail. Note that τ here is analogous to the inverse standard deviation $\sqrt{\tau}$ of the normal, not the inverse variance τ.

Student's *t* x $\sim$ dt(mu,tau,k)

Density $p(x|\mu,\tau,k) \quad = \quad \frac{\Gamma((k+1)/2)}{\Gamma(k/2)}\sqrt{\frac{\tau}{k\pi}}\left\{1 + \frac{\tau}{k}(x-\mu)^2\right\}^{-(k+1)/2}$ for $-\infty < x < \infty$, $\tau > 0$, $k \geq 2$ (WinBUGS), $k \geq 1$ (OpenBUGS), or $k \geq 0$ (JAGS).

Mean μ **Variance** $\frac{k}{\tau(k-2)}$ if $k \geq 2$, otherwise infinite or undefined

Used in place of the commoner normal distribution when heavier tails are required. As $k \to \infty$, the tails become thinner and this tends to the normal distribution with mean μ and precision τ. The "standard" t distribution, familiar in classical hypothesis tests, has $\mu = 0$, $\tau = 1$. Given $X \sim N(0,1)$ and $V \sim \chi_k^2$, then $X/\sqrt{V/k}$ has the standard t_k distribution.

Due to this relation with the χ^2 distribution, the "degrees of freedom" k is conventionally an integer. BUGS allows non-integer k, through generalising the χ_k^2 as Gamma$(\frac{k}{2}, \frac{1}{2})$. WinBUGS is restricted to $k \geq 2$, OpenBUGS to $k \geq 1$, while JAGS allows all $k \geq 0$.

The t distribution with $k = 1$ is called the Cauchy distribution, whose mean and variance are undefined. The half-Cauchy distribution has an application as a prior for standard deviations in hierarchical models (§10.2.3). The Cauchy and other t distributions may be implemented in all varieties of BUGS using the above definition in terms of the normal and gamma, as illustrated in §5.5, §10.2.3.

Double exponential   x ~ ddexp(mu,tau)

Density $p(x|\mu,\tau) = \frac{\tau}{2}e^{-\tau|x-\mu|}$ for $-\infty < x < \infty, \tau > 0$

Mean μ **Variance** $2/\tau^2$

A symmetric distribution whose density is formed by coupling the density of the exponential distribution with its reflection in the y-axis, then shifting and scaling. Sometimes used in place of a normal distribution when heavier tails are required, for example, in modelling spatial correlation (§11.3). Also known as a Laplace distribution.

Improper uniform   x ~ dflat() (WinBUGS + OpenBUGS)

Density $p(x) = 1$ for $-\infty < x < \infty$

An "improper" distribution does not integrate to 1. This improper distribution may be used as a prior, but the responsibility is on the user to ensure the implied posterior distribution is proper, without which inferences will not make sense. A typical use of dflat() is as a minimally informative prior based on Jeffreys' principle (§ 5.2.3). Another example of a valid use is in spatial modelling (§11.3).

C.2   Continuous univariate, restricted to be positive

Exponential   x ~ dexp(theta)

Density $p(x|\theta) = \theta e^{-\theta x}$ for $x > 0, \theta > 0$

Mean $1/\theta$ **Variance** $1/\theta^2$

A simple distribution for a positive quantity, typically the time to an event, where the risk or rate θ of the event is constant through time.

Gamma   x ~ dgamma(a,b)

Density $p(x|a,b) = b^a x^{a-1} e^{-bx}/\Gamma(a)$ for $x > 0, a,b > 0$

Mean a/b **Variance** a/b^2

A Gamma$(1,b)$ distribution is exponential with mean $1/b$, and Gamma$(\frac{v}{2}, \frac{1}{2})$ is a chi-squared χ^2_v distribution on v degrees of freedom. Gamma distributions are used as sampling distributions for positive and skewed data, such as costs or times to events. If a is an integer, then the Gamma(a,b) is the distribution of the sum of a independent exponential variables with rate b. Note that the gamma is sometimes parameterised in terms of the *scale* $1/b$ instead of the rate b.

The gamma is commonly used as a conjugate prior distribution for inverse-scale parameters — see §5.2.7 and Table 3.1. However, the gamma with small a, b is not recommended as a vague prior for the precision of random effects in a hierarchical model. See §10.2.3 for discussion of appropriate alternatives.

Chi-squared   x ~ dchisqr(k)

Density $p(x|k) = 2^{-\frac{k}{2}} x^{\frac{k}{2}-1} e^{-\frac{x}{2}} / \Gamma(k/2)$ for $x > 0, k \geq 0$

Mean k **Variance** $2k$

The distribution of the sum of the squares of k independent standard normal random variables. More commonly used in classical hypothesis testing than in Bayesian applications.

Noncentral chi-squared   x ~ dnchisqr(k,delta)   (JAGS only)

Density $p(x|k, \delta) = \exp(-\frac{\delta}{2}) \sum_{r=0}^{\infty} \frac{(\delta/2)^r}{r!} p_{\chi^2}(x|k + 2r)$ for $x > 0, k \geq 0, \delta \geq 0$, where $p_{\chi^2}(x|k)$ is the density $p(x|k)$ of the χ^2 distribution above

Mean $k + \delta$ **Variance** $2(k + 2\delta)$

The distribution the sum of squares of k independent normal random variables each with variance one, where δ is the sum of squares of the normal means.

Weibull   x ~ dweib(a,b)

Density $p(x|a, b) = a\,b\,x^{a-1} e^{-bx^a}$ for $x > 0, \ a, b > 0$

Mean $b^{-1/a}\Gamma(1 + 1/a)$ **Variance** $b^{-2/a}(\Gamma(1 + 2/a) - \Gamma(1 + 1/a)^2)$

A common model for times x to events (see §11.1). The *hazard* or instantaneous risk of the event is $h(x) = abx^{a-1}$. For $a < 1$ the hazard decreases with x; for $a > 1$ it increases. $a = 1$ is the exponential distribution dexp(b) with constant hazard.

Be careful of alternative parameterisations in different software; for example, dweibull() in R uses $p(x|a, b_2) = (a/b_2)(x/b_2)^{a-1} \exp(-(x/b_2)^a)$, where $b_2 = b^{-1/a}$, and survreg() in the R package survival (Therneau, 2010) reports estimates of $\log(b_2)$ and $1/a$.

Log-normal   x ~ dlnorm(mu,tau)

Density $p(x|\mu, \tau) = \sqrt{\frac{\tau}{2\pi}} \frac{1}{x} e^{-\frac{\tau}{2}(\log x - \mu)^2}$ for $x > 0, \ \tau > 0$

Mean $\exp(\mu + 1/(2\tau))$ **Variance** $\exp(2\mu + 1/\tau)(\exp(1/\tau) - 1)$

The distribution of the log of a normally distributed random variable with mean μ and *precision* (not variance) τ. The median is $\exp(\mu)$.

Generalised gamma $\quad$ x ~ gen.gamma(a,b,c)

Density $\quad p(x|a,b,c) = c\, b^{ca} x^{ca-1} e^{-(bx)^c}/\Gamma(a) \quad$ for $\quad x > 0,\ a,b,c > 0$

Mean $\mu = \frac{\Gamma(a+1/c)}{b\Gamma(a)}$ $\quad$ **Variance** $\frac{\Gamma(a+2/c)}{b^2\Gamma(a)} - \mu^2$

Called `dggamma` in OpenBUGS and `dgen.gamma` in JAGS, with the same parameters.

If $W \sim \text{Gamma}(a,1)$ then $X = W^{1/c}/b$ has the generalised gamma distribution. With $a = 1$ this is equivalent to `dweibull(c, c2)` where $c2 = (1/b)^{-c}$, with $c = 1$ it is equivalent to `dgamma(a, b)`, and with $a = c = 1$ to `dexp(b)`. As $a \to \infty$ it tends to `dlnorm(mu, tau)` where `mu` is $\log(1/b) + \log(a)/c$ and the precision `tau` is $c^2 a$ (Lawless, 1980).

Again there are many different parameterisations of this distribution. This one is based on Stacy (1962). A more flexible parameterisation was developed by Prentice (1974), in which the parameter called a here is transformed and unbounded. This avoids problems with estimation when the supported values of a are close to the log-normal boundary. WinBUGS modules for this alternative, and the even more flexible four-parameter generalised F distributions, are provided by Jackson et al. (2010b).

Pareto $\quad$ x ~ dpar(a,b)

Density $\quad p(x|a,b) = ab^a x^{-(a+1)} \quad$ for $x > b,\ a,b > 0$

Mean $\frac{ba}{a-1}$ (undefined $a \leq 1$) $\quad$ **Variance** $\frac{b^2 a}{(a-1)^2(a-2)}$ (undefined $a \leq 2$)

The distribution function is $\Pr(Y < y) = 1 - b^a/y^k$, that is, the tail probability is a "power law" function of the tail length. The probability density is a decreasing function, like the density of the exponential distribution.

The "Pareto principle" is that for many events, a proportion p of the effects comes from roughly $1 - p$ of the causes. For example, a common observation is that 20% of all individuals own 80% of a society's wealth. Furthermore, for all n, $(1-p)^n$ will own p^n of the wealth; for example, 20% of that richest 20% will own 80% of that richest 20%'s share. This model holds if individual wealth has a Pareto distribution with $a = \log(1-p)/\log(\frac{1-p}{p}) > 1$. The *Gini coefficient* commonly used to measure income inequality is $\frac{1}{2a-1}$ under this model, ranging between 0 ($a \to \infty, p = 0.5$, perfect equality) and 1 (or 100%, $a = 1, p = 1$, complete inequality).

The Pareto is also a natural model for Bayesian inference about extreme values, since it is the conjugate distribution for the upper limit of a uniform distribution with a fixed lower limit (Table 3.1).

If $X \sim \text{Pareto}(a,b)$, then $\log(X/b)$ is exponential with rate a.

Generalised extreme value   x ~ dgev(mu,sigma,eta)
(OpenBUGS only)

Density $p(x|\mu,\sigma,\eta) =$
$$\frac{1}{\sigma}\left(1+\frac{\eta}{\sigma}(x-\mu)\right)^{-(1+1/\eta)}\exp\left\{-(1+\frac{\eta}{\sigma}(x-\mu))^{-1/\eta}\right\}\quad\text{for }\frac{\eta}{\sigma}(x-\mu)\geq-1$$

Mean $\begin{cases}\mu+\sigma\frac{\Gamma(1-\eta)-1}{\eta} & \text{if }\eta\neq0,\eta<1,\\\mu+\sigma\gamma & \text{if }\eta=0,\\\text{indeterminate} & \text{if }\eta\geq1\end{cases}$ **Variance** $\begin{cases}\sigma^2(g_2-g_1^2)/\eta^2 & \text{if }\eta\neq0,\eta<\frac{1}{2},\\\sigma^2\frac{\pi^2}{6} & \text{if }\eta=0,\\\infty & \text{if }\eta\geq\frac{1}{2}\end{cases}$

γ is Euler's constant (0.5772 to four decimal places) and $g_k=\Gamma(1-k\eta)$.

This distribution arises through the extreme value theorem as the distribution of the normalised maximum of a sequence of random variables. When $\sigma=\mu\eta$ with $\mu,\eta<0$, it reduces to a Weibull distribution for $-x$, with $a=-1/\eta$, $b=(-\eta/\sigma)^{-1/\eta}$ as parameterised above.

Generalised Pareto   x ~ dgpar(mu,sigma,eta)
(OpenBUGS only)

Density $p(x|\mu,\sigma,\eta)=\frac{1}{\sigma}\left(1+\frac{\eta}{\sigma}(x-\mu)\right)^{-(1+1/\eta)}$
for $\frac{\eta}{\sigma}(x-\mu)\geq-1$, $x\geq\mu$

Mean $\mu+\frac{\sigma}{1-\eta}$ ($\eta<1$, otherwise indeterminate) **Variance** $\frac{\sigma^2}{(1-\eta)^2(1-2\eta)}$ ($\eta<\frac{1}{2}$, otherwise indeterminate)

When $\eta=\sigma/\mu$, this reduces to the Pareto distribution with parameters $a=\mu/\sigma$ and $b=\sigma/\eta$. As $\eta\to0$ with $\mu=0$, this tends to the exponential distribution with rate $1/\sigma$. Used for modelling events which exceed some extreme threshold.

F   x ~ df(n,m,mu,tau) (JAGS and OpenBUGS only)

Density $p(x|n,m,\mu,\tau)=$
$$\frac{\Gamma(\frac{n+m}{2})}{\Gamma(\frac{n}{2})\Gamma(\frac{m}{2})}\left(\frac{n}{m}\right)^{\frac{n}{2}}\sqrt{\tau}(\sqrt{\tau}(x-\mu))^{\frac{n}{2}-1}\left\{1+\frac{n\sqrt{\tau}(x-\mu)}{m}\right\}^{-\frac{(n+m)}{2}}$$
for $x>0,n,m>0$

Mean $\frac{m}{m-2}$ for $m>2$ **Variance** $\frac{2m^2(n+m-2)}{n(m-2)^2(m-4)}$ for $m>4$

The standard F distribution, familiar from classical analysis of variance, is the distribution of the ratio of the mean squares of n and m independent standard normal variates, hence of the ratio of two independent chi-squared variates each divided by its degrees of freedom.

The F distribution in OpenBUGS is a *generalised* F distribution which also includes location and inverse scale parameters μ and τ. The F distribution in JAGS is the standard one restricted to $\mu=0$, $\tau=1$. WinBUGS modules for the generalised F distribution with alternative parameterisations described by Prentice (1975) and Cox (2008) are provided by Jackson et al. (2010b).

Other positive distributions

The `Reliability` subsystem in OpenBUGS (Kumar et al., 2010) implements several positive distributions which are used to model failure times in engineering. These include Birnbaum–Saunders, Burr X, Burr XII, exponential power, exponentiated Weibull, extended exponential, extended Weibull, flexible Weibull, generalised exponential, generalised power Weibull, Gompertz, Gumbel, inverse Gaussian, inverse Weibull, linear failure rate, logistic exponential, log-logistic, log-Weibull, and modified Weibull. See its manual for further details.

C.3 Continuous univariate, restricted to a finite interval

Uniform $\quad$ x ~ dunif(a,b)

Density $\quad p(x|a,b) = 1/(b-a) \quad$ for $x \in [a,b],\ b > a$

Mean $\frac{a+b}{2}$ $\quad$ **Variance** $\frac{(b-a)^2}{12}$

The uniform is common in Bayesian applications as an intuitively vague prior distribution. When using it in this way, make sure that the limits really cover all possible values for the parameter. Also, as discussed in §5.2.3, it is only "uninformative" in this way for one choice of scale; for example, if logit(p) is uniform then p is not uniform.

Beta $\quad$ x ~ dbeta(a,b)

Density $\quad p(x|a,b) = \frac{\Gamma(a+b)}{\Gamma(a)\Gamma(b)} x^{a-1}(1-x)^{b-1} \quad$ for $0 < x < 1,\ a,b > 0$

Mean $a/(a+b)$ $\quad$ **Variance** $ab/((a+b)^2(a+b+1))$

The beta distribution is most commonly used as a prior for proportions (§5.2.5), since it is the conjugate prior for the probability in a binomial model (Table 3.1). If $a,b > 1$, the beta has a single mode at $(a-1)/(a+b-2)$. With $a = b = 1$ this is simply the Uniform(0,1) distribution. Otherwise, if $a \geq 1, b \leq 1$, the density is increasing and skewed towards one; if $a \leq 1, b \geq 1$, the density is decreasing and skewed towards zero; or if $a, b < 1$, the distribution is U-shaped. See §5.3.1 for discussion of how to express or elicit prior beliefs as a beta distribution.

C.4   Continuous multivariate distributions

Multivariate normal   x[1:d] ~ dmnorm(mu[],T[,])

Density $p(x|\mu,T) = (2\pi)^{-\frac{d}{2}}|T|^{\frac{1}{2}}e^{-\frac{1}{2}(x-\mu)'T(x-\mu)}$ for T symmetric and positive definite, $d = dim(x)$, $\infty < x_i < \infty$

Analogously to the univariate normal, this is parameterised by the *precision* matrix T, which is the inverse of the covariance matrix Σ.

For smaller-dimensional x it is often preferable to specify a multivariate normal model or prior as a sequence of conditional univariate normal distributions (see, e.g., § 10.2.3). This gives the parameters a more intuitive meaning. For example, if X and Y are jointly bivariate normal with marginal means μ_X, μ_Y, marginal variances σ_X^2, σ_Y^2, respectively, and correlation $\rho = Cov(X,Y)/(\sigma_X\sigma_Y)$, then the conditional distribution $(Y|X = x)$ is normal with mean $\mu_Y + \frac{\sigma_Y}{\sigma_X}\rho(x - \mu_X)$ and variance $(1 - \rho^2)\sigma_Y^2$. Together with an independent univariate normal model for X, this fully defines the joint distribution of X and Y in terms of their marginal properties and their correlation, an understandable quantity between 0 and 1.

In higher dimensions, given a pair of vectors x_1, x_2, where the concatenated pair is multivariate normal, the general formula for the conditional distribution of $(x_1|x_2 = a)$ is multivariate normal with mean $\mu_1 + \Sigma_{12}\Sigma_{22}^{-1}(a - \mu_2)$ and covariance matrix $\Sigma_{11} - \Sigma_{12}\Sigma_{22}^{-1}\Sigma_{21}$, where μ_r is the marginal mean of x_r and $\Sigma_{rs} = Cov(x_r, x_s)$.

Multivariate t   x[1:d] ~ dmt(mu[],T[,],k)

Density $p(x|\mu,T,k) = \Gamma\left((k+d)/2\right)(k\pi)^{-\frac{d}{2}}|T|^{\frac{1}{2}} \times$
$$\{1 + (x - \mu)'T(x - \mu)/k\}^{-(k+d)/2}\,/\Gamma(k/2)$$
for T symmetric and positive definite, $d = dim(x)$, $k \geq 2$, $-\infty < x_i < \infty$

Analogously to the multivariate normal, this reduces to the univariate Student t distribution when $dim(x) = 1$ and can be applied in place of the multivariate normal when a heavier-tailed distribution is needed.

Wishart   x[1:d,1:d] ~ dwish(R[,],k)

Density $p(x|R,k) = |R|^{\frac{k}{2}}|x|^{\frac{k-d-1}{2}}e^{-\frac{1}{2}\text{tr}(Rx)}/2^{dk/2}\Gamma_d(k/2)$ for R, x symmetric and positive definite, $d = dim(x)$, $k \geq d$

The Wishart is a distribution for a matrix which is restricted to be positive-definite. It is a multivariate generalisation of the gamma distribution and is conjugate prior for the precision matrix of a multivariate normal distribution. The mean is kR^{-1}, and when k is lower, the distribution is less informative. As discussed in §10.2.3, since the parameters of the Wishart are difficult to interpret, multivariate normal distributions may be better specified as sequences of conditional distributions, where dependence can be expressed by bivariate correlations instead of covariance matrices.

This may only be used as a conjugate prior or for forward sampling, and the parameters must be specified as constants and cannot be estimated.

Dirichlet   x[1:d] ~ ddirch(theta[])

Density $p(x|\theta) = \frac{\Gamma(\sum_i \theta_i)}{\prod_i \Gamma(\theta_i)} \prod_i x_i^{\theta_i - 1}$ for $\theta_i > 0$, $x_i \in [0, 1]$, $\sum_i x_i = 1$

In OpenBUGS and JAGS this may also be spelt ddirch.

A distribution for a vector where all elements are constrained in [0,1] and add up to one, such as a vector of probabilities for mutually exclusive events. The Dirichlet is a multivariate generalisation of the beta distribution and the conjugate distribution for the probabilities governing a multinomial or categorical model. In JAGS, but not in OpenBUGS or WinBUGS, *structural zeros* are allowed, so that if some of the elements of theta are zero, then the corresponding element of x is fixed to zero.

ddirch can be used as a prior but not as a likelihood in any variety of BUGS. In other words, the parameters of the distribution cannot be estimated and must be supplied as constants. However, if $Y_1, \ldots, Y_n \sim$ Gamma(a_i, b) independently, then $V = \sum_{i=1}^{n} Y_i \sim$ Gamma$(\sum a_i, b)$ and $(Y_1/V, \ldots, Y_n/V) \sim$ Dirichlet(a), where $a = (a_1, \ldots, a_n)$. Since the parameters of the gamma can be estimated in BUGS, this enables the Dirichlet distribution to be fitted to data, for example, or to be used in a hierarchical model where the random effects consist of vectors of probabilities (Example 10.3.4).

Additionally, in WinBUGS, ddirch can only be used as a conjugate prior in a model with a multinomial (dmulti) or categorical outcome (dcat), or for forward sampling. OpenBUGS and JAGS do not have this restriction.

Spatial distributions

The GeoBUGS facilities of WinBUGS and OpenBUGS provide several distributions for modelling sets of spatially correlated quantities. These are described in §11.3.

C.5   Discrete univariate distributions

Bernoulli   x ~ dbern(theta)

Density $p(x|\theta) = \theta^x (1-\theta)^{1-x}$ for $x = 0, 1$, $\theta \in [0, 1]$

Mean θ **Variance** $\theta(1-\theta)$

The distribution of a single event which occurs with probability θ.

Binomial   x ~ dbin(theta,n)

Density $p(x|\theta, n) = \frac{n!}{x!(n-x)!}\theta^x (1-\theta)^{n-x}$ for $\theta \in [0, 1]$, $n \in \mathbb{Z}^+$, $x = 0, \ldots, n$, where $\mathbb{Z}^+$ denotes the set of all positive integers.

Mean $n\theta$ **Variance** $n\theta(1-\theta)$

The distribution of the number of "successes" x out of n independent Bernoulli trials with probability θ. By generalising $x! = \Gamma(x+1)$, non-integer x may be modelled by the binomial distribution in BUGS.

Categorical   x ~ dcat(theta[])

Density $p(x|\theta) = \theta_x$ for $x = 1, 2, \ldots, n$, $\theta_i \in [0, 1]$, $\sum_i \theta_i = 1$

The distribution of an event which has one of n mutually exclusive outcomes with probabilities $\theta_1, \ldots, \theta_n$. The mean and variance are not defined, as the outcomes are not necessarily quantitative. In JAGS, the probabilities for a categorical (or multinomial) distribution may be any positive quantities, which are normalised internally to sum to 1. In WinBUGS or OpenBUGS, the elements of theta must be between 0 and 1 and sum to 1.

Poisson   x ~ dpois(theta)

Density $p(x|\theta) = \frac{\theta^x}{x!}e^{-\theta}$ for $x = 0, 1, \ldots$, $\theta > 0$

Mean θ **Variance** θ

A simple distribution for count data. It models the number of independent events in a fixed interval, when the expected number of events is θ and the event rate is constant. Unlike the binomial distribution, the number of events is theoretically unbounded. For rare events, where θ is low, the Binomial(n, θ) distribution is approximately equivalent to the Poisson$(n\theta)$.

dpois in BUGS has an unexpected use in the "zeros trick" for defining a new distribution; see §9.5.

Geometric   x ~ dgeom(theta) (OpenBUGS only)

Density $p(x|\theta) = \theta(1-\theta)^{x-1}$ for $x = 0, 1, \ldots$, $\theta \in [0, 1]$

Mean θ **Variance** θ

The distribution of the number x of Bernoulli trials required for one success to occur; where the success probability is θ.

Negative binomial   x ~ dnegbin(theta,n)

Density $p(x|\theta, n) = \frac{(x+n-1)!}{x!(n-1)!}\theta^n(1-\theta)^x$ for $\theta \in [0,1]$, $n \in \mathbb{Z}^+$, $x = 0, 1, 2, \ldots$

Mean $\frac{(1-\theta)n}{\theta}$ **Variance** $\frac{(1-\theta)n}{\theta^2}$

The negative binomial is the distribution of the number of failures in a sequence of Bernoulli events with success probability θ before n successes occur, see Example 5.2.2.

The negative binomial also arises as a generalisation of the Poisson distribution where the variance is greater than the mean. Thus it is commonly used for modelling overdispersed count data. If $(Y|W) \sim \text{Poisson}(W)$, and the Poisson rate is random with $W \sim \text{Gamma}(n, \frac{\theta}{1-\theta})$, then the distribution of Y marginalised over W is negative binomial with parameters θ, n. When modelling count data in BUGS, explicitly using dpois and dgamma is a more flexible and clearer alternative to using dnegbin, although it does not lend itself to regression situations, where the response mean is modelled as a function of covariates (as in Example 6.5.2). The parameters of the gamma may both be continuous, whereas n in dnegbin is discrete.

Beta-binomial   x ~ dbetabin(a,b,n) (JAGS only)

Density $p(x|a, b, n) = \binom{a+x-1}{x}\binom{b+n-x-1}{n-x}\binom{a+b+n-1}{n}^{-1}$ for $a, b > 0, n \in \mathbb{Z}^+$

Mean $\frac{na}{a+b}$ **Variance** $\frac{nab(a+b+n)}{(a+b)^2(a+b+1)}$

The beta-binomial is an overdispersed version of the binomial distribution, where the success probability is random. If $(Y|p) \sim \text{Binomial}(n, p)$, and $p \sim \text{Beta}(a, b)$, then the marginal distribution of Y is beta-binomial with parameters a, b. Thus it can be used in place of the binomial as a more flexible model for bounded count data where the variance is not defined entirely by the mean. Unlike the negative binomial, the parameters of dbetabin are the same as the underlying beta distribution; therefore, there is little clarity gained by explicitly using dbinom and dbeta.

It reduces to the Bernoulli distribution when $n = 1$ and to a discrete uniform distribution when $a = b = 1$.

Non-central hypergeometric   x ~ dhyper(n1,n2,m1,psi)
(JAGS and OpenBUGS only)

Density $p(x|n_1, n_2, m_1, \psi) = \frac{\binom{n_1}{x}\binom{n_2}{m_1-x}\psi^x}{\sum_{i=max(0,m_1-n_2)}^{min(n_1,m_1)}\binom{n_1}{i}\binom{n_2}{m_1-i}\psi^i}$ for $n_i \geq 0$, $0 < m_1 \leq n_1 + n_2$, $\max(0, m_1 - n_2) \leq x \leq \min(n_1, m_1)$

Mean m **Variance** v

The JAGS parameterisation is given above. OpenBUGS parameterises it slightly differently as x $\sim$ dhyper(n,m,N,psi), where $n = n_1$, $m = m_1$, $N = n_1 + n_2$, and ψ is unchanged.

The non-central hypergeometric distribution is used for sampling without replacement. Consider an urn with N balls, n of which are white, and the remainder are black. The standard hypergeometric distribution (with $\phi = 1$) governs the number of white balls in a sample of m balls drawn from the urn. The non-central hypergeometric distribution applies when the probability that an individual ball is drawn is different between a black (p_b) and a white (p_w) ball.* The odds ratio for drawing a white ball is $\psi = \frac{p_w(1-p_b)}{p_b(1-p_w)}$.

It also has an application to inference for a 2×2 table when the margins are known — either as a likelihood when the cell counts are known (§7.1.3) or as a prior when these are unknown ("ecological" inference: Wakefield (2004)). Consider a table of the number of individuals with and without a disease who are either unexposed or exposed to a risk factor. For a fixed population size N, number of exposed individuals n, diseased individuals m, and a fixed odds ratio of ϕ, then the number of unexposed individuals with the disease follows the non-central hypergeometric distribution.

It is also known as Fisher's non-central hypergeometric distribution, or the extended hypergeometric.

C.6   Discrete multivariate distributions

Multinomial   x[1:R] $\sim$ dmulti(theta[],n)

Density $p(x|\theta, n) = \frac{n!}{\prod_r x_r!} \prod_r \theta_r^{x_r}$ for $\theta_r \in [0,1]$, $\sum_r \theta_r = 1$, $\sum_r x_r = n$, $n \in \mathbb{Z}^+$

The distribution of a set of n events, each of which can have one of R mutually exclusive outcomes with probabilities $\theta_1, \ldots, \theta_R$. It generalises the categorical distribution to more than one event and the binomial to more than two outcomes.

JAGS allows unnormalised probabilities in the multinomial (as in the categorical) and internally sums them to one. In OpenBUGS and WinBUGS, theta must consist of proper probabilities.

In *multinomial logistic regression*, a multinomial outcome is modelled as a function of covariates. The probability of the rth of R categories given

---

*Perhaps if their weights or temperatures are different.

covariate vector x_i (changing notation so x_i is a covariate) is

$$p(r|x_i) = \frac{\exp(\beta_r' x_i)}{\sum_{r=1}^{R} \exp(\beta_r' x_i)}$$

where $\beta_1 = 0$ for identifiability. Let the observed counts for all data with the ith covariate pattern be $y_{i1}, ..., y_{iR}, \sum_{r=1}^{R} y_{ir} = n_i$. Then the likelihood is

$$\prod_i \frac{\exp(\sum_{r=1}^{R} y_{ir}\, \beta_r' x_i)}{\left[\sum_{r=1}^{R} \exp(\beta_r' x_i)\right]^{n_i}}$$

This likelihood can be handled in BUGS using `dmulti` (§6.4). However, in many circumstances it is more efficient to specify

$$y_{ir} \sim \text{Poisson}(\mu_{ir}), \qquad \log(\mu_{ir}) = \lambda_i + \beta_r' x_i.$$

With a Gamma(ϵ,ϵ) prior on λ_i as $\epsilon \to 0$, equivalent to a uniform prior on $\log(\lambda_i)$, integrating over λ_i produces the same likelihood for the β_r as the multinomial model (proof as exercise). A BUGS example is in §7.2.4.

Bibliography

Akaike, H. (1979). A Bayesian extension of the minimum AIC procedure of autoregressive model fitting. *Biometrika*, **66**, (2), 237.

Altham, P. (1969). Exact Bayesian analysis of a 2 × 2 contingency table, and Fisher's "exact" significance test. *Journal of the Royal Statistical Society. Series B (Methodological)*, **31**, (2), 261–9.

Andersen, P. K., Borgan, O., Gill, R. D., and Keiding, N. (1993). *Statistical models based on counting processes*. Springer, New York.

Anderson, T. W. (1971). *The statistical analysis of time series*. John Wiley & Sons, New York..

Andrieu, C., Doucet, A., and Holenstein, R. (2010). Particle Markov chain Monte Carlo methods. *Journal of the Royal Statistical Society: Series B (Statistical Methodology)*, **72**, (3), 269–342.

Asmussen, S. and Glynn, P. W. (2011). A new proof of convergence of MCMC via the ergodic theorem. *Statistics and Probability Letters*, **81**, 1482–5.

Barnard, J., McCulloch, R., and Meng, X. (2000). Modeling covariance matrices in terms of standard deviations and correlations, with application to shrinkage. *Statistica Sinica*, **10**, (4), 1281–312.

BaSiS (2001). Standards for Reporting of Bayesian Analyses in the Scientific Literature. `http://lib.stat.cmu.edu/bayesworkshop/2001/BaSis.html`.

Bayes, T. (1763). An essay towards solving a problem in the doctrine of chances. *Philosophical Transactions of the Royal Society*, **53**, 370–418.

Beaumont, M. (2010). Approximate Bayesian computation in evolution and ecology. *Annual Review of Ecology, Evolution and Systematics*, **41**, 379–406.

Beaumont, M. A., Zhang, W., and Balding, D. J. (2002). Approximate Bayesian computation in population genetics. *Genetics*, **162**, 2025–35.

Berger, J. (1985). *Statistical decision theory and Bayesian analysis*. Springer.

Bernardo, J. M. and Smith, A. F. M. (1994). *Bayesian theory*. John Wiley & Sons, New York.

Berry, S. M., Carlin, B. P., Lee, J. J., and Müller, P. (2010). *Bayesian adaptive methods for clinical trials*. CRC Press, Boca Raton, FL.

Besag, J., York, J., and Mollié, A. (1991). Bayesian image restoration, with two applications in spatial statistics. *Annals of the Institute of Statistical Mathematics*, **43**, (1), 1–20.

Best, N., Cockings, S., Bennett, J., Wakefield, J., and Elliott, P. (2001). Ecological regression analysis of environmental benzene exposure and childhood leukaemia: sensitivity to data inaccuracies, geographical scale and ecological bias. *Journal of the Royal Statistical Society, Series A*, **164**, (1), 155–74.

Best, N., Ickstadt, K., and Wolpert, R. (2000a). Spatial Poisson regression for health and exposure data measured at disparate resolutions. *Journal of the American Statistical Association*, **95**, (452), 1076–1088.

Best, N., Ickstadt, K., Wolpert, R., and Briggs, D. (2000b). Combining models of health and exposure data: the SAVIAH study. In *Spatial epidemiology: Methods and applications* (ed. P. Elliott, J. C. Wakefield, N. G. Best, and D. J. Briggs), pp. 393–414. Oxford University Press, Oxford, UK.

Best, N. G., Cowles, M. K., and Vines, S. K. (1995). *CODA: Convergence Diagnosis and Output Analysis software for Gibbs Sampler output: Version 0.3*. Medical Research Council Biostatistics Unit, Cambridge, UK.

Best, N. G., Spiegelhalter, D. J., Thomas, A., and Brayne, C. E. G. (1996). Bayesian analysis of realistically complex models. *Journal of the Royal Statistical Society, Series A*, **159**, 323–42.

Bishop, C. M. (2006). *Pattern recognition and machine learning*. Springer, London, UK.

Bivand, R. S., Pebesma, E. J., and Gómez-Rubio, V. (2008). *Applied spatial data analysis with R*. Springer, London, UK.

Bowmaker, J. K., Jacobs, G. H., Spiegelhalter, D. J., and Mollon, J. D. (1985). Two types of trichromatic squirrel monkey share a pigment in the red-green spectral region. *Vision Research*, **25**, (12), 1937–46.

Box, G. E. P. and Tiao, G. C. (1973). *Bayesian inference in statistical analysis*. John Wiley & Sons, New York.

Box, G. E. P. (1980). Sampling and Bayes inference in scientific modelling and robustness. *Journal of Royal Statistical Society*, Series A (General), 383–430.

Breiman, L. (1992). The little bootstrap and other methods for dimensionality selection in regression: X-fixed prediction error. *Journal of the American Statistical Association*, **87**, (419), 738–54.

Breslow, N. (1984). Extra-Poisson variation in log-linear models. *Applied Statistics*, **33**, (1), 38–44.

Breslow, N. E. and Clayton, D. G. (1993). Approximate inference in generalized linear mixed models. *Journal of the American Statistical Association*, **88**, 9–25.

Briggs, A.H., Ades, A.E., and Price, M.J. (2003). Probabilistic sensitivity analysis for decision trees with use of the Dirichlet distribution in a Bayesian framework, *Medical Decision Making*, **23**, (4), 341–350.

Brooks, S., Gelman, A., Jones, G. L., and Meng, X.-L. (ed.) (2011). *Handbook of Markov chain Monte Carlo*. CRC Press, Boca Raton, FL.

Brooks, S. P. and Gelman, A. (1998). General methods for monitoring convergence of iterative simulations. *Journal of Computational and Graphical Statistics*, **7**, 434–55.

Browne, W. (2009). *MCMC estimation in MLwiN, v2.10*. Centre for Multilevel Modelling, University of Bristol.

Buckland, S. T., Burnham, K. P., and Augustin, N. H. (1997). Model selection: an integral part of inference. *Biometrics*, **53**, (2), 603–18.

Burnham, K. P. and Anderson, D. R. (2002). *Model selection and multi-model inference: a practical information-theoretic approach*. Springer, New York.

Caldwell, D. M., Ades, A. E., and Higgins, J. P. T. (2005). Simultaneous comparison of multiple treatments: combining direct and indirect evidence. *British Medical Journal*, **331**, 897–900.

Carlin, B. and Gelfand, A. (1991). An iterative Monte Carlo method for nonconjugate Bayesian analysis. *Statistics and Computing*, **1**, (2), 119–28.

Carlin, B., Gelfand, A., and Smith, A. (1992). Hierarchical Bayesian analysis of changepoint problems. *Applied Statistics*, **41**, (2), 389–405.

Carlin, B. P. and Chib, S. (1995). Bayesian model choice via Markov chain Monte Carlo methods. *Journal of the Royal Statistical Society, Series B*, **57**, (3), 473–84.

Carlin, B. P. and Louis, T. A. (2008). *Bayesian methods for data analysis, third edition*. CRC Press, Boca Raton, FL.

Carroll, R. J., Gail, M. H., and Lubin, J. H. (1993). Case-control studies with errors in covariates. *Journal of the American Statistical Association*, **88**, (421), 185–99.

Casella, G. and George, E. I. (1992). Explaining the Gibbs sampler. *The American Statistician*, **46**, 167–74.

Celeux, G., Forbes, F., Robert, C., and Titterington, D. M. (2006). Deviance information criteria for missing data models (with discussion). *Bayesian Analysis*, **1**, (4), 651–706.

Celeux, G., Hurn, M., and Robert, C. (2000). Computational and inferential difficulties with mixture posterior distributions. *Journal of the American Statistical Association*, **95**, (451), 957–70.

Chase, M. and Dummer, G. (1992). The role of sports as a social status determinant for children. *Research Quarterly for Exercise and Sport*, **63**, (4), 418–24.

Chib, S. (1995). Marginal likelihood from the Gibbs output. *Journal of the American Statistical Association*, **90**, (432), 1313–21.

Chib, S. and Greenberg, E. (1998). Analysis of multivariate probit models. *Biometrika*, **85**, (2), 347–61.

Chib, S. and Jeliazkov, I. (2001). Marginal likelihood from the Metropolis-Hastings output. *Journal of the American Statistical Association*, **96**, (453), 270–81.

Chien, C. H. (1988). Small sample theory for steady state confidence intervals. In *Proceedings of the Winter Simulation Conference* (ed. M. Abrams, P. Haigh, and J. Comfort), pp. 408–13.

Clayton, D. G. and Kaldor, J. (1987). Empirical Bayes estimates of age-standardized relative risks for use in disease mapping. *Biometrics*, **43**, 671–681.

Congdon, P. (2003). *Applied Bayesian modelling*, John Wiley & Sons, New York.

Congdon, P. (2005). *Bayesian models for categorical data*. John Wiley & Sons, New York.

Congdon, P. (2006). *Bayesian statistical modelling* (2nd edn). John Wiley & Sons, New York.

Congdon, P. (2010). *Applied Bayesian hierarchical methods*. John Wiley & Sons, New York.

Coursaget, P., Yvonnet, B., Chiron, J. P., Gilks, W. R., Day, N. E., Wang, C. C., et al. (1991). Scheduling of revaccination against hepatitis B virus. *The Lancet*, **337**, (8751), 1180–3.

Cowles, M. K. and Carlin, B. P. (1996). Markov chain Monte Carlo convergence diagnostics: a comparative review. *Journal of the American Statistical Association*, **91**, 883–904.

Cox, C. (2008). The generalized F distribution: An umbrella for parametric survival analysis. *Statistics in Medicine*, **27**, 4301–12.

Cox, D. R. and Hinkley, D. V. (1974). *Theoretical statistics*. Chapman and Hall, London.

Cox, D. R. and Miller, H. D. (1965). *The theory of stochastic processes*. Chapman and Hall, London.

Crowder, M. J. (1978). Beta-binomial ANOVA for proportions. *Applied Statistics*, **27**, 34–7.

Daniels, M. J. and Hogan, J. W. (2008). *Missing data in longitudinal studies: Strategies for Bayesian modeling and sensitivity analysis*. Chapman & Hall, Boca Raton, FL.

Davis, P. and Rabinowitz, P. (1975). *Methods of numerical integration*. Academic Press, Waltham, MA.

Dawid, A. (1973). Posterior expectations for large observations. *Biometrika*, **60**, (3), 664–7.

de Finetti, B. (1931). *Funzione caratteristica di un fenomeno aleatorio*. Academia Nazionale del Linceo.

Del Moral, P., Doucet, A., and Jasra, A. (2006). Sequential Monte Carlo samplers. *Journal of the Royal Statistical Society: Series B (Statistical Methodology)*, **68**, 411–36.

Dellaportas, P., Forster, J., and Ntzoufras, I. (2002). On Bayesian model and variable selection using MCMC. *Statistics and Computing*, **12**, (1), 27–36.

Demiris, N. and Sharples, L. D. (2006). Bayesian evidence synthesis to extrapolate survival estimates in cost-effectiveness studies. *Statistics in Medicine*, **25**, 1960–75.

DerSimonian, R. and Laird, N. (1986). Meta-analysis in clinical trials. *Controlled clinical trials*, **7**, (3), 177–88.

Diebolt, J. and Robert, C. (1994). Estimation of finite mixture distributions through Bayesian sampling. *Journal of the Royal Statistical Society. Series B (Methodological)*, **56**, (2), 363–75.

Diggle, P. J., Tawn, J. A., and Moyeed, R. A. (1998). Model-based geostatistics (with discussion). *Journal of the Royal Statistical Society: Series C (Applied Statistics)*, **47**, (3), 299–350.

Dobson, A. (1983). *Introduction to statistical modelling*. Chapman & Hall, Boca Raton, FL.

Doucet, A., de Freitas, N., and Gordon, N. (ed.) (2001). *Sequential Monte Carlo methods in practice*. Springer, London, UK.

Draper, D. (1995). Assessment and propagation of model uncertainty (with discussion). *Journal of the Royal Statistical Society, Series B*, **57**, (1), 45–97.

Duane, S., Kennedy, A. D., Pendleton, B. J., and Roweth, D. (1987). Hybrid Monte Carlo. *Physics Letters B*, **195**, 216–22.

Elliott, P., Wakefield, J. C., Best, N. G., and Briggs, D. J. (ed.) (2000). *Spatial epidemiology: Methods and applications*. Oxford University Press, Oxford, UK.

Elston, R. and Grizzle, J. (1962). Estimation of time-response curves and their confidence bands. *Biometrics*, **18**, (2), 148–59.

Escobar, M. D. and West, M. (1995). Bayesian density estimation and inference using mixtures. *Journal of the American Statistical Association*, **90**, (430), 577–588.

Gamerman, D. and Lopes, H. F. (2006). *Markov chain Monte Carlo: Stochastic simulation for Bayesian inference* (2nd edn). Taylor & Francis, Boca Raton, FL.

Gehan, E. (1965). A generalized Wilcoxon test for comparing arbitrarily singly-censored samples. *Biometrika*, **52**, (1-2), 203–223.

Geisser, S. and Eddy, W. (1979). A predictive approach to model selection. *Journal of the American Statistical Association*, **74**, 153–60.

Gelfand, A. and Dey, D. (1994). Bayesian model choice: asymptotics and exact calculations. *Journal of the Royal Statistical Society, Series B*, **56**, (3), 501–14.

Gelfand, A., Hills, S., Racine-Poon, A., and Smith, A. (1990). Illustration of Bayesian inference in normal data models using Gibbs sampling. *Journal of the American Statistical Association*, **85**, (412), 972–85.

Gelfand, A., Sahu, S., and Carlin, B. (1995). Efficient parametrisations for normal linear mixed models. *Biometrika*, **82**, (3), 479.

Gelfand, A. E. and Smith, A. F. M. (1990). Sampling-based approaches to calculating marginal densities. *Journal of the American Statistical Association*, **85**, 398–409.

Gelman, A. (2006). Prior distributions for variance parameters in hierarchical models. *Bayesian Analysis*, **1**, (3), 515–33.

Gelman, A., Carlin, J. B., Stern, H. S., and Rubin, D. B. (2004). *Bayesian data analysis, second edition*. Chapman & Hall/CRC, London, UK.

Gelman, A. and Hill, J. (2007). *Data analysis using regression and multi-level/hierarchical models*. Cambridge University Press, New York.

Gelman, A., Jakulin, A., Pittau, M., and Su, Y. (2008). A weakly informative default prior distribution for logistic and other regression models. *The Annals of Applied Statistics*, **2**, (4), 1360–83.

Gelman, A. and Meng, X. (1998). Simulating normalizing constants: From importance sampling to bridge sampling to path sampling. *Statistical Science*, **13**, (2), 163–85.

Gelman, A. and Rubin, D. B. (1992). Inference from iterative simulation using multiple sequences (with discussion). *Statistical Science*, **7**, 457–511.

Geman, S. and Geman, D. (1984). Stochastic relaxation, Gibbs distributions and the Bayesian restoration of images. *IEEE Transactions on Pattern Analysis and Machine Intelligence*, **6**, 721–41.

George, E. and McCulloch, R. (1993). Variable selection via Gibbs sampling. *Journal of the American Statistical Association*, **88**, (423), 881–9.

Geweke, J. (1992). Evaluating the accuracy of sampling based approaches to the calculation of posterior moments. In *Bayesian statistics 4*, (ed. J. O. Bernardo, J. M. Berger, A. P. Dawid, and A. F. M. Smith), pp. 169–94.

Gilks, W. (1992). Derivative-free adaptive rejection sampling for Gibbs sampling. In *Bayesian statistics 4*, (ed. J. M. Bernardo, J. O. Berger, A. P. Dawid, and A. F. M. Smith), pp. 641–65. Oxford University Press, Oxford, UK.

Gilks, W. R., Richardson, S., and Spiegelhalter, D. J. (ed.) (1996). *Markov chain Monte Carlo in practice*. Chapman & Hall/CRC, Boca Raton, FL.

Gilks, W. R. and Wild, P. (1992). Adaptive rejection sampling for Gibbs sampling. *Applied Statistics*, **41**, (2), 337–48.

Girolami, M. and Calderhead, B. (2011). Riemann manifold Langevin and Hamiltonian Monte Carlo methods. *Journal of the Royal Statistical Society: Series B (Statistical Methodology)*, **73**, 123–214.

Goldstein, H. (2010). *Multilevel statistical models* (4th edn). John Wiley & Sons, New York.

Goubar, A., Ades, A. E., De Angelis, D., McGarrigle, C. A., Mercer, C. H., Tookey, P. A., Fenton, K., and Gill, O. N. (2008). Estimates of human immunodeficiency virus prevalence and proportion diagnosed based on Bayesian multiparameter synthesis of surveillance data. *Journal of the Royal Statistical Society: Series A (Statistics in Society)*, **171**, (3), 541–80.

Green, P. J. (1995). Reversible jump Markov chain Monte Carlo computation and Bayesian model determination. *Biometrika*, **82**, (4), 711–32.

Green, P. J. and Mira, A. (2001). Delayed rejection in reversible jump Metropolis-Hastings. *Biometrika*, **88**, 1035–53.

Han, C. and Carlin, B. P. (2001). Markov chain Monte Carlo methods for computing Bayes factors: a comparative review. *Journal of the American Statistical Association*, **96**, (455), 1122–32.

Hanson, K. (2001). Markov chain Monte Carlo posterior sampling with the Hamiltonian method. *Proc. SPIE*, 4322, pp. 456–67.

Hastings, W. K. (1970). Monte Carlo sampling-based methods using Markov chains and their applications. *Biometrika*, **57**, 97–109.

Hjort, N. L., Holmes, C., Müller, P., and Walker, S. G. (ed.) (2010). *Bayesian nonparametrics*. Cambridge University Press, Cambridge.

Holsinger, K. (2001–2010). Lecture Notes in Population Genetics. University of Connecticut, Storrs, CT.

Howard, J. V. (1998). The 2×2 table: A discussion from a Bayesian viewpoint. *Statistical Science*, **13**, (4), 351–67.

Ibrahim, J. G. and Chen, M.-H. (2000). Power prior distributions for regression models. *Statistical Science*, **15**, (1), 46–60.

Ickstadt, K. and Wolpert, R. L. (1998). Multiresolution assessment of forest inhomogeneity. In *Case studies in Bayesian statistics, Volume 3. Lecture notes in statistics* (ed. C. Gatsonis, J. S. Hodges, R. E. Kass, R. McCulloch, P. Rossi, and N. D. Singpurwalla), pp. 371–86. Springer-Verlag, New York.

Jackman, S. (2009). *Bayesian analysis for the social sciences*. John Wiley & Sons, New York.

Jackson, C. H., Best, N. G., and Richardson, S. (2006). Improving ecological inference using individual-level data. *Statistics in Medicine*, **25**, (12), 2136–59.

Jackson, C. H., Sharples, L. D., and Thompson, S. G. (2010a). Structural and parameter uncertainty in Bayesian cost-effectiveness models. *Applied Statistics*, **59**, (2), 233–53.

Jackson, C. H., Sharples, L. D., and Thompson, S. G. (2010b). Survival models in health economic evaluations: balancing fit and parsimony to improve prediction. *International Journal of Biostatistics*, **6**, (1). Article 34.

Jara, A., Hanson, T., Quintana, F., Müller, P., and Rosner, G. (2011). DP-package: Bayesian semi and nonparametric modeling in R. *Journal of Statistical Software*, **40**, (5), 1–30.

Jeffreys, H. (1939). *Theory of probability*. Oxford University Press, Oxford, UK.

Johnson, S. R. (2011). Bayesian inference: Statistical gimmick or added value? *The Journal of Rheumatology*, **38**, (5), 794.

Jones, G. L. (2004). On the Markov chain central limit theorem. *Probability Surveys*, **1**, 299–320.

Kadane, J. and Wolfson, L. J. (1998). Experiences in elicitation. *Journal of the Royal Statistical Society: Series D (The Statistician)*, **47**, (1), 3–19.

Kalbfleisch, J. D. and Prentice, R. L. (2002). *The statistical analysis of failure time data* (second edn). John Wiley & Sons, New York.

Kass, R. E. and Wasserman, L. (1995). A reference Bayesian test for nested hypotheses with large samples. *Journal of the American Statistical Association*, **90**, 928–34.

Kelsall, J. E. and Wakefield, J. C. (1999). Discussion of "Bayesian models for spatially correlated disease and exposure data" by Best et al. In *Bayesian statistics 6*, p. 151. Oxford University Press, Oxford, UK.

Kéry, M. (2010). *Introduction to WinBUGS for ecologists: Bayesian approach to regression, ANOVA, mixed models and related analyses*. Academic Press, Waltham, MA.

Kéry, M. and Schaub, M. (2011). *Bayesian population analysis using WinBUGS: A hierarchical perspective*. Academic Press, Waltham, MA.

Kruschke, J. K. (2010). *Doing Bayesian data analysis: A tutorial with R and BUGS*. Academic Press, Waltham, MA.

Kumar, V., Ligges, U., and Thomas, A. (2010). *ReliaBUGS user manual, version 1.0*. Available online: http://www.openbugs.info/ Manuals/\breakReliability/Contents.html.

Kynn, M. (2005). *Eliciting expert knowledge for Bayesian logistic regression in species habitat modelling*. PhD thesis, Queensland University of Technology, Brisbane, Australia.

Laplace, P. S. (1774). *Mémoire sur la probabilité des causes par les évènemens*. De l'Imprimerie Royale. Translated and discussed by S. M. Stigler (1986) in *Statistical Science*, **1**, (3), 359–378.

Lauritzen, S. L., Dawid, A. P., Larsen, B. N., and Leimer, H. G. (1990). Independence properties of directed Markov fields. *Networks*, **20**, 491–505.

Lawless, J. F. (1980). Inference in the generalized gamma and log gamma distributions. *Technometrics*, **22**, (3), 409–19.

Lawson, A. B., Browne, W. J., and Rodeiro, C. L. V. (2003). *Disease mapping with WinBUGS and MLwiN*. Wiley-Blackwell, New York.

Lee, P. (2004). *Bayesian statistics: An introduction*. John Wiley & Sons, New York.

Lindley, D. V. (1984). A Bayesian lady tasting tea. In *Statistics: An appraisal* (ed. H. A. David and H. T. David). Iowa State University Press, Ames, IA.

Little, R. J. A. and Rubin, D. B. (2002). *Statistical analysis with missing data*. John Wiley & Sons, New York.

Lu, G. and Ades, A. (2004). Combination of direct and indirect evidence in mixed treatment comparisons. *Statistics in Medicine*, **23**, (20), 3105–24.

Lunn, D., Best, N., Spiegelhalter, D., Graham, G., and Neuenschwander, B. (2009a). Combining MCMC with "sequential" PKPD modelling. *Journal of Pharmacokinetics and Pharmacodynamics*, **36**, (1), 19–38.

Lunn, D., Spiegelhalter, D., Thomas, A., and Best, N. (2009b). The BUGS project: Evolution, critique and future directions. *Statistics in Medicine*, **28**, (25), 3049–67.

Lunn, D. J. (2003). WinBUGS development interface (WBDev). *ISBA Bulletin*, **10**, (3), 10–1.

Lunn, D. J., Best, N. G., Thomas, A., Wakefield, J., and Spiegelhalter, D. (2002). Bayesian analysis of population PK/PD models: general concepts and software. *Journal of Pharmacokinetics and Pharmacodynamics*, **29**, (3), 271–307.

Lunn, D. J., Best, N. G., and Whittaker, J. C. (2009c). Generic reversible jump MCMC using graphical models. *Statistics and Computing*, **19**, (4), 395–408.

Lunn, D. J., Thomas, A., Best, N., and Spiegelhalter, D. (2000). WinBUGS — a Bayesian modelling framework: concepts, structure, and extensibility. *Statistics and Computing*, **10**, 325–37.

Mackay, D. J. C. (2003). *Information theory, inference, and learning algorithms*. Cambridge University Press, Cambridge, UK.

Marshall, E. and Spiegelhalter, D. (2007). Identifying outliers in Bayesian hierarchical models: a simulation-based approach. *Bayesian Analysis*, **2**, (2), 409–44.

Mason, A., Richardson, S., Plewis, I., and Best, N. (2012). Strategy for modelling non-random missing data mechanisms in observational studies using Bayesian methods. *Journal of Official Statistics* (forthcoming).

Matthews, R. A. J. (2001). Methods for assessing the credibility of clinical trial outcomes. *Drug Information Journal*, **35**, (4), 1469–78.

McCullagh, P. and Nelder, J. (1989). *Generalized linear models*. Chapman & Hall/CRC, Boca Raton, FL.

Mengersen, K. L., Robert, C. P., and Guihenneuc-Jouyaux, C. (1999). MCMC convergence diagnostics: a reviewww. In *Bayesian Statistics 6* (ed. J. M. Bernardo, J. O. Berger, A. P. Dawid, and A. F. M. Smith), pp. 415–40. Oxford University Press, Oxford, UK.

Metropolis, N., Rosenbluth, A. W., Rosenbluth, M. N., Teller, A. H., and Teller, E. (1953). Equations of state calculations by fast computing machines. *Journal of Chemical Physics*, **21**, 1087–91.

Meyer, R. and Millar, R. B. (1999). BUGS in Bayesian stock assessments. *Canadian Journal of Fisheries and Aquatic Sciences*, **56**, 1078–86.

Michael, J. and Schucany, W. (2002). The mixture approach for simulating bivariate distributions with specified correlations. *The American Statistician*, **56**, (1), 48–54.

Minka, T., Winn, J., Guiver, J., and Kannan, A. (2011). Infer.NET version 2.4. Microsoft Research, Cambridge.

Mitchell, T. and Beauchamp, J. (1988). Bayesian variable selection in linear regression. *Journal of the American Statistical Association*, **83**, (404), 1023–32.

Molenberghs, G. and Kenward, M. G. (2007). *Missing data in clinical studies*. John Wiley & Sons, New York.

Molitor, N., Best, N., Jackson, C., and Richardson, S. (2009). Using Bayesian graphical models to model biases in observational studies and to combine multiple data sources: Application to low birth-weight and water disinfection by-products. *Journal of the Royal Statistical Society, Series A*, **172**, (3), 615–37.

Neal, R. (1996). Sampling from multimodal distributions using tempered transitions. *Statistics and Computing*, **6**, (4), 353–66.

Neal, R. (1998). *Learning in graphical models*, chapter Suppressing random walks in Markov chain Monte Carlo using ordered over-relaxation, pp. 205–39. Kluwer Academic Publishers, Dordrecht.

Neal, R. M. (2003). Slice sampling. *Annals of Statistics*, **31**, (3), 705–41.

Neal, R. M. (2008). The harmonic mean of the likelihood: worst Monte Carlo method ever. Radford Neal's blog, August 17. http://radfordneal.wordpress.com/2008/08/17/the-harmonic-mean-of-the-likelihood-worst-monte-carlo-method-ever/.

Neal, R. M. (2010). MCMC using Hamiltonian dynamics. In *Handbook of Markov chain Monte Carlo* (ed. S. Brooks, A. Gelman, G. Jones, and X. L. Meng). Chapman & Hall–CRC Press, Boca Raton, FL.

Ntzoufras, I. (2009). *Bayesian modeling using WinBUGS*. John Wiley & Sons, New York.

O'Hagan, A. (2003). HSSS model criticism (with discussion). In *Highly structured stochastic systems* (ed. P. J. Green, N. L. Hjort, and S. T. Richardson), pp. 423–53. Oxford University Press, Oxford, UK.

O'Hagan, A., Buck, C., Daneshkhah, A., Eiser, J., Garthwaite, P., Jenkinson, D., Oakley, J., and Rakow, T. (2006). *Uncertain judgements: Eliciting experts' probabilities*, Statistics in practice. John Wiley & Sons, New York.

O'Hara, R. B. and Sillanpää, M. J. (2009). A review of Bayesian variable selection methods: what, how and which. *Bayesian Analysis*, **4**, (1), 85–118.

Ohlssen, D. I., Sharples, L. D., and Spiegelhalter, D. (2007). Flexible random-effects models using Bayesian semi-parametric models: applications to institutional comparisons. *Statistics in Medicine*, **26**, 2088–112.

O'Malley, A. J. and Zaslavsky, A. M. (2005). Cluster-level covariance analysis for survey data with structured nonresponse. *Technical Report*, Department of Health Care Policy, Harvard Medical School, Boston, MA.

Pearl, J. (1988). *Probabilistic reasoning in intelligent systems: Networks of plausible inference*. Morgan Kaufmann.

Penny, W. D., Friston, K. J., Ashburner, J. T., Kiebel, S. J., and Nichols, T. E. (ed.) (2006). *Statistical parametric mapping: The analysis of functional brain images*. Academic Press, Waltham, MA.

Plummer, M. (2003). JAGS: A program for analysis of Bayesian graphical models using Gibbs sampling. In *Proceedings of the 3rd International Workshop on Distributed Statistical Computing*, Vienna, Austria, pp. 20–2.

Plummer, M. (2008). Penalized loss functions for Bayesian model comparison. *Biostatistics*, **9**, (3), 523–39.

Plummer, M. (2011). *rjags: Bayesian graphical models using MCMC*. R package version 3-5, http://CRAN.R-project.org/package=rjags.

Plummer, M., Best, N., Cowles, K., and Vines, K. (2006). CODA: Convergence diagnosis and output analysis for MCMC. *R News*, **6**, (1), 7–11.

Pocock, S. and Spiegelhalter, D. (1992). Domiciliary thrombolysis by general practitioners. *British Medical Journal*, **305**, (6860), 1015.

Prentice, R. L. (1974). A log gamma model and its maximum likelihood estimation. *Biometrika*, **61**, (3), 539–44.

Prentice, R. L. (1975). Discrimination among some parametric models. *Biometrika*, **62**, (3), 607–14.

Presanis, A. M., De Angelis, D., Spiegelhalter, D. J., Seaman, S., Goubar, A., and Ades, A. E. (2008). Conflicting evidence in a Bayesian synthesis of surveillance data to estimate human immunodeficiency virus prevalence.

Journal of the Royal Statistical Society: Series A (Statistics in Society), **171**, (4), 915–37.

Press, S. J. (1971). *Some effects of an increase in police manpower in the 20th precinct of New York City.* RAND Corporation, New York.

Press, W. H., Teukolsky, S. A., Vetterling, W. T., and Flannery, B. P. (2002). *Numerical recipes in C++: The art of scientific computing* (2nd edn). Cambridge University Press, Cambridge, UK.

Pritchard, J. K., Seielstad, M. T., Perez-Lezaun, A., and Feldman, M. T. (1999). Population growth of human Y chromosomes: A study of Y chromosome microsatellites. *Molecular Biology and Evolution*, **16**, 1791–8.

R Development Core Team (2011). *R: A language and environment for statistical computing.* R Foundation for Statistical Computing, Vienna, Austria.

Raftery, A. and Lewis, S. (1992). How many iterations in the Gibbs sampler? In *Bayesian statistics 4*, pp. 763–73. Oxford University Press, Oxford, UK.

Rasbash, J., Charlton, C., Browne, W., Healy, M., and Cameron, B. (2009). *MLwiN version 2.1.* Centre for Multilevel Modelling, University of Bristol.

Reid, A. W. N., Harper, S., Jackson, C. H., Wells, A. C., Summers, D. M., Gjorgjimajkoska, O., Sharples, L. D., Bradley, J. A., and Pettigrew, G. J. (2011). Expansion of the kidney donor pool by using cardiac death donors with prolonged time to cardiorespiratory arrest. *American Journal of Transplantation*, **11**, (5), 995–1005.

Richardson, S. and Best, N. (2003). Bayesian hierarchical models in ecological studies of health-environment effects. *Environmetrics*, **14**, (2), 129–47.

Richardson, S. and Green, P. (1997). On Bayesian analysis of mixtures with an unknown number of components (with discussion). *Journal of the Royal Statistical Society: Series B (Statistical Methodology)*, **59**, (4), 731–92.

Riley, R., Lambert, P., Staessen, J., Wang, J., Gueyffier, F., Thijs, L., and Boutitie, F. (2008). Meta-analysis of continuous outcomes combining individual patient data and aggregate data. *Statistics in Medicine*, **27**, (11), 1870–93.

Ripley, B. D. (1987). *Stochastic simulation.* John Wiley & Sons, New York.

Ripley, B. D. (2004). *Spatial statistics.* Wiley-Blackwell, New York.

Robert, C. and Casella, G. (2004). *Monte Carlo statistical methods*, (2nd edn). Springer-Verlag, London, UK.

Robert, C., Cornuet, J.-M., Marin, J.-M., and Pillai, N. S. (2011). Lack of confidence in approximate Bayesian computational (ABC) model choice. *PNAS*, **108**, 15112–7.

Roberts, G. and Rosenthal, J. (2004). General state space Markov chains and MCMC algorithms. *Probability Surveys*, **1**, 20–71.

Rockova, V., Lesaffre, E., Luime, J., and Löwenberg, B. (2012). Hierarchical Bayesian formulations for selecting variables in regression models. *Statistics in Medicine* (early view, doi:10.1002/sim.4439).

Roeder, K. (1990). Density estimation with confidence sets exemplified by superclusters and voids in the galaxies. *Journal of the American Statistical Association*, **85**, (411), 617–24.

Rubin, D. (1987). *Multiple imputation for nonresponse in surveys*. John Wiley & Sons, New York.

Rubin, D. B. (1976). Inference and missing data. *Biometrika*, **63**, (3), 581–92.

Rubin, D. B. (1981). The Bayesian bootstrap. *Annals of Statistics*, **9**, (1), 130–4.

Rue, H., Martino, S., and Chopin, N. (2009). Approximate Bayesian inference for latent Gaussian models by using integrated nested Laplace approximations (with discussion). *Journal of the Royal Statistical Society, Series B*, **71**, 319–92.

Salway, R. and Wakefield, J. (2005). Sources of bias in ecological studies of non-rare events. *Environmental and Ecological Statistics*, **12**, (3), 321–47.

Schwartz, G. (1978). Estimating the dimension of a model. *Annals of Statistics*, 6, 342.

Senn, S. (1997). *Statistical issues in drug development*. Wiley Interscience, New York.

Smith, B. J. (2000). *Bayesian output analysis program (BOA) version 0.5.0 user manual*. Department of Biostatistics, University of Iowa College of Public Health, Iowa City, IA.

Spiegelhalter, D. J., Abrams, K. R., and Myles, J. P. (2004). *Bayesian approaches to clinical trials and health-care evaluation*. Wiley, Chichester, UK.

Spiegelhalter, D. J. and Best, N. G. (2003). Bayesian approaches to multiple sources of evidence and uncertainty in complex cost-effectiveness modelling. *Statistics in Medicine*, **22**, (23), 3687–709.

Spiegelhalter, D. J., Best, N. G., Carlin, B. P., and van der Linde, A. (2002). Bayesian measures of model complexity and fit (with discussion). *Journal of the Royal Statistical Society, Series B*, **64**, (4), 583–639.

Stacy, E. W. (1962). A generalization of the gamma distribution. *Annals of Mathematical Statistics*, **33**, 1187–92.

Stephens, M. (2000). Dealing with label switching in mixture models. *Journal of the Royal Statistical Society: Series B (Statistical Methodology)*, **62**, (4), 795–809.

Stigler, S. (1977). Do robust estimators work with real data? *The Annals of Statistics*, **5**, (6), 1055–98.

Stigler, S.M. (1986). Laplace's 1774 memoir on inverse probability, *Statistical Science* , Vol. 1, Number 3, 359–363.

Stone, M. (1977). An asymptotic equivalence of choice of model by cross-validation and Akaike's criterion. *Journal of the Royal Statistical Society, Series B*, **39**, (1), 44–7.

Sturtz, S., Ligges, U., and Gelman, A. (2005). R2WinBUGS: a package for running WinBUGS from R. *Journal of Statistical Software*, **12**, (3), 1–16.

Su, Y.-S. and Yajima, M. (2011). *R2jags: A package for running jags from R*. R package version 0.03-02, http://CRAN.R-project.org/package=R2jags.

Sun, D. and Berger, J. (1994). Bayesian sequential reliability for Weibull and related distributions. *Annals of the Institute of Statistical Mathematics*, **46**, (2), 221–49.

Sung, L., Hayden, J., Greenberg, M., Koren, G., Feldman, B., and Tomlinson, G. (2005). Seven items were identified for inclusion when reporting a Bayesian analysis of a clinical study. *Journal of Clinical Epidemiology*, **58**, (3), 261–8.

Ter Braak, C. (2006). A Markov chain Monte Carlo version of the genetic algorithm differential evolution: easy Bayesian computing for real parameter spaces. *Statistics and Computing*, **16**, (3), 239–49.

Therneau, T. (2010). *Survival: Survival analysis, including penalised likelihood*. R package version 2.36-1. Original Splus→R port by Thomas Lumley. Available online: http://CRAN.R-project.org/package=survival

Tierney, L. (1994). Markov chains for exploring posterior distributions (with discussion). *Annals of Statistics*, **22**, 1701–86.

UCL Institute for Risk and Disaster Reduction (2010). *Volcanic hazard from Iceland: analysis and implications of the Eyjafjallajökull eruption*. University College, London, UK.

U.S. Department of Health and Human Services (2010). Guidance for the use of Bayesian statistics in medical device clinical trials. Food and Drug Administration, Center for Devices and Radiological Health.

van der Linde, A. (2005). DIC in variable selection. *Statistica Neerlandica*, **59**, (1), 45–56.

Vehtari, A. and Lampinen, J. (2002). Bayesian model assessment and comparison using cross-validation predictive densities. *Neural Computation*, **14**, (10), 2439–68.

Wakefield, J. (2004). Ecological inference for 2×2 tables (with discussion). *Journal of the Royal Statistical Society, Series A*, **167**, (3), 385–445.

Wakefield, J., Haneuse, S., Dobra, A., and Teeple, E. (2011). Bayes computation for ecological inference. *Statistics in Medicine*, **30**, (12), 1381–96.

Whittemore, A.S., and Keller, J.B. (1988). Approximations for errors in variables regression. *Journal of the American Statistical Association*, 83, 1057–1066.

Woodward, P. (2011). *Bayesian analysis made simple: An Excel GUI for WinBUGS*, CRC Biostatistics Series. Chapman & Hall, Boca Raton, FL.

Index

2 × 2 tables, 97–100, **121–126**
  both margins fixed, 121–122, 126
  case-control studies, 125
  neither margin fixed, 122, 132
  one margin fixed, 121–125

adapting phase, 69, 306
aggregate data, 275–278, 322
AIC, 159, 165, 169, 175, 177
air pollution, 200
analysis of variance/covariance, 104, 119
asthma, 127, 130
asymptotics, 53, 160
autocorrelation, 72

Bayes factors, 148, **169–172**
Bayes' theorem, 33–36
Bayes, Thomas, 1, 33
Bayesian model averaging, 173–175
Bayesian modelling and reporting strategies, xvi, 5
Bayesian paradigm, 1–2
  advantages and disadvantages, 3
beetles, 115, 118
Bernoulli distribution, 351
  priors for, 84
  use in the "ones" trick, 205
beta distribution, 349
  as a conjugate prior, 37–41, 46, 84–85
  eliciting, 90
  simple usage, 6, 30
beta-binomial distribution, 353

as a predictive distribution, 30, 38–41, 179
bias modelling, 93
BIC, 170, 172, 175
binomial distribution, 352
  Bayes' paper on, 1
  conjugacy, 37–41, 46
  prediction for, 30
  priors for, 84
  simple example of, 9–10
Blackbox, 300, 301, 314, 316
BOA, 73
bootstrap, 118–119, 214–215
  Bayesian, 175–176
Bristol surgery mortality, 141, 146, 149, 221, 223, 241, 243, 244
Brooks, Gelman, and Rubins' convergence diagnostic, **73–76**, 312
BRugs, 303, 315, **318–319**
BUGS
  Classic BUGS, 299
  engines and interfaces, 297–298
  implementations of, 297–327
  running models, 16–17
BUGS language, 15–21, 26, 39, 48, 323
  distributions, 342–355
  functions, 337–342
  syntax, **327–336**
BUGS project, 13, 299
burn-in, 72

cadralazine, 238, 279
case-control studies, 125, 195